Annals of Mathematics Studies

Number 174

Introduction to Ramsey Spaces

Stevo Todorcevic

PRINCETON UNIVERSITY PRESS

PRINCETON AND OXFORD

2010

Published by Princeton University Press
41 William Street, Princeton, New Jersey 08540

In the United Kingdom: Princeton University Press
6 Oxford Street, Woodstock, Oxfordshire OX20 1TW

Library of Congress Cataloging-in-Publication Data

Todorcevic, Stevo
    Introduction to Ramsey Spaces / Stevo Todorcevic.
        p. cm. (Annals of mathematics studies ; no. 174)
    Includes bibliographical references and index.
    ISBN 978-0-691-14541-9 (hardcover : alk. paper)
    ISBN 978-0-691-14542-6 (pbk. : alk. paper)
    1. Ramsey theory. 2. Algebraic spaces. I. Title.
    QA166.T635 2010
    511'.5–dc22   2009036738

British Library Cataloging-in-Publication Data is available

This book has been composed in LaTeX

The publisher would like to acknowledge the author of this volume for
providing the camera-ready copy from which this book was printed.

Printed on acid-free paper. ∞

press.princeton.edu

Printed in the United States of America

10 9 8 7 6 5 4 3 2 1

Contents

Introduction to Ramsey Spaces

Introduction

This book is intended to be an introduction to a rich and elegant area of Ramsey theory that concerns itself with coloring infinite sequences of objects and which is for this reason sometimes called infinite-dimensional Ramsey theory. Transferring basic pigeon hole principles to their higher dimensional versions to increase their applicability is thus the subject matter of this theory. In fact, this tendency in Ramsey theory could be traced back to the invention of the original Ramsey theorem, which is nothing other than a higher dimensional version of the principle that says that a finite coloring of an infinite set must involve at least one infinite monochromatic subset. Ramsey's original application of the finite-dimensional Ramsey theorem was to obtain a rough classification of relational structures on the set $\mathbb{N}$ of natural numbers that he needed for a decision procedure that would test the validity of a certain kind of logical sentence. This original application of the finite-dimensional Ramsey theorem was matched in depth only forty years later by the Brunel-Sucheston use of this theorem in showing the existence of the so-called spreading model of a given Banach space, a notion that has eventually triggered important developments in that area of mathematics. The infinite-dimensional extension was also done for utilitarian reasons. It was initiated forty years ago by Nash-Williams in the course of developing his theory of better-quasi-ordered sets that eventually led him to the proof of that trees are well-quasi-ordered under the embedability relation. The full statement of the infinite-dimensional Ramsey theorem came, however, only through the work of Galvin-Prikry, Silver, Mathias, and especially Ellentuck, who was the first to use topological notions to describe what is today generally considered the optimal form of this result. In this book we present a general procedure to transfer any other Ramsey theoretic principles to higher and especially infinite dimensions trying to match the clarity of the Ellentuck result, but going beyond his topological Ramsey theory. As seen in the prototype example of the Ramsey space of infinite sequences of words and variable words over a fixed finite alphabet, topological Ramsey theory fails to capture the situation in which the objects that generate combinatorial subspaces are not the objects that one colors. For this, one needs the new theory of Ramsey spaces in which there are no natural topologies to describe the complexity of the allowed colorings. In other words, the new theory addresses not only the challenging problem of finding the right hypothesis for the colorings but also the problem of whether such a hypothesis can be preserved under classical operations such as the Souslin operation. The topological Ramsey theory of Ellentuck relies at this point on the classical result of Nikodym asserting that the Baire property relative to an arbitrary topology is preserved under the Souslin operation, while general Ramsey space theory requires a special proof of the corresponding fact. The abstract infinite-dimensional Ramsey theorem that we prove in Chapter Four leads to many interesting examples of Ramsey spaces. We had to be quite selective when choosing which of these Ramsey spaces to present in some detail and which not, and our choices are all made on the basis of known applications of these spaces. It is expected that in the years to come many other Ramsey spaces will find similar applications explaining our main motivation for writing this book.

The book is organized as follows. The Appendix gathers some special notation and supplementary material to help the reader in following the book. Preliminaries about the Ramsey theorem are given in Chapter One. This chapter is intended to serve as an indication of the general high-dimensional Ramsey theory that will be developed from Chapter Four on. So the reader who is encountering this material for the first time is kindly asked to patiently wait for a more thorough understanding until the abstract theory is developed in later chapters. As it will be seen, however, we assume nothing more from the reader than the familiarity with the general mathematical culture. The basic pigeon hole principles used in the rest of the book are briefly presented in Chapters Two and Three. The reader is, however, advised to skip these two chapters on first reading and instead return to a particular pigeon hole principle when needed. The reason for being brief here is that all of these pigeon hole principles are well known, and have already whole monographs devoted to them and so we found no reason to reproduce more than is needed to make this book partially self-contained. The abstract Ramsey theorem is given in Chapter Four. The two chapters that follow Chapter Four are devoted to analysis of this theorem based on one of the basic principles given in Chapters Two and Three. Local Ramsey theory is presented in Chapter Seven. The optimal parametrization of the infinite-dimensional Ramsey theorem is given in Chapter Nine, a chapter which is in part based on Chapter Eight and deals with the Ramsey theory of products of finite sets. Historical notes, remarks, and suggestions for further reading and information about related developments are given at the end of each chapter.

Chapter One

Ramsey Theory: Preliminaries

1.1 COIDEALS

Recall the notion of a *coideal* on some index set S, a collection $\mathcal{H}$ of subsets of S with the following properties:

(1) $\emptyset \notin \mathcal{H}$ but $S \in \mathcal{H}$,

(2) $M \subseteq N$ and $M \in \mathcal{H}$ imply $N \in \mathcal{H}$,

(3) $M = N_0 \cup N_1$ and $M \in \mathcal{H}$ imply $N_i \in \mathcal{H}$ for some $i = 0, 1$.

We shall be concerned here only with infinite index sets S and we shall always make the implicit assumption that the coideal $\mathcal{H}$ is *nonprincipal* which means that we shall consider coideals with the first condition strengthened as follows:

(1') $S \in \mathcal{H}$, but $\{x\} \notin \mathcal{H}$ for all $x \in S$.

Thus, coideals are notions of largeness for subsets of various sets S that typically carry some structure, and the purpose of Ramsey theory is to discover and organize them, as well as to lift them to higher dimensions. Consider the following four examples of families of subsets of some infinite index sets S.

Example 1.1.1 (Ramsey) *Let $S = \mathbb{N}$ be the set of natural numbers and let $\mathcal{H}$ be the collection of all infinite subsets of $\mathbb{N}$.*

While it is difficult to imagine a more simple fact than that $\mathcal{H}$ of Example 1.1.1 is indeed a coideal on $\mathbb{N}$, the corresponding fact in the following example is a major result of Ramsey theory, which we introduce in Sections 2.3 and 2.6 below (Hindman's theorem).

Example 1.1.2 (Hindman) *Let $S = \mathrm{FIN}$ be the collection of all nonempty finite subsets of $\mathbb{N}$ and let $\mathcal{H}$ be the collection of all subsets M of FIN for which we can find an infinite sequence (x_n) of pairwise disjoint elements of FIN such that $x_{n_0} \cup \ldots \cup x_{n_k} \in M$ for every finite increasing sequence $n_0 < \ldots < n_k$ of integers.*

The fact that the family $\mathcal{H}$ of the following example is a coideal is also a major result of Ramsey theory, which we treat briefly in Section 2.5 below (The Hales-Jewett Theorem).

Example 1.1.3 (Hales-Jewett) *Let $S = W_L$ be the semigroup of words over some fixed finite alphabet L and let $\mathcal{H}$ be the collection of all subsets M of W_L for which one can find an infinite sequence (x_n) of variable-words[1] over L such that $x_{n_0}[\lambda_0]^\frown \ldots ^\frown x_{n_k}[\lambda_k] \in M$ for every finite sequence $n_0 < \ldots < n_k$ of indexes and every choice $\lambda_0, \ldots, \lambda_k$ of letters from L to substitute all the occurrences of the variable in variable-words $x_{n_0}, \ldots, x_{n_k}$ that they are assigned to.*

Finally, the fact that the family $\mathcal{H}$ of the following example is indeed an example of a coideal on its index set is one of the finest results of Ramsey theory, the Halpern-Läuchli theorem (see Sections 3.1 and 3.2 below).

Example 1.1.4 (Halpern-Läuchli) *Let $S = \prod_{i<d} T_i$ be the product of a finite sequence T_i ($i < d$) of finitely branching rooted trees of height ω with no terminal nodes and let $\mathcal{H}$ be the collection of all subsets D of S for which we can find $t_i \in T_i$ ($i < d$) such that for every integer n the set D contains a product $\prod_{i<d} X_i$, where for each $i < d$ the set $X_i \subseteq T_i$ dominates every node of T_i at level n comparable with t_i.*

Remark 1.1 As already noted in the preface, one purpose of this book is to develop methods that would step up such basic Ramsey theoretic principles to higher dimensions. Perhaps it should also be noted that this is a nontrivial matter even in the case of Example 1.1.1. In fact, this was the original contribution by F. P. Ramsey himself. Perhaps at this point it is not even clear what "higher dimension" in each of these four examples means. It is for this reason that we reproduce Ramsey's original contribution below not only as a way of getting a feeling for what "dimension" means, but also as a way of showing why there is a need for developing this higher-dimensional Ramsey theory at all.

The notion of a coideal has an important extremal case worth pointing out. Recall, thus, that an *ultrafilter* on some index set S is a coideal $\mathcal{U}$ of subsets of S which in addition to (1), (2) and (3), or (1'), (2) and (3), has the following property that makes them *minimal* coideals on the given index set S :

(4) $M \in \mathcal{U}$ and $N \in \mathcal{U}$ imply $M \cap N \in \mathcal{U}$.

A given coideal $\mathcal{H}$ on S can be identified with a quantifier $(\mathcal{H}x)$, where the variable x ranges over elements of S. Thus, for a given property $\varphi(x)$ of elements of S, we write $(\mathcal{H}x)\,\varphi(x)$ to mean that

$$\{x \in S : \varphi(x)\} \in \mathcal{H}.$$

Analyzing such quantifiers is what usually lies behind the discovery and proof of basic Ramsey theoretic results such as these given in the four examples

---

[1] Here, a *variable-word* is simply a word x over the extended alphabet $L \cup \{v\}$ for some symbol ("variable") $v \notin L$ such that v occurs in x at least once. Given a variable-word x and $\lambda \in L$, we denote by $x[\lambda]$, the word obtained from x substituting by λ any occurrence of v in x.

above. Particularly useful are quantifiers $(\mathcal{U}x)$ that correspond to ultrafilters $\mathcal{U}$ on S. In this case, the quantifier $(\mathcal{U}x)$ has the following pleasant properties.

(a) $(\mathcal{U}x)\ \varphi_0(x) \wedge (\mathcal{U}x)\ \varphi_1(x)$ is equivalent to $(\mathcal{U}x)\ (\varphi_0(x) \wedge \varphi_1(x))$.

(b) $(\mathcal{U}x)\ \varphi_0(x) \vee (\mathcal{U}x)\ \varphi_1(x)$ is equivalent to $(\mathcal{U}x)\ (\varphi_0(x) \vee \varphi_1(x))$.

(c) $\neg(\mathcal{U}x)\ \varphi(x)$ is equivalent to $(\mathcal{U}x)\ \neg\varphi(x)$.

If the index set S carries some (partial) semigroup operation, the quantifier notation suggests the way to extend this operation on the space of all ultrafilters on S. We shall exploit this in the next chapter, which is devoted to semigroup colorings.

Remark 1.2 Readers not familiar with the notion of ultrafilter and the corresponding quantifier are kindly urged to check the properties (a), (b) and (c) of the quantifier. For these readers we have also reproduced below proofs of some of the standard results from this theory that use this quantifier which they can then compare with the proofs that avoid these quantifiers. These comparisons are suggested only as a way of getting familiar with ultrafilters as tools in this area. While ultrafilters as tools for achieving the main goals of this book are secondary, there is a growing number of important basic Ramsey theoretic principles (even on the level of importance of those presented above in Examples 1–4) whose only known proofs use ultrafilters. This situation is indeed unsatisfactory, but the objection that these proofs fail to be satisfactory because they use the Axiom of Choice is not formally valid. Namely, for given a Ramsey theoretic problem[2] P, we consider the field $\mathcal{G}_P$ of subsets of S that are Gödel-constructible from P. As Gödel has shown (without using AC!), the field of sets $\mathcal{G}_P$ does admit a rich family of ultrafilters suitable for any argument one ever wishes to pursue in this area. This is much like the way the field of Lebesgue-measurable sets works for analysts.

1.2 DIMENSIONS IN RAMSEY THEORY

The concept of dimension in Ramsey theory comes naturally when one encounters a problem that involves colorings of subsets of some structure S, all of some fixed cardinality k that is typically assumed to be either finite or countably infinite. Given a k-element subset F of S, one colors it with the isomorphism type of the structure induced from S. It is for this reason that one usually fixes the type τ and concentrates on solving the Ramsey theoretic problem that involves colorings of subsets of S of type τ. Having fixed one such τ, one uses the notation $S^{[k]}$ for the family of all k-element subsets of S of type τ. For example, if S is equal to the set $\mathbb{N}$ of natural numbers

---

[2]The P could be seen as a particular subset of S (a "coloring") we wish to consider, a set of integers of positive upper density, etc.

with the usual ordering, $\mathbb{N}^{[k]}$ is simply the set of all k-element subsets of $\mathbb{N}$ that can also be viewed as the set of all increasing k-sequences of elements of $\mathbb{N}$. So the k-dimensional Ramsey theoretic result in this context is the following famous theorem.

Theorem 1.3 (Ramsey) *For every positive integer k and every finite coloring of the family $\mathbb{N}^{[k]}$ of all k-element subsets of $\mathbb{N}$, there is an infinite subset M of $\mathbb{N}$ such that the set $M^{[k]}$ of all k-element subsets of M is monochromatic.*

Corollary 1.4 (Ramsey) *For all positive integers k, l, and m, there is a positive integer n such that for every n-element set N and every l-coloring of $N^{[k]}$, there is $M \subseteq N$ of cardinality m such that $M^{[k]}$ is monochromatic.*

It follows that the family $\mathcal{H}^{[k]}$ of subsets X of $\mathbb{N}^{[k]}$ that include some set of the form $M^{[k]}$ for infinite $M \subseteq \mathbb{N}$ forms a coideal on $\mathbb{N}^{[k]}$, a fact that is considerably deeper than the fact that the family $\mathcal{H}$ of Example 1.1.1 forms a coideal. Thus, $\mathcal{H}^{[k]}$ is some sort of k-power of $\mathcal{H} = \mathcal{H}^{[1]}$. It turns out that the power operation is considerably easier to visualize when applying it to ultrafilters rather than coideals. To see this, let $\mathcal{U}$ be a given nonprincipal ultrafilter[3] on $\mathbb{N}$. For a positive integer k, let $\mathcal{U}^k$ be a family of subsets of $\mathbb{N}^{[k]}$ defined as follows:

$$A \in \mathcal{U}^k \quad \text{iff} \quad (\mathcal{U}x_0)(\mathcal{U}x_1) \dots (\mathcal{U}x_{k-1}) \; \{x_0, x_1, \dots, x_{k-1}\} \in A.$$

Then $\mathcal{U}^k$ is an ultrafilter on $\mathbb{N}^{[k]}$. Equivalently, one can define a quantifier $\mathcal{U}^k \vec{x}$ on increasing k-sequences $\vec{x} = (x_0, x_1, \dots, x_{k-1})$ of elements of $\mathbb{N}$ as follows:

$$(\mathcal{U}^k \vec{x}) \; \varphi(\vec{x}) \quad \text{iff} \quad (\mathcal{U}x_0)(\mathcal{U}x_1) \dots (\mathcal{U}x_{k-1}) \; \varphi(x_0, x_1, \dots, x_{k-1}).$$

Theorem 1.3 is proved once we establish the following fact.

Lemma 1.5 *For every $A \in \mathcal{U}^k$, there is infinite $M \subseteq \mathbb{N}$ such that $A \supseteq M^{[k]}$.*

Proof. Clearly, we may assume $k > 1$. We shall think of A as a set of increasing k-sequences $\vec{x}$ of elements of $\mathbb{N}$, we shall think of $\mathcal{U}^k$ as living on the set of all increasing k-sequences, and we shall use also its lower-dimensional versions $\mathcal{U}^l$ $(1 \leq l \leq k)$. The set $M = (m_i)_{i=0}^{\infty}$ will be picked recursively according to its increasing enumeration. Since

$$(\mathcal{U}^k \vec{x}) \; \vec{x} \in A,$$

there must be an integer m_0 such that

$$(\mathcal{U}^{k-1}\vec{y}) \; (m_0){}^\frown\vec{y} \in A.$$

Suppose we have already chosen $m_0, \dots, m_p$ such that

$$(\forall l \leq k)(\forall \vec{x} \in \{m_0, \dots, m_p\}^{[l]})(\mathcal{U}^{k-l}\vec{y}) \; \vec{x}{}^\frown\vec{y} \in A,$$

---

[3] Recall that an ultrafilter on $\mathbb{N}$ (or any other index set, for that matter) is *nonprincipal* if it is not of the form $\{X \subseteq \mathbb{N} : k \in X\}$ for some $k \in \mathbb{N}$.

where $\mathcal{U}^0$ is to be interpreted as containing the single set, the singleton over the empty sequence. It follows that

$$(\forall l < k)(\forall \vec{x} \in \{m_0, \ldots, m_p\}^{[l]})(\mathcal{U}m)(\mathcal{U}^{k-l-1}\vec{y}) \quad \vec{x}^\frown(m)^\frown \vec{y} \in A.$$

Since $\mathcal{U}$ is nonprincipal there is $m_{p+1} > m_p$ such that the inductive hypothesis

$$(\forall l \leq k)(\forall \vec{x} \in \{m_0, \ldots, m_p, m_{p+1}\}^{[l]})(\mathcal{U}^{k-l}\vec{y}) \quad \vec{x}^\frown \vec{y} \in A$$

remains preserved. Once the set $M = (m_i)_{i=0}^\infty$ has been constructed, it is clear that we will have the inclusion $M^{[k]} \subseteq A$. □

A nonprincipal ultrafilter $\mathcal{U}$ on $\mathbb{N}$ with the property that for every positive integer k, its k-power $\mathcal{U}^k$ is generated by sets of the form $M^{[k]}$ ($M \in \mathcal{U}$), i.e., if $\mathcal{U}^k = \mathcal{U}^{[k]}$, is called a *selective ultrafilter*, or *Ramsey ultrafilter*. Thus a Ramsey ultrafilter $\mathcal{U}$ has the property that for every finite coloring of some finite power $\mathbb{N}^{[k]}$, there is $M \in \mathcal{U}$ such that $M^{[k]}$ is monochromatic.

Let us now give Ramsey's original application of Theorem 1.3. It is a good example of how one applies a Ramsey theoretic result to a given problem. It could also serve as a good indicator of which kind of problems such Ramsey theoretic results could be relevant to.

Definition 1.6 *Fix a positive integer k. For a sequence $\vec{\rho} \in \{<, =, >\}^{k \times k}$, define a relation $R_{\vec{\rho}} \subseteq \mathbb{N}^k$ by*

$$R_{\vec{\rho}}(\vec{x}) \quad \text{iff} \quad (\forall(i,j) \in k \times k) \; x_i \; \vec{\rho}(i,j) \; x_j.$$

Relations of the form $R_{\vec{\rho}}$ are called atomic canonical k-ary relations *on $\mathbb{N}$. We call a relation $R \subseteq \mathbb{N}^k$ a* canonical k-ary relation *on $\mathbb{N}$ if R is equal to the disjunction of a set of atomic canonical k-ary relations on $\mathbb{N}$.*

The point here is that there are no more than $2^{3^{k^2}}$ (a list computable from k)[4] canonical k-ary relations on $\mathbb{N}$ and that if R is one such relation then the structure $\langle \mathbb{N}, R \rangle$ is isomorphic to any of its restrictions $\langle M, R \rangle$ for M an infinite subset of $\mathbb{N}$. On the other hand, note that when $k \geq 2$ there exist continuum many nonisomorphic structures of the form $\langle \mathbb{N}, R \rangle$ for R a k-ary relation on $\mathbb{N}$. So the following "rough classification" result is quite interesting.

Theorem 1.7 (Ramsey) *For every positive integer k and every relation $S \subseteq \mathbb{N}^k$ there is an infinite subset M of $\mathbb{N}$ and a canonical k-ary relation R on $\mathbb{N}$ such that $S \cap M^k = R \cap M^k$.*

Proof. Given a relation $S \subseteq \mathbb{N}^k$ consider the following equivalence relation E on $\mathbb{N}^{[k]} = \{a \subseteq \mathbb{N} : |a| = k\}$. Two k-element subsets $a = \{x_0, \ldots, x_{k-1}\}$ and $b = \{y_0, \ldots, y_{k-1}\}$, enumerated increasingly, are equivalent if

$$(\forall \iota \in k^k) \; [S(x_{\iota(0)}, \ldots, x_{\iota(k-1)}) \leftrightarrow S(y_{\iota(0)}, \ldots, y_{\iota(k-1)})].$$

---

[4] The reader is indeed invited to make a list of all different canonical relations for small dimensions k. For $k = 2$, the list is $\{\bot, =, <, \leq, >, \geq, \neq, \top\}$.

Note that E has no more than 2^{k^k} equivalence classes, so by Theorem 1.3 there exists an infinite subset M of $\mathbb{N}$ such that $M^{[k]}$ is included in one of the classes; i.e., every two k-element subsets of M are E-equivalent. Let $\{m_0, \ldots, m_{k-1}\}$ be the increasing enumeration of the first k members of M. Let

$$\Sigma = \{\iota \in k^k : S(m_{\iota(0)}, \ldots, m_{\iota(k-1)}) \text{ holds}\}.$$

For $\iota \in \Sigma$, define $\vec{\rho}_\iota \in \{<, =, >\}^{k \times k}$ by letting

$$\vec{\rho}_\iota(i, j) = \rho \text{ iff } (\iota(i), \iota(j)) \in \rho,$$

for $(i, j) \in k \times k$ and $\rho \in \{<, =, >\}$. Let R be the disjunction of the atomic canonical relations $R_{\vec{\rho}_\iota} (\iota \in \Sigma)$. Then it is straightforward to check that $S \cap M^k = R \cap M^k$. $\qquad \square$

In the particular case of equivalence relations restricted to $\mathbb{N}^{[k]}$ viewed as the subset of $\mathbb{N}^k$ consisting of increasing k-tuples, one has a very clear picture of the canonical form. The canonical equivalence relations are determined by subsets $I \subseteq \{0, \ldots, k-1\}$ as follows

$$(x_0, \ldots, x_{k-1}) E_I (y_0, \ldots, y_{k-1}) \text{ iff } (\forall i \in I) \ x_i = y_i,$$

where the k-tuples $(x_0, \ldots, x_{k-1})$ and $(y_0, \ldots, y_{k-1})$ are taken to be increasing according the order of $\mathbb{N}$. This gives us the following well-known result,[5] which we give as an application of Theorem 1.7 and therefore ultimately as an application of the original high-dimensional Ramsey theorem 1.3.

Theorem 1.8 (Erdös-Rado) *For every equivalence relation E on $\mathbb{N}^{[k]}$ there is an infinite subset M of $\mathbb{N}$ and an index set $I \subseteq \{0, \ldots, k-1\}$ such that $E \upharpoonright M^{[k]} = E_I \upharpoonright M^{[k]}$.*

Proof. Let

$$R_E = \{(x_0, \ldots, x_{2k-1}) \in \mathbb{N}^{2k} : \{x_0, \ldots, x_{k-1}\} E \{x_k, \ldots, x_{2k-1}\}\}.$$

By Theorem 1.7, we get an infinite subset M of $\mathbb{N}$ and $\Sigma \subseteq \{<, =, >\}^{2k \times 2k}$ such that R_E is equal to the disjunction of $R_{\vec{\rho}}$ ($\vec{\rho} \in \Sigma$). Let

$$I = \{i < k : (\forall \vec{\rho} \in \Sigma) \ \vec{\rho}(i, k+i) ==\}.$$

Choose an infinite subset N of M which has the property that between every two integers of N there is at least one integer of M. We shall show that

$$E \upharpoonright N^{[k]} = E_I \upharpoonright N^{[k]}.$$

Suppose $s, t \in N^{[k]}$ when enumerated increasingly as $\{s_0, \ldots, s_{k-1}\}$ and $\{t_0, \ldots, t_{k-1}\}$, respectively, agree on indices from I. Let us show that s and t are E-equivalent. This is done by induction on the cardinality of the set

$$D(s, t) = \{i < k : s_i \neq t_i\},$$

---

[5]Since we are still at the very basic level, the reader for whom all this is very much new may wish to prove directly the case $k = 2$ of Theorem 1.8.

which is by our assumption disjoint from the set I. If $D(s,t) = \emptyset$, then s and t are equal and therefore E-equivalent. Suppose now that $|D(s,t)| = 1$, or in other words that $D(s,t)$ is equal to a singleton $\{i\}$. Since $i \notin I$, there is $\vec{\rho} \in \Sigma$ such that $\vec{\rho}(i, k+i) \neq\!=$. By the assumption that between every two integers from N there is one from M, we can find $u \in M^{[k]}$ such that

$$(s_0, \ldots, s_{k-1}, u_0, \ldots, u_{k-1}) \in R_{\vec{\rho}} \text{ and } (t_0, \ldots, t_{k-1}, u_0, \ldots, u_{k-1}) \in R_{\vec{\rho}}.$$

It follows that sEu and tEu and therefore sEt. Consider now the case $|D(s,t)| > 2$ and let i be the minimal member of $D(s,t)$. Let t' be obtained from t by replacing its ith member with s_i. Then $|D(s,t')| < |D(s,t)|$ and s and t' agree on I, so sEt' by the induction hypothesis. By the transitivity of E, we would get the desired conclusion sEt, provided we show that tEt'. However, note that $D(t,t') = \{i\}$, so the desired conclusion tEt' follows from the case $D(s,t) = 1$.

Conversely, suppose that pairs $s = \{s_0, \ldots, s_{k-1}\}$ and $t = \{t_0, \ldots, t_{k-1}\}$ of k-element subsets of N are E-equivalent. Let us show that $s_i = t_i$ for all $i \in I$. Pick a $\vec{\rho} \in \Sigma$ such that $(s_0, \ldots, s_{k-1}, t_0, \ldots, t_{k-1}) \in R_{\vec{\rho}}$. Then $\vec{\rho}(i, k+i) ==$ for all $i \in I$, and therefore $s_i = t_i$ for all $i \in I$, as required. This finishes the proof.   □

Corollary 1.9 *For all positive integers k and m there is an integer n such that for every equivalence relation E on $\{0, 1, \ldots, n\}^{[k]}$, there is a set $M \subset \{0, 1, \ldots, n\}$ of cardinality m such that the restriction $E \upharpoonright M^{[k]}$ is equal to one of the 2^k canonical equivalence relations on $M^{[k]}$.*

Corollary 1.10 *Given an integer $k \geq 1$ and regressive[6] map $f : \mathbb{N}^{[k]} \to \mathbb{N}$, there is an infinite $M \subseteq \mathbb{N}$ such that $f(s) = f(t)$ for all $s, t \in M^{[k]}$ with the property $\min(s) = \min(t)$.*

Proof. Apply Erdös-Rado theorem to the equivalence relation E_f on $\mathbb{N}^{[k]}$ induced by f, i.e., sE_ft iff $f(s) = f(t)$, and get infinite $M \subseteq \mathbb{N}$ and $I \subseteq \{0, 1, \ldots, k-1\}$ such that $E_f \upharpoonright M^{[k]} = E_I \upharpoonright M^{[k]}$. Since f is regressive and since M is infinite, it must be that either $I = \emptyset$, or $I = \{0\}$, as required.   □

Corollary 1.11 *For all positive integers k and m there is an integer n such that for every regressive mapping $f : \{0, 1, \ldots, n\}^{[k]} \to \mathbb{N}$, there is $M \subseteq \{0, 1, \ldots, n\}$ of cardinality m such that $f(s) = f(t)$ for all $s, t \in M^{[k]}$ with the property $\min(s) = \min(t)$.*   □

Note that Theorem 1.8 is a strengthening of Theorem 1.3 as it applies to colorings of $\mathbb{N}^{[k]}$ into any number of colors, not just finite. In fact, this result suggests that many Ramsey theoretic facts have "canonical versions" that apply to an unrestricted number of colors. Indeed, this is an important and deep line of investigation that has already reached some maturity and that typically also involves methods from other areas of combinatorics, such as various methods of enumerating combinatorial configurations. However, in order to keep this book to a reasonable length, we shall mention here very few results of this sort.

---

[6] A map $f : \mathbb{N}^{[k]} \to \mathbb{N}$ is *regressive* if $f(s) < \min(s)$ for all $s \in \mathbb{N}^{[k]}$ such that $\min(s) \neq 0$.

1.3 HIGHER DIMENSIONS IN RAMSEY THEORY

The purpose of this section is to isolate the property of a family $\mathcal{F}$ of finite subsets of $\mathbb{N}$ that permits us to state and prove the analog of the Ramsey theorem for $\mathcal{F}$. The analog should of course apply to the case

$$\mathcal{F} = \mathbb{N}^{[k]} = \{s \subseteq \mathbb{N} : |s| = k\}$$

giving us back the original Ramsey theorem, but the point here is that $\mathcal{F}$ could be of a considerably higher complexity we could precisely measure. Before we proceed further, we fix some notation. Let

$$\mathbb{N}^{[<\infty]} = \{s \subseteq \mathbb{N} : s \text{ is finite}\}$$

denote the family of *all* finite subsets of $\mathbb{N}$ including the empty set $\emptyset$.[7] For a given family $\mathcal{F}$ of finite subsets of $\mathbb{N}$ and for a (typically infinite) subset M of $\mathbb{N}$, we shall consider the following kind of *restriction* of $\mathcal{F}$ on M,

$$\mathcal{F}|M = \{s \in \mathcal{F} : s \subseteq M\}.$$

This is of course not the only kind of restriction one can make. For example, we can take the *trace* $\{s \cap M : s \in \mathcal{F}\}$, which is actually a quite different kind of restriction of $\mathcal{F}$ on M. While in general carelessness about the distinctions between the restrictions can lead to confusion, the notion of *barrier* that we are about to define in this section will, among other things, help us avoid these traps. Finally, we let $\sqsubseteq$ denote the initial segment relation on $\mathbb{N}^{[<\infty]}$, or more precisely,

$$s \sqsubseteq t \text{ iff } s = t \text{ or } s = \{m \in t : m < n\} \text{ for some } n \in t.$$

Let $\sqsubset$ denote the strict version of $\sqsubseteq$. We are now ready to define the three properties of families of finite subsets of $\mathbb{N}$ that are the subject matter of this section.

Definition 1.12 *For a family $\mathcal{F}$ of finite subsets of $\mathbb{N}$, we say that*

(1) *$\mathcal{F}$ is* Ramsey *if for every finite partition $\mathcal{F} = \mathcal{F}_0 \cup \ldots \cup \mathcal{F}_k$ and for every infinite set $N \subseteq \mathbb{N}$ there is an infinite set $M \subseteq N$ such that at most one of the restrictions $\mathcal{F}_0|M, \ldots, \mathcal{F}_k|M$ is nonempty;*

(2) *$\mathcal{F}$ is* Nash-Williams *if $s \not\sqsubseteq t$ for every pair $s \neq t \in \mathcal{F}$;*

(3) *$\mathcal{F}$ is* Sperner *if $s \not\subseteq t$ for every pair $s \neq t \in \mathcal{F}$.*

Lemma 1.13 *For every Ramsey family $\mathcal{F}$ of finite subsets of $\mathbb{N}$ and every infinite subset N of $\mathbb{N}$, there is an infinite set $M \subseteq N$ such that the restriction $\mathcal{F}|M$ is Sperner and therefore Nash-Williams.*

Proof. Apply the assumption that $\mathcal{F}$ is a Ramsey family to the partition $\mathcal{F} = \mathcal{F}_0 \cup \mathcal{F}_1$, where $\mathcal{F}_0$ is the family of all $\subseteq$-minimal elements of $\mathcal{F}$. $\square$

The following result shows that some sort of converse to this is true.

---

[7] Recall that in Example 1.1.2 we use the notation FIN for the family of all *nonempty* subsets of $\mathbb{N}$, an important difference, which is going to become more clear as we go on.

Theorem 1.14 (Nash-Williams) *Every Nash-Williams family is Ramsey.*

Corollary 1.15 (Ramsey) *For every positive integer k, the family $\mathbb{N}^{[k]}$ of k-element subsets of $\mathbb{N}$ is Ramsey.*

Proof. Clearly, $\mathbb{N}^{[k]}$ is Sperner and, therefore, Nash-Williams. □

The original proof of Theorem 1.14 which we now present is the first use of combinatorial forcing a technique that plays a quite prominent role in the rest of the book. We shall, however, not follow the original terminology and notation. To avoid repeating certain phrases, we shall let the variable M, N, P, ... run over infinite subsets of $\mathbb{N}$, the variables $s, t, u, \ldots$ over finite subsets, and $m, n, p, \ldots$ over elements of $\mathbb{N}$.

For the purpose of the next definition of combinatorial forcing and the three lemmas that follow, we fix a Nash-Williams family $\mathcal{F}$ of finite subsets of $\mathbb{N}$.

Definition 1.16 *We say that M accepts s if s is $\sqsubseteq$-comparable[8] to some $t \in \mathcal{F}$ such that $t \subseteq s \cup M$. We say that M rejects s if there is no $t \subseteq s \cup M$ in $\mathcal{F}$ that is $\sqsubseteq$-comparable to s. We say that M strongly accepts s if every $P \subseteq M$ accepts s. Finally, we say that M decides s if either M rejects s or if M strongly accepts s.*

Note the following immediate properties of these notions.

Lemma 1.17 (1) *For every M and s, there is $N \subseteq M$ that decides s.*

(2) *If M strongly accepts (respectively, rejects) s then every infinite subset of M strongly accepts (respectively, rejects) s.*

Using this and a simple diagonalization procedure we get the following.

Lemma 1.18 *Every infinite subset of $\mathbb{N}$ can be refined to an infinite set that decides all of its finite subsets.*

Fix an M that decides all of its finite subsets.

Lemma 1.19 *If M strongly accepts some subset s, then M strongly accepts $s \cup \{n\}$ for all but finitely many n in M.*

Proof. Otherwise, let

$$N = \{n \in M : n > \max(s) \text{ and } M \text{ rejects } s \cup \{n\}\}.$$

Then N is an infinite subset of M that rejects s, contradicting the assumption that M strongly accepts s. □

We are now ready to complete the proof of Nash-Williams's theorem. So fix a partition $\mathcal{F} = \mathcal{F}_0 \cup \mathcal{F}_1$ of some Nash-Williams family $\mathcal{F}$ and fix some infinite subset M of $\mathbb{N}$. We need to find an infinite subset P of M such

---

[8]That is, $s \sqsubseteq t$ or $t \sqsubseteq s$.

that one of the restrictions $\mathcal{F}_0|P$ or $\mathcal{F}_1|P$ is empty. Applying the lemmas to the first piece $\mathcal{F}_0$ of the partition and shrinking $\mathcal{F}$, we may assume that M $\mathcal{F}_0$-decides all of its finite subsets and, moreover, that if M $\mathcal{F}_0$-strongly accepts one of its finite subsets, then M $\mathcal{F}_0$-strongly accepts $s \cup \{n\}$ for all $n \in M$ such that $n > \max(s)$. If M $\mathcal{F}_0$-rejects $\emptyset$, then clearly $\mathcal{F}_0|M = \emptyset$. So, suppose that M strongly accepts $\emptyset$. Then by induction on the cardinality $|s|$ of a subset s of M, one shows that M strongly accepts all of its finite subsets. Since no two distinct elements of the family $\mathcal{F} = \mathcal{F}_0 \cup \mathcal{F}_1$ are $\sqsubseteq$-comparable, it follows that M does not contain an element of the family $\mathcal{F}_1$, which finishes the proof.

Theorem 1.14 was originally invented to facilitate the following notion which has found many uses, not only in the original theory of well-quasi-orderings, but also in such diverse areas as Banach space geometry.

Definition 1.20 *A family $\mathcal{F}$ of finite subsets of $\mathbb{N}$ is a* front *on some infinite subset M of $\mathbb{N}$ if $\mathcal{F}$ is a Nash-Williams family and if every infinite subset of M has an initial segment in $\mathcal{F}$. If, moreover, $\mathcal{F}$ is a Sperner family then we say that $\mathcal{F}$ is a* barrier *on M.*

Corollary 1.21 *If $\mathcal{F}$ is a front on M, then there is an infinite subset P of M such that the restriction $\mathcal{F}|P$ is a barrier on P.*

Proof. Apply Theorem 1.14 to the partition $\mathcal{F} = \mathcal{F}_0 \cup \mathcal{F}_1$, where $\mathcal{F}_0$ is the set of all $\subseteq$-minimal elements of $\mathcal{F}$. □

Corollary 1.22 *For every finite partition $\mathcal{F} = \mathcal{F}_0 \cup ... \cup \mathcal{F}_k$ of a family $\mathcal{F}$ that is a barrier on some infinite subset N of $\mathbb{N}$, there is infinite $M \subseteq N$ and $1 \le i \le k$ such that the restriction $\mathcal{F}_i|M$ is a barrier on M.*

The point of introducing fronts as well as barriers is that, while one usually works with barriers, the natural recursive constructions will give us families that are only fronts rather than barriers. For example, if for every integer $n \in M$ we fix a front $\mathcal{F}_n$ on the tail $M/n = \{m \in M : m > n\}$ then

$$\mathcal{F} = \{\{n\} \cup s : n \in M \text{ and } s \in \mathcal{F}_n\}$$

is a front on M. Naturally, concrete examples of fronts like $\mathcal{F} = \mathbb{N}^{[k]}$ or

$$\mathcal{S} = \{s \subseteq \mathbb{N} : |s| = \min(s) + 1\}$$

are already barriers. The family $\mathcal{S}$ is the famous *Schreier barrier*, which plays an important role in the Banach space geometry. It is in some sense the minimal barrier of *infinite rank*. So let us give some information about the *rank* of a barrier that will be useful in some places later in the book. The notion of rank is facilitated by the following immediate property of barriers.

Lemma 1.23 *Suppose $\mathcal{F}$ is a barrier on $\mathbb{N}$. Letting $\overline{\mathcal{F}}$ be the topological closure of $\mathcal{F}$ inside the Cantor set $2^{\mathbb{N}}$,[9] we have the following equalities:*

$$\overline{\mathcal{F}} = \{t : (\exists s \in \mathcal{F}) \ t \subseteq s\} = \{t : (\exists s \in \mathcal{F}) \ t \sqsubseteq s\}.$$

---

[9]We are using here the standard identification of sets and characteristic functions.

Lemma 1.23 gives us two equivalent ways to define the rank of a barrier. For example, it gives us that the topological closure $\overline{\mathcal{F}}$ is a countable compactum, and therefore, $K = \overline{\mathcal{F}}$ is a compact scattered space, and so it makes sense to talk about its Cantor-Bendixson index, the minimal ordinal α with the property that the $(\alpha + 1)$'st Cantor-Bendixson derivative $K^{(\alpha+1)} = \emptyset$, or equivalently, the uniquely determined ordinal α with the property $K^{(\alpha)} = \{\emptyset\}$.[10] The same index can be obtained in a purely combinatorial way by observing that

$$T(\mathcal{F}) = \{s : (\exists t \in \mathcal{F}) \, s \sqsubseteq t\}$$

considered as a tree ordered by the relation $\sqsubseteq$ of end-extension has no infinite branches. This allows us to define recursively a strictly decreasing map $\rho = \rho_{T(\mathcal{F})}$ from $T(\mathcal{F})$ into the ordinals by the following rule:

$$\rho(s) = \sup\{\rho(t) + 1 : t \in T(\mathcal{F}) \text{ and } t \sqsupset s\}. \tag{1.1}$$

It is easily seen that $\alpha = \rho_{T(\mathcal{F})}(\emptyset)$ is the maximal ordinal with the property $K^{(\alpha)} \neq \emptyset$ for $K = \overline{\mathcal{F}}$, so we can make the following two equivalent definitions of rank of a barrier.

Definition 1.24 *If a family $\mathcal{F}$ of finite subsets of $\mathbb{N}$ is a barrier on some infinite set $M \subseteq \mathbb{N}$, then its* rank *on M, denoted by $\mathrm{rk}_M(\mathcal{F})$ is defined to be the Cantor-Bendixson index of the countable compactum $\overline{\mathcal{F}}|M$. Equivalently, $\mathrm{rk}_M(\mathcal{F}) = \rho_{T(\mathcal{F}|M)}(\emptyset)$, where $\rho_{T(\mathcal{F}|M)}$ is the rank function on the well-founded tree $T(\mathcal{F}|M)$.*

We shall suppress the index M in the notation for the rank when this is clear from the context, for example, when we are working with a barrier $\mathcal{F}$ on $\mathbb{N}$. To express better our next information about the rank of some barrier $\mathcal{F}$, we need a notation for a section of a given barrier $\mathcal{F}$ over an integer n,

$$\mathcal{F}_{\{n\}} = \{s : \min(s) > n \text{ and } \{n\} \cup s \in \mathcal{F}\}.$$

Note that if $\mathcal{F}$ is a barrier on some infinite subset M of $\mathbb{N}$, then for each integer $n \in M$, if we let M/n denote the tail set $\{m \in M : m > n\}$ of M, the restriction $\mathcal{F}_{\{n\}}|(M/n)$ is a barrier on M/n. Note also that for $n \in M$ the tree $T(\mathcal{F}_{\{n\}}|(M/n))$ is naturally isomorphic to the cone subtree

$$\{s \in T(\mathcal{F}|M) : n = \min(s)\} = \{s \in T(\mathcal{F}|M) : \{n\} \sqsubseteq s\}$$

of the tree $T(\mathcal{F}|M)$, so we have that

$$\rho_{T(\mathcal{F}|M)}(\{n\}) = \rho_{T(\mathcal{F}_{\{n\}}|(M/n))}(\emptyset) \text{ for all } n \in M. \tag{1.2}$$

On the other hand, from the recursive definition given in Equation (1.1), we infer that

$$\rho_{T(\mathcal{F}|M)}(\emptyset) = \sup\{\rho_{T(\mathcal{F}|M)}(\{n\}) + 1 : n \in M\}. \tag{1.3}$$

This establishes the following information about the rank of a given barrier $\mathcal{F}$, which will be quite useful in some later chapters of this book.

---

[10]Recall the definition of the standard Cantor-Bendixson derivation: $K^{(0)} = K$, $K^{(\alpha+1)} = K^{(\alpha)} \setminus \{x \in K^{(\alpha)} : x \text{ is isolated in } K^{(\alpha)}\}$, and $K^{(\lambda)} = \bigcap_{\alpha < \lambda} K^{(\alpha)}$ for a limit ordinal $\lambda > 0$.

Lemma 1.25 *Suppose that a family $\mathcal{F}$ of finite subsets of $\mathbb{N}$ is a barrier on some infinite subset M of $\mathbb{N}$. Then*

$$\mathrm{rk}_M(\mathcal{F}) = \sup\{\mathrm{rk}_{M/n}(\mathcal{F}_{\{n\}}) + 1 : n \in M\}.$$

It follows that $\mathrm{rk}(\mathbb{N}^{[k]}) = k$ for every positive integer k and that $\mathrm{rk}(\mathcal{S}) = \omega$ for the Schreier barrier

$$\mathcal{S} = \{s \subseteq \mathbb{N} : |s| = \min(s) + 1\}.$$

Lemma 1.25 allows us to prove results about barriers by induction on their ranks. For example, we suggest the reader examine the corresponding natural inductive proof of the Nash-Williams Theorem and compare it with the standard proof of the Ramsey Theorem which uses induction on the dimension k.

We finish this section with an important result that relates an *arbitrary* family of finite subsets to Nash-Williams notions of blocks and barriers.

Theorem 1.26 (Galvin) *For every family $\mathcal{F}$ of finite subsets of $\mathbb{N}$, there is an infinite subset M of $\mathbb{N}$ such that either $\mathcal{F}|M = \emptyset$, or else every infinite subset of M has an initial segment in $\mathcal{F}$, or in other words, the restriction $\mathcal{F}|M$ contains a barrier on M.*

The proof again uses combinatorial forcing, this time even more relevant to the subject matter of this book. We adopt the previous convention about variables and define the following notion of combinatorial forcing relative to some fixed family $\mathcal{F}$ of finite subsets of $\mathbb{N}$.

Definition 1.27 *We say that M accepts s if every infinite subset P of $s \cup M$ that has s as an initial segment also has an initial segment that belongs to $\mathcal{F}$. It there is no infinite subset of M that accepts s, then we say that M rejects s. We say that M decides s if M either accepts or rejects s.*

Note that this notion of "accepts" corresponds to "strongly accepts" in the earlier version of the combinatorial forcing. We again have the following two immediate properties.

Lemma 1.28   (1) *For every M and s, there is $N \subseteq M$ which decides s.*

(2) *If M accepts(rejects) s, then every $N \subseteq M$ accepts(rejects) s.*

Corollary 1.29 *Every infinite subset N of $\mathbb{N}$ can be refined to an infinite subset M that decides all of its finite subsets.*

Proof. We construct recursively an infinite sequence (M_k) of infinite subsets of N as follows. By Lemma 1.28 (1) find an infinite set $M_0 \subseteq N$ that decides $\emptyset$. Suppose we have constructed a decreasing sequence $M_0 \supseteq M_1 \supseteq \cdots \supseteq M_k$ of infinite sets such that $m_0 = \min(M_0) < m_1 = \min(M_1) < \cdots < m_k = \min(M_k)$. Apply Lemma 1.28 and get an infinite subset M_{k+1} of M_k such that $m_{k+1} = \min(M_{k+1}) > m_k$ and such that M_{k+1} decides all

$s \subseteq \{m_0, m_1, \ldots, m_k\}$ such that $\max(s) = m_k$. This gives us the inductive step. Let $M = \{m_k : k \in \mathbb{N}\}$. Then M decides all of its finite subsets. $\square$

Note also the following immediate property of acceptance.

Lemma 1.30 M accepts s if and only if M accepts $s \cup \{n\}$ for all $n \in M$ such that $n > \max(s)$.

Fix an infinite subset M of $\mathbb{N}$ that decides all of its finite subsets.

Lemma 1.31 Suppose that M rejects one of its finite subsets s. Then M rejects $s \cup \{n\}$ for all but finitely many $n \in M$, $n > \max(s)$.

Proof. Otherwise, the set

$$N = \{n \in M : n > \max(s) \text{ and } M \text{ accepts } s \cup \{n\}\}$$

is an infinite subset of M that by Lemma 1.30 accepts s, contradicting the assumption that M rejects s and the fact that rejection is monotone. $\square$

We are now ready to finish the proof of Theorem 1.26. Given a family $\mathcal{F}$ of finite subsets of $\mathbb{N}$, we consider the corresponding notion of combinatorial forcing. By Corollary 1.29 we fix an infinite set M that decides all of its finite subsets. If M accepts $\emptyset$, the second alternative of Theorem 1.26 is true. So suppose that M rejects $\emptyset$. By Lemma 1.31 we can fix $n_0 \in M$ such that M rejects $\{n\}$ for all $n \geq n_0$. Having defined an increasing sequence $n_0 < \ldots < n_k$ of elements of M such that M rejects all $s \subseteq \{n_0, \ldots, n_k\}$, by Lemma 1.31, we can find $n_{k+1} > n_k$ in M such that M rejects $s \cup \{n\}$ for all $s \subseteq \{n_0, \ldots, n_k\}$ and $n \geq n_{k+1}$. Finally, let $N = \{n_0, n_1, \ldots, n_k, \ldots\}$. Then N is an infinite subset of $\mathbb{N}$ such that $\mathcal{F}|N = \emptyset$. This finishes the proof.

The following reformulation of Theorem 1.26 is worth pointing out.

Corollary 1.32 Let $\mathcal{F}$ be an arbitrary family of finite subsets of $\mathbb{N}$ and let $\mathcal{F}_0$ be the collection of all $\subseteq$-minimal members of $\mathcal{F}$. Then either there is an infinite subset M of $\mathbb{N}$ such that $\mathcal{F}|M = \emptyset$, or else there is an infinite subset M of $\mathbb{N}$ such that the restriction $\mathcal{F}_0|M$ is a barrier on M.

We finish this section with some applications of this result. For this we need the following definition.

Definition 1.33 A family $\mathcal{F}$ of finite subsets of $\mathbb{N}$ is

(1) dense *if every infinite subset of $\mathbb{N}$ contains an element of $\mathcal{F}$,*

(2) hereditary *if $s \subseteq t$ and $t \in \mathcal{F}$ imply $s \in \mathcal{F}$,*

(3) relatively compact *if the topological closure $\overline{\mathcal{F}}$ of $\mathcal{F}$ viewed as a subset[11] of $2^{\mathbb{N}}$ contains only finite sets,*

(4) extensible *if $\{n : s \cup \{n\} \notin \mathcal{F}\}$ is finite for all $s \in \mathcal{F}$.*

---

[11] We are identifying here subsets of $\mathbb{N}$ with their characteristics functions.

Recall that Lemma 1.23 above says that if $\mathcal{F}$ is a family of finite subsets of some infinite set $M \subseteq \mathbb{N}$, which, moreover, is a barrier on M, then the topological closure of $\mathcal{F}$ is equal to both versions of its downward closures, or more precisely,

$$\overline{\mathcal{F}} = \{s \subseteq M : (\exists t \in \mathcal{F})\, s \subseteq t\} = \{s \subseteq M : (\exists t \in \mathcal{F})\, s \sqsubseteq t\}. \qquad (1.4)$$

Corollary 1.34 *Suppose $\mathcal{F}$ and $\mathcal{G}$ are two barriers on the same infinite set $N \subseteq \mathbb{N}$. Then there is an infinite $M \subseteq N$ such that $\overline{\mathcal{F}|M} \subseteq \overline{\mathcal{G}|M}$, or, vice versa, $\overline{\mathcal{G}|M} \subseteq \overline{\mathcal{F}|M}$.*

Proof. Color a given element s from $\mathcal{F}$ in two colors according to whether or not it has an initial segment in $\mathcal{G}$. Similarly, color a given element t from $\mathcal{G}$ in two colors according to whether or not it has an initial segment in $\mathcal{F}$. By Theorem 1.14(Nash-Williams), find infinite $M \subseteq N$ such that the restrictions $\mathcal{F}|M$ and $\mathcal{G}|M$ are both monochromatic. Since $\mathcal{F}$ and $\mathcal{G}$ are barriers on M, the set M itself has an initial segment $s \in \mathcal{F}$ and another initial segment $t \in \mathcal{G}$. Then either $s \sqsubseteq t$ or else $t \sqsubseteq s$. By symmetry, we may assume that $s \sqsubseteq t$. It then follows from the general Equation (1.4) for the closures of barriers that $\overline{\mathcal{F}|M} \subseteq \overline{\mathcal{G}|M}$. $\qquad\square$

Note that from Equation (1.4) it follows, in particular, that the closure of a barrier is a *compact hereditary* family of finite subsets of $\mathbb{N}$. The following result is some sort of converse of this.

Lemma 1.35 *For every relatively compact family $\mathcal{S}$ of finite subsets of $\mathbb{N}$ there is an infinite set M and a barrier $\mathcal{F}$ on M such that the trace of $\mathcal{S}$ on M is equal to the closure of $\mathcal{F}$, or more precisely, $\{s \cap M : s \in \mathcal{S}\} = \overline{\mathcal{F}}$.*

Proof. As before, we will recursively construct an infinite decreasing sequence

$$M_0 \supseteq M_1 \supseteq \cdots \supseteq M_k \supseteq \cdots \qquad (1.5)$$

of infinite subsets of $\mathbb{N}$ such that the corresponding sequence of integers $m_i = \min(M_i)$ is strictly increasing and gives us eventually the desired set M and barrier $\mathcal{F}$ on M. We start by choosing an infinite set $M_0 \subseteq \mathbb{N}$ that has one of the following two properties,

$$(\forall s \in \mathcal{S})\, s \cap M_0 = \emptyset \text{ or } (\forall n \in M_0)(\exists s \in \mathcal{S})\, s \cap M_0 = \{n\}. \qquad (1.6)$$

If M_k is defined, we choose infinite $M_{k+1} \subseteq M_k$ such that $\min(M_{k+1}) > m_k$ and such that for all $t \subseteq \{m_0, \ldots, m_k\}$ with the property $\max(t) = m_k$, we have that either

$$(\forall s \in \mathcal{S})\, s \sqsupseteq t \rightarrow s \cap M_{k+1} = \emptyset \qquad (1.7)$$

or else

$$(\forall n \in M_{k+1})(\exists s \in \mathcal{S})\, s \sqsupseteq t \text{ and } s \cap M_{k+1} = \{n\}. \qquad (1.8)$$

Let $M = \{m_k : k \in \mathbb{N}\}$ and let $\mathcal{F}$ be the collection of all $\sqsubseteq$-maximal elements of the trace $\{s \cap M : s \in \mathcal{S}\}$. Then it follows easily from our construction that $\mathcal{F}$ is a front on the set M, i.e., that every infinite subset of M contains an initial segment in $\mathcal{F}$. Let $\mathcal{F}_0$ be the collection of all $\subseteq$-minimal elements of $\mathcal{F}$. Applying Theorem 1.14 (Nash-Williams) to the coloring

$$\mathcal{F} = \mathcal{F}_0 \cup (\mathcal{F} \setminus \mathcal{F}_0),$$

we find an infinite subset N of M such that $\mathcal{F}|N = \mathcal{F}_0|N$. It follows that $\mathcal{F}_0|N$ is a barrier on N and that the trace $\{s \cap N : s \in \mathcal{S}\}$ is equal to the closure $\overline{\mathcal{F}_0|N}$. This finishes the proof. $\qquad\square$

One may think of the following corollary as a simultaneous version of Galvin's lemma (Theorem 1.26), which admittedly works only in the realm of extensible families of finite subsets of $\mathbb{N}$.

Corollary 1.36 *For every finite sequence $\mathcal{S}_0, \mathcal{S}_1, ..., \mathcal{S}_k$ of dense extensible families of finite subsets of $\mathbb{N}$, there is an infinite set M such that the intersection $\bigcap_{i=0}^{k}(\mathcal{S}_i|M)$ contains a barrier on M.*

Proof. Applying Galvin's Lemma (Theorem 1.26), we first find an infinite set $N \subseteq \mathbb{N}$ such that for every $i \leq k$, the restriction $\mathcal{S}_i|N$ contains a barrier $\mathcal{F}_i$ on N. Applying Corollary 1.34 and reindexing if necessary, we may assume that we have an infinite set $M \subseteq N$ such that

$$\overline{\mathcal{F}_0|M} \subseteq \overline{\mathcal{F}_1|M} \subseteq \cdots \subseteq \overline{\mathcal{F}_k|M}.$$

Using the assumption that $\mathcal{S}_i$ is a sequence of extensible families and using a sufficiently thin subset of M, we may further assume that for every $i \leq k$, every $s \in \mathcal{S}_i|M$, and every $n \in M$ such that $n > \max(s)$, we have that $s \cup \{n\} \in \mathcal{S}_i$. It follows then easily that the intersection $\bigcap_{i=0}^{k}(\mathcal{S}_i|M)$ contains the barrier $\mathcal{F}_k|M$, as required. $\qquad\square$

Remark 1.37 Note that Corollary 1.36 has Theorem 1.14(Nash-Williams) as an immediate consequence. To see this, let $\mathcal{F} = \mathcal{F}_0 \cup \mathcal{F}_1$ be a given coloring of some Nash-Williams family $\mathcal{F}$. For $i = 0, 1$, let $\mathcal{S}_i$ be the collection of all finite subsets of $\mathbb{N}$ that have initial segment in $\mathcal{F}_i$. If the conclusion of Theorem 1.14 (Nash-Williams) fails, $\mathcal{S}_0$ and $\mathcal{S}_1$ would be two dense extensible families. Note however that these two families do not intersect, so this would contradict Corollary 1.36.

In the rest of this section we present results surrounding an extension of Theorem 1.8 of Erdős-Rado. It is a Ramsey classification result for equivalence relations defined on barriers. First of all, note that an arbitrary equivalence relation defined on some barrier $\mathcal{F}$ has the form E_φ for some mapping $\varphi : \mathcal{F} \to \mathbb{N}^{[<\infty]}$, where

$$sE_\varphi t \text{ if and only if } \varphi(s) = \varphi(t).$$

When $E = E_\varphi$, we shall say that E is *represented* by φ. As we shall see, the following definition isolates the class of mappings that represent canonical equivalence relations defined on barriers.

Definition 1.38 *For a mapping of* $\varphi : \mathcal{F} \to \mathbb{N}^{[<\infty]}$ *defined on some barrier* $\mathcal{F}$, *we say that*

(a) φ *is* inner *if* $\varphi(s) \subseteq s$ *for all* $s, t \in \mathcal{F}$,

(b) φ *is* Nash-Williams *if* $\varphi(s) \not\sqsubseteq \varphi(t)$ *whenever* $\varphi(s) \neq \varphi(t)$,

(c) φ *is* Sperner *if* $\varphi(s) \not\subseteq \varphi(t)$ *whenever* $\varphi(s) \neq \varphi(t)$,

(d) φ *is* irreducible *if* φ *is inner and Nash-Williams.*

Note that by applying Theorem 1.14(Nash-Williams), for every irreducible map $\varphi : \mathcal{F} \to \mathbb{N}^{[<\infty]}$, we can find an infinite set $M \subseteq \mathbb{N}$ such that the restriction $\varphi \upharpoonright \mathcal{F}|M$ is Sperner. So we could have defined irreducible maps as inner Sperner rather than Nash-Williams maps. However, in the infinite-dimensional case, the inner Sperner maps could not serve our purpose here. Note that by Galvin's lemma (modulo restricting to an infinite set) the range of an irreducible mapping is a barrier as well. Thus, irreducible mappings are a special kind of mappings from one barrier to another. The following fact, whose proof is given after Corollary 1.42 below, explains the reason behind the name.

Lemma 1.39 *Suppose* φ_0 *and* φ_1 *are two irreducible mappings defined on the same barrier* $\mathcal{F}$ *that represent the same equivalence relation on* $\mathcal{F}$. *Then there is an infinite set* $M \subseteq \mathbb{N}$ *such that* $\varphi_0 \upharpoonright (\mathcal{F}|M) = \varphi_1 \upharpoonright (\mathcal{F}|M)$.

Theorem 1.40 (Pudlak-Rödl) *For every barrier* $\mathcal{F}$ *on* $\mathbb{N}$ *and every equivalence relation* E *on* $\mathcal{F}$, *there is an infinite* $M \subseteq \mathbb{N}$ *such that the restriction of* E *to* $\mathcal{F}|M$ *is represented by an irreducible mapping* φ *defined on* $\mathcal{F}|M$.

This result suggests that solving a mathematical problem that can be reformulated as a problem involving barriers on $\mathbb{N}^{12}$ will likely involve an analysis of mappings from one barrier to another. The following is one example of a useful result of this sort.

Lemma 1.41 *Let* $\varphi : \mathcal{F} \to \mathbb{N}^{[<\infty]}$ *be a mapping whose domain is a barrier* $\mathcal{F}$ *on* $\mathbb{N}$ *and whose range is a precompact family of finite subsets of* $\mathbb{N}$. *Suppose that* $\varphi(s) \cap s = \emptyset$ *for all* $s \in \mathcal{F}$. *Then there is infinite* $M \subseteq \mathbb{N}$ *such that* $\varphi(s) \cap M = \emptyset$ *for all* $s \in \mathcal{F}|M$.

Proof. The proof is by induction on the rank of the barrier $\mathcal{F}$. First of all, note that the conclusion is true for the barrier $\mathcal{F} = \mathbb{N}^{[1]}$ of rank 1. To see this, using that the range of φ is precompact we first go to an infinite subset N of $\mathbb{N}$ such that the sequence $(\varphi(\{n\}))_{n \in N}$ converges to some finite set s. Find now an infinite set $M \subseteq N$ whose minimum is above s such that for all $m < n$ in M,

$$\varphi(\{n\}) \cap \{0, 1, \ldots, m\} = s \text{ and } n > \max(\varphi(\{m\})).$$

---

[12]Indeed, there a growing number of such problems especially in the infinite-dimensional geometry of normed spaces.

Then $\varphi(\{n\}) \cap M = \emptyset$ for all $n \in M$. Suppose now that $\mathcal{F}$ is a barrier of some rank > 1 and that the conclusion is true for its sections $\mathcal{F}_{\{n\}}$ $(n \in \mathbb{N})$. Then using the inductive hypothesis and the Ramsey property of barriers $\mathcal{F}_{\{n\}}$, we can build a decreasing sequence $M_0 \supseteq M_1 \supseteq \cdots \supseteq M_k \supseteq \cdots$ of infinite subsets of $\mathbb{N}$ such that the corresponding sequence $m_i = \min(M_i)$ of minimums is strictly increasing and such that for all i we can find a single set $t_i \subseteq \{0, 1, \ldots, m_i - 1\}$ such that

$$\varphi(s) \cap (\{0, 1, \ldots, m_i - 1\} \cup M_i) = t_i \text{ for all } s \in \mathcal{F}|M_i \text{ with } \min(s) = m_i.$$

Let $M_\infty = \{m_i : i \in \mathbb{N}\}$. Applying the rank 1 case of the lemma to the mapping $\{m_i\} \mapsto t_i$, we can find an infinite set $M \subseteq M_\infty$ such that $M \cap t_i = \emptyset$ for all i such that $m_i \in M$. Then $\varphi(s) \cap M = \emptyset$ for all $s \in \mathcal{F}|M$. This finishes the proof. □

Lemma 1.41 is saying in particular that maps $\psi : \mathcal{F} \to \mathbb{N}^{[<\infty]}$ whose domains are barriers and ranges precompact families of finite sets are essentially inner, or in other words, we can always restrict to an infinite set $M \subseteq \mathbb{N}$ such that $\psi(s) \cap M \subseteq s$ for all $s \in \mathcal{F}|M$. (To see this, apply Lemma 1.41 to the map $\varphi(s) = \psi(s) \setminus s$.)

Corollary 1.42 *For every irreducible mapping φ defined on some barrier $\mathcal{F}$ there is an infinite set $M \subseteq \mathbb{N}$ such that for all $s \in \mathcal{F}|M$ there is $t \in \mathcal{F}$ such that*

$$\varphi(s) = \varphi(t) = s \cap t.$$

Proof. Let $\mathcal{G}$ be the range of φ. We have already observed that $\mathcal{G}$ is a barrier when restricted to some infinite subset of $\mathbb{N}$. Choose a mapping $\psi : \mathcal{G} \to \mathcal{F}$ such that $\varphi(\psi(v)) = v$ for all $v \in \mathcal{G}$. Apply Lemma 1.41 to the mapping $v \mapsto \psi(v) \setminus v$ to obtain an infinite set $M \subseteq \mathbb{N}$ such that

$$(\psi(v) \setminus v) \cap M = \emptyset \text{ for all } v \in \mathcal{G}|M.$$

To verify that M satisfies the conclusion, consider an $s \in \mathcal{F}|M$ and let $v = \varphi(s)$. Let $t = \psi(v)$. Then $v = \varphi(t)$ and $v = t \cap M$. Since $v \subseteq s \subseteq M$, we get the remaining conclusion, $s \cap t = v$. □

Proof of Lemma 1.39. Suppose that the conclusion of the lemma is false. Applying the Ramsey property of $\mathcal{F}$ we can find an infinite set $N \subseteq \mathbb{N}$ such that

$$\varphi_0(s) \neq \varphi_1(s) \text{ for all } s \in \mathcal{F}|N. \tag{1.9}$$

Applying Corollary 1.34 (and 1.4) and shrinking further, we may assume that, say,

$$(\forall s \in \mathcal{G}_0)(\exists t \in \mathcal{G}_1) \ s \subseteq t. \tag{1.10}$$

By Corollary 1.42 we can find infinite $M \subseteq N$ and a pair of finite sets $s \in \mathcal{F}|M$ and $t \in \mathcal{F}$ such that $\varphi_0(s) = \varphi_0(t) = s \cap t$. Since φ_0 and φ_1 represent the same equivalence relation (and since φ_1 is inner), we have that

$\varphi_1(s) = \varphi_1(t) \subseteq s \cap t$. Using Equation (1.10) and the fact that $\mathcal{G}_1$ is a Sperner family, we conclude that $\varphi_1(s) = \varphi_1(t) = s \cap t = \varphi_0(s) = \varphi_0(t)$, contradicting Equation (1.9). This finishes the proof. $\qquad\square$

The following result extends Corollary 1.42 to the realm of arbitrary inner maps.

Theorem 1.43 *For every inner mapping φ defined on some barrier $\mathcal{F}$ there is an infinite set $M \subseteq \mathbb{N}$ such that for all $s \in \mathcal{F}|M$ there exist $t, u \in \mathcal{F}$ such that*

$$s \cap t = \varphi(t) \sqsubseteq \varphi(s) \text{ and } s \cap u = \varphi(s) \sqsubseteq \varphi(u).$$

1.4 RAMSEY PROPERTY AND BAIRE PROPERTY

Let us now discuss the infinite-dimensional extensions of Theorem 1.3. There are some new phenomena that show up in the dimension $k = \infty$. In particular, one quickly learns that there is a need to add some restrictions on the colorings of the infinite-dimensional space $\mathbb{N}^{[\infty]}$ as the following example shows.

Example 1.4.1 *Given a nonprincipal ultrafilter $\mathcal{U}$ on $\mathbb{N}$ define $c : \mathbb{N}^{[\infty]} \to \{-1, 1\}$ as follows:*

$$c(M) = \lim_{n \to \mathcal{U}} \ (-1)^{|M \cap \{0, \dots, n-1\}|}.$$

Then

$$c(X) \neq c(X \setminus \{\min(X)\}) \text{ for every } X \in \mathbb{N}^{[\infty]},$$

so, in particular, the coloring c is not constant on any set of the form $M^{[\infty]}$ for an infinite set $M \subseteq \mathbb{N}$.

An analysis of this example shows that c is neither Lebesgue nor Baire-measurable relative to the standard measure and topology of the Cantor space $2^{\mathbb{N}}$. So it is natural to ask whether imposing Lebesgue or Baire measurability on a given coloring of $\mathbb{N}^{[\infty]}$ would guarantee us the existence of a monochromatic set of the form $M^{[\infty]}$ for $M \subseteq \mathbb{N}$ infinite. Unfortunately, even this is not possible, as the following example shows.

Example 1.4.2 *Given the coloring c of Example 1.4.1, define $b : \mathbb{N}^{[\infty]} \to \{-1, 1\}$ as follows,*

$$b(M) = \min\{c(M), \ \min_{m,n \in M} (-1)^{|m-n|}\}.$$

Note that the set $\mathcal{O} = \{M \in \mathbb{N}^{[\infty]} : \min_{m,n \in M} (-1)^{|m-n|} = -1\}$ is a dense open subset of $\mathbb{N}^{[\infty]}$ of full measure,[13] so the coloring b is both Lebesgue- and Baire-measurable. On the other hand, note that for every infinite set $N \subseteq \mathbb{N}$, there is an infinite set $M \subseteq N$ such that $M^{[\infty]} \cap \mathcal{O} = \emptyset$, and

---

[13] We look here at $\mathbb{N}^{[\infty]}$ as a subset of the Cantor space $2^{\mathbb{N}}$ equipped with the standard topology and measure.

therefore, $b \upharpoonright M^{[\infty]} = c \upharpoonright M^{[\infty]}$. *Therefore, as in Example 1.4.1, we conclude that there is no infinite* $M \subseteq \mathbb{N}$ *such that b is constant on* $M^{[\infty]}$.

So in order to capture the Ramsey property one has to look for some other restriction. It turns out that the right restriction is the Baire measurability, not relative to the usual metric topology of $\mathbb{N}^{[\infty]}$ but relative to the *exponential topology* or *Ellentuck topology* on this set. The exponential topology, also known under the name of *Vietoris topology*, in this particular case has its basic open sets of the form[14]

$$[s, M] = \{N \in \mathbb{N}^{[\infty]} : s \sqsubseteq N \text{ and } N/s \subseteq M\},$$

where s is an arbitrary finite and M an arbitrary infinite subset of $\mathbb{N}$. Let us recall also the basic Baire notions.

Definition 1.44 *A subset S of a topological space X is* nowhere dense *if every nonempty open subset includes a nonempty open subset that avoids S. A subset T of X is* meager *if it can be covered by countably many sets that are nowhere dense in X.*

Definition 1.45 *A subset Z of a topological space X has the* Baire property *in X if it is equal to an open subset of X modulo the ideal of meager sets, or in other words, can be written as the symmetric difference of an open set and a meager set. A mapping $f : X \to Y$ is said to be* Baire measurable *if $f^{-1}(U)$ has the property of Baire for every open subset U of the topological space Y.*

Having these notions at hand, we can state the infinite-dimensional Ramsey theorem, which forms a basis of a whole branch of Ramsey theory.

Theorem 1.46 (Ellentuck) *Every finite coloring of $\mathbb{N}^{[\infty]}$ that is Baire-measurable relative to the exponential topology of $\mathbb{N}^{[\infty]}$ is constant on a set of the form $M^{[\infty]}$ for some infinite $M \subseteq \mathbb{N}$.*

In fact, the restriction on colorings given by this theorem is optimal in the sense that if one introduces the natural notion of a *Ramsey property* for subsets of $\mathbb{N}^{[\infty]}$, one immediately realizes that it coincides with the Baire property relative to the exponential topology. Since the proof of this theorem also inspires many of the other proofs in this theory, we give it here with some details.

The basic ingredient of the proof of Theorem 1.46 is contained in the following notion of combinatorial forcing which is given relative to some *fixed* subset $\mathcal{X}$ of $\mathbb{N}^{[\infty]}$ for which we would like to construct a Baire-measurable envelope of $\mathcal{X}$, much in the spirit of the classical theory of measure and category.

Definition 1.47 *Fixing the set $\mathcal{X} \subseteq \mathbb{N}^{[\infty]}$, we say that an $M \in \mathbb{N}^{[\infty]}$* accepts *an $s \in \mathbb{N}^{[<\infty]}$ if $[s, M] \subseteq \mathcal{X}$. We say that M* rejects *s if there is no infinite*

---

[14]Notation: $N/s = \{n \in N : n > \max(s)\}$.

$N \subseteq M$ that accepts s. We say that M decides s if M either accepts or rejects s.

Note the following immediate properties of these notions.

Lemma 1.48 (a) If M accepts (rejects) s, then every infinite subset N of M accepts (rejects) s.

(b) For every s and M, there is infinite $N \subseteq M$ such that N decides s. $\square$

Lemma 1.49 Every infinite subset M of $\mathbb{N}$ can be refined to an infinite subset N such that every finite $s \subseteq N$ is decided by N.

Proof. Choose infinite $M_0 \subseteq M$ that decides $\emptyset$ and let $n_0 = \min M_0$. Suppose we have chosen $M_0 \supseteq \ldots \supseteq M_k$ with $n_0 = \min M_0 < \ldots < n_k = \min M_k$. Choose $M_{k+1} \subseteq M_k$ that decides all $s \subseteq \{n_0, \ldots, n_k\}$ such that $n_{k+1} = \min M_{k+1} > n_k$. Finally, let $N = \{n_0, \ldots, n_k, \ldots\}$. Then N is as required. $\square$

Lemma 1.50 Suppose M decides all of its finite subsets. If M rejects its finite subset s, then M rejects also $s \cup \{n\}$ for all but finitely many $n \in M$.

Proof. Suppose the set N of all $n \in M$ for which M does not reject, and therefore accepts, $s \cup \{n\}$ is infinite. Then $[s, N] \subseteq \mathcal{X}$, and so N accepts s, contradicting the assumption that M rejects s. $\square$

Lemma 1.51 Suppose M decides all of its finite subsets. If M rejects $\emptyset$, then there is an infinite $N \subseteq M$ such that N rejects all of its finite subsets.

Proof. Let $M_0 = M$ and $n_0 = \min M_0$. Suppose that for some integer $k > 0$ we have chosen $M_0 \supseteq \ldots \supseteq M_{k-1}$ with $n_0 = \min M_0 < \ldots < n_{k-1} = \min M_{k-1}$ such that M_{k-1} rejects every $s \subseteq \{n_0, \ldots, n_{k-2}\}$. By Lemma 1.50, there is infinite $M_k \subseteq M_{k-1}$ such that $n_k = \min M_k > n_{k-1}$ and such that M rejects $s \cup \{n\}$ for all $s \subseteq \{n_0, \ldots, n_{k-1}\}$ and $n \in M_k$. Let $N = \{n_0, \ldots, n_k, \ldots\}$. Then N is as required. $\square$

This finishes the series of lemmas about combinatorial forcing relative to a fixed set $\mathcal{X} \subseteq \mathbb{N}^{[\infty]}$.

Lemma 1.52 Let $\mathcal{O}$ be an exponentially open subset of $\mathbb{N}^{[\infty]}$. Then for every basic open set $[s, M]$, there is $N \in [s, M]$ such that $[s, N]$ is either included or is disjoint from $\mathcal{O}$.

Proof. We shall use the already established facts about the combinatorial forcing applied to the set $\mathcal{X} = \mathcal{O}$, and we shall use the forcing lemmas relativized to the basic set $[s, M]$ in place of $[\emptyset, \mathbb{N}] = \mathbb{N}^{[\infty]}$. Choose $N \in [s, M]$ that $\mathcal{O}$-decides all sets of the form $s \cup t$, where t is a finite subset of $N/s = \{n \in N : n > s\}$. If N $\mathcal{O}$-accepts s, then we are done so let us assume it rejects it. By Lemma 1.51 we can find $P \in [s, N]$ that $\mathcal{O}$-rejects all finite sets of the form $s \cup t$, where t is a subset of P/s. Since $\mathcal{O}$ is exponentially open, this means that $[s, P] \cap \mathcal{O} = \emptyset$. $\square$

Lemma 1.53 *Let M be a subset that is meager relative to the exponential topology. Then for every basic open set $[s, M]$, there is $N \in [s, M]$ such that $[s, N]$ is disjoint from M.*

Proof. First of all, note that the conclusion of the lemma is true under the stronger assumption that the set M is nowhere dense, since by applying Lemma 1.52 to the closure $\overline{M}$, the alternative $[s, N] \subseteq \overline{M}$ is impossible. Let $M = \bigcup_{k=0}^{\infty} M_k$ be a decomposition of M into nowhere dense sets. Let $[s, M]$ be a given basic open set. Relativizing the argument, we may assume that in fact $s = \emptyset$. Using the fact that the conclusion of the lemma is true for the nowhere dense sets M_k, we build a decreasing sequence $M \supseteq M_0 \supseteq \ldots \supseteq M_k \supseteq \ldots$ such that $n_0 = \min M_0 < \ldots < n_k = \min M_k < \ldots$ and such that $[s, M_k] \cap M_n = \emptyset$ for all $n \leq k$ and $s \subseteq \{n_0, \ldots, n_{k-1}\}$. Let $N = \{n_0, \ldots, n_k, \ldots\}$. Then $[s, N] \cap M = \emptyset$. $\square$

We are now ready to prove Theorem 1.46 in the following equivalent formulation.

Theorem 1.54 (Ellentuck) *Suppose $\mathcal{X}$ is a subset of $\mathbb{N}^{[\infty]}$ that has the Baire property relative to the exponential topology of $\mathbb{N}^{[\infty]}$. Then for every basic open set $[s, M]$, there is $N \in [s, M]$ such that $[s, N]$ is either included in or is disjoint from $\mathcal{X}$.*

Proof. Choose an open set $\mathcal{O}$ and a meager set M such that $\mathcal{X} \triangle \mathcal{O} = M$. By Lemma 1.53, we can choose $N \in [s, M]$ such that $[s, N]$ is disjoint from M. By lemma 1.52, we can choose $P \in [s, N]$ such that $[s, P]$ is either included in, or is disjoint from, $\mathcal{O}$. It follows that $[s, P]$ is either included in, or disjoint from, $\mathcal{X}$. $\square$

Let us say that a subset $\mathcal{X}$ of $\mathbb{N}^{[\infty]}$ has the *Ramsey property* if it satisfies the conclusion of Theorem 1.54. Clearly, every subset of $\mathbb{N}^{[\infty]}$ that has the Ramsey property also has the Baire property, so Theorem 1.54 says that these two properties are in fact equivalent.

Corollary 1.55 (Silver) *Every analytic subset of $\mathbb{N}^{[\infty]}$ has the Ramsey property.*

Proof. This follows from the standard fact that the property of Baire relative to any topological space is closed under the Souslin operation (see Section 4.1 below). $\square$

Corollary 1.56 (Galvin-Prikry) *Every Borel subset of $\mathbb{N}^{[\infty]}$ has the Ramsey property.*

It turns out that there are a number of weaker restrictions that one can impose on colorings of $\mathbb{N}^{[\infty]}$ guaranteeing the conclusion of the infinite-dimensional Ramsey theorem. In particular, the exponential topology is not the only topology on $\mathbb{N}^{[\infty]}$ whose Baire measurability will give us a sufficient restriction for the infinite-dimensional Ramsey theorem, but it is the only topology that gives us the *characterization* of the Ramsey property. The power behind any result of this sort is hidden in the fact that the Baire property relative to any topology on $\mathbb{N}^{[\infty]}$ that is finer than the usual product topology on that set is a considerably weaker restriction than the classical descriptive requirements, such as Borel or Souslin measurability. In fact, a large body of this book is concerned with finding an abstract notion of Baire measurability that would work in many contexts and, in particular, in contexts where no topological approach could used.

Let us now turn to the infinite-dimensional interpretation of the Pudlak-Rödl theorem (Theorem 1.40 above).

Theorem 1.57 (Pudlak-Rödl) *For every Borel equivalence relation E on $\mathbb{N}^{[\infty]}$ with countably many classes, there is infinite $M \subseteq \mathbb{N}$ such that the restriction of E to $M^{[\infty]}$ is represented by an irreducible*[15] *1-Lipschitz map*[16] $\varphi : M^{[\infty]} \to M^{[<\infty]}$.

The Lipschitz condition here is really only to ensure the continuity of the mapping φ. The reader will have no difficulty showing that for every irreducible map φ defined on a barrier $\mathcal{F}$, there is an infinite set $M \subseteq \mathbb{N}$ such that the restriction of φ to $\mathcal{F}|M$ is 1-Lipschitz. So this new condition makes no difference in the old formulation of the Pudlak-Rödl theorem. In the new formulation, however, the Pudlak-Rödl theorem naturally leads to an analogous result that applies to a larger class of Borel equivalence relations. Before we state this result, let us recall that a Borel equivalence relation E on a Polish space X is *smooth* if there is Borel map f from X into some other Polish space Y, such that f indices E, or in other words, for every $x, y \in X$, we have xEy if and only if $f(x) = f(y)$.

Theorem 1.58 (Mathias, Prömel-Voigt) *For every smooth Borel equivalence relation E on $\mathbb{N}^{[\infty]}$, there is infinite $M \subseteq \mathbb{N}$ such that the restriction of E to $M^{[\infty]}$ is induced by an 1-Lipschitz irreducible map $\varphi : M^{[\infty]} \to \mathcal{P}(M)$.*

Note that the second condition of irreducibility, condition (b) of Definition 1.38, plays little or no role in the Mathias-Prömel-Voigt theorem unless E has countably many classes on $M^{[\infty]}$, in which case the result adds nothing more to the Pudlak-Rödl theorem. Namely, assuming that E has uncountably many classes on any symmetric cube $N^{[\infty]}$ over an infinite set $N \subseteq M$ and applying the Galvin-Prikry theorem, we can find infinite $N \subseteq M$ such that the $\varphi(X)$ is infinite for every $X \in N^{[\infty]}$. So, in particular, for no $X, Y \in N^{[\infty]}$, $\varphi(X)$ can be a strict initial segment of $\varphi(Y)$. Thus, $\varphi \upharpoonright N^{[\infty]}$ automatically satisfies the second condition of Definition 1.38. Thus unless Theorem 1.58 reduces to Theorem 1.57, its conclusion is really saying that E on $M^{[\infty]}$ is represented by 1-Lipschitz inner map.

---

[15] In the sense of Definition 1.38, which works equally well for arbitrary maps of the form $\varphi : \mathcal{P}(\mathbb{N}) \to \mathcal{P}(\mathbb{N})$.

[16] That is, if $\min(\varphi(X) \triangle \varphi(Y)) \geq \min(X \triangle Y)$, or equivalently, if for all n, $X \cap n = Y \cap n$ implies $\varphi(X) \cap n = \varphi(Y) \cap n$.

We finish this section with a result that shows that the field of Ramsey sets behaves quite analogously to the classical fields of sets such as, for example, the field of Lebesgue measurable sets of reals or the field of sets of reals with the property of Baire.

Theorem 1.59 (Ramsey Uniformization Theorem) *Suppose X is a Polish space and R is a coanalytic subset of the product $\mathbb{N}^{[\infty]} \times X$ with the property that for all $M \in \mathbb{N}^{[\infty]}$ there is $x \in X$ such that $R(M, x)$ holds. Then there is an infinite subset M of $\mathbb{N}$ and a continuous map $F : M^{[\infty]} \to X$ such that $R(N, F(N))$ holds for all $N \in M^{[\infty]}$.*

The proof of this theorem is closely related to Silver's original proof of his theorem and it is beyond the scope of this book. We mention it here because it exposes a rather important phenomenon true in all other infinite-dimensional Ramsey theoretic contexts encountered here. We shall, however, explicitly mention this phenomenon only when we consider it useful.

NOTES TO CHAPTER ONE

Theorems 1.3 and 1.7 both appear in the original paper [93] of F. P. Ramsey. The idea behind Theorem 1.7 has led to important developments in practically every area of modern mathematical logic (see, for example, [26]). The canonical version of Ramsey's theorem, Theorem 1.8, appears in the paper of Erdös-Rado [30]. Similar results about unrestricted number of colors are valid in many other Ramsey theoretic contexts that we study in the following chapters of this book. Note that to each infinitary Ramsey theoretic result corresponds its finitary form deduced from it using the standard compactness argument. It should be mentioned, however, that finitary Ramsey theoretic results are of independent interest and are studied as a separate subject (see, for example, [41]). It is interesting that sometimes substantial infinitary methods are necessary when proving finitary Ramsey theoretic results. The first such phenomenon was that of Paris and Harrington (see, for example, [55]), but another one is the result of Corollary 1.11, which requires a theory stronger than Peano arithmetic for its proof (see [52]). Regarding the comments in Remark 1.2, we mention monograph [36] as the original reference for Gödel constructibility, but the reader will find its exposition in almost every standard text in axiomatic set theory. The generalization of Ramsey's theorem due to Nash-Williams [82] is the first result of infinite-dimensional Ramsey theory. Its extension due to Galvin appears originally in his announcement [33] and is in some sense more relevant to the infinite-dimensional theory, since it is equivalent to the fact that open subsets of $\mathbb{N}^{[\infty]}$ are Ramsey. The notion of barrier, due to Nash-Williams [82], has seen many applications far beyond its original use in the development of the theory of well-quasi-orderings. The notion makes perfect sense for every other Ramsey space that we develop later in this book, although, we hope, applications of comparable wealth are still to come. Corollary 1.36 appears in the Galvin-Prikry paper [35], where it is deduced from the main result of the paper, which says that the infinite-dimensional Ramsey theorem is true for Borel colorings. The Pudlak-Rödl theorem appears in their paper [92], and the reader can find its proof also in [3]. Theorem 1.43 appears in the paper of Lopez-Abad and Todorcevic [66], where we refer the reader for a wealth of information on uses of barriers in problems about unconditional convergence in Banach spaces. The first counterexample to the unrestricted infinite-dimensional Ramsey theorem is due to R. Rado (see, e.g., [29]) who used a well ordering of the continuum for its construction. In [73] Mathias shows that the existence of a q-point ultrafilter on $\mathbb{N}$ also leads to a counterexample to the unrestricted infinite-dimensional Ramsey theorem. It turns out that an arbitrary nonprincipal ultrafilter on $\mathbb{N}$ is sufficient for this. This was shown by Baumgartner (see [73]) and it is his proof that we reproduce above in Example 1.4.1. Example 1.4.2 is due to Galvin-Prikry [35]. Baire category notions for sets of reals were originally introduced in Baire's thesis [4]. Ellentuck [27] proved his theorem in order to supply a proof of Silver's Theorem [99] stating that analytic sets have the Ramsey property thus starting the whole area of topological Ramsey theory. He seems to have rediscovered the Vietoris topology of $\mathbb{N}^{[\infty]}$ although he does have in his reference list Kuratowski's book [58], whose Section I.17 is devoted to this topology. As can be seen from the proofs we present, the combinatorial forcing encountered above in the proof of Ellentuck's Theorem is due to Galvin and Prikry [35] using some ideas of Nash-Williams [82]. As stated, Theorem 1.58 appears in the paper of Prömel and Voigt [91] although an equivalent formulation (to the effect that the restriction of the equivalence relation is represented by a 1-Lipschitz inner map) appears in the much earlier paper of Mathias [74].

Chapter Two

Semigroup Colorings

2.1 IDEMPOTENTS IN COMPACT SEMIGROUPS

A *compact semigroup* S is a nonempty semigroup with a compact Hausdorff topology for which

$$x \mapsto xs$$

is a continuous map for all $s \in S$. The reader should be warned that this terminology is nonstandard, since usually "compact semigroup" means that the semigroup operation is jointly continuous in both factors. We chose this terminology only to avoid the somewhat awkward "compact semitopological semigroup" that corresponds better to the notion we study here.

Example 2.1.1 *If X is a compact Hausdorff space, then the Tychonov cube X^X is a compact semigroup with the composition operation, since for each g the map $f \mapsto f \circ g$ is continuous on X^X. Note that in general the operation of composition from the right $f \mapsto g \circ f$ is not necessarily continuous on X^X unless $g : X \to X$ is continuous.*

An element x of a (compact) semigroup S is *idempotent* if $x^2 = x$. We shall need the following important fact about this notion

Lemma 2.1 (Ellis) *Every compact semigroup S has an idempotent.*

Proof. Pick by Zorn's Lemma a minimal compact subsemigroup $R \subseteq S$ and an arbitrary $s \in R$. Then Rs is also a compact subsemigroup and $Rs \subseteq R$. Hence $Rs = R$. Let $P = \{x \in R : xs = s\}$. Then $P \neq \emptyset$, since $s \in Rs$. Note that P is also a compact subsemigroup of S. Hence $P = R$ and therefore $s^2 = s$. □

Fix from now on a compact semigroup S. A *left-ideal* of S is a nonempty subset I of S such that $SI \subseteq I$. A *right-ideal* of S is a nonempty subset I of S such that $IS \subseteq I$. A *two-sided ideal* of S is a nonempty subset of S which is both left and right ideal. In this context, left-ideals seem to be richer in properties than right-ideals. For example, note that for every $x \in S$, Sx is a closed left-ideal, so every minimal left-ideal is closed and if a left-ideal is minimal among all closed left-ideals, then it is also minimal among all left-ideals. Clearly every closed ideal (one-sided or two-sided) of S is a compact subsemigroup of S, so by Lemma 2.1 it contains idempotents. Idempotents belonging to minimal left ideals are rather special. To state this property, we need the following important relation on S :

$$x \leq y \text{ iff } xy = yx = x.$$

Note that $\leq$ is transitive and antisymmetric on S. Note also that $x \leq x$ only when $x^2 = x$. Thus $\leq$ is a partial order on the idempotents of S.

Lemma 2.2 *An idempotent belonging to a minimal left ideal is minimal in the ordering $\leq$.*

Proof. Let y be an idempotent belonging to a minimal left-ideal I of S and let $x \leq y$ be a given idempotent. Since $xy = x$ we conclude that $x \in I$ and therefore $Ix = I$ by the minimality of I. Choose $z \in I$ such that $y = zx$. Then $yx = zx^2 = zx = y$. From $x \leq y$, we have that $yx = x$. Hence $x = y$. $\square$

Example 2.1.2 *Note that an element f of the compact semigroup X^X is idempotent if and only if it is a retraction, i.e., if the restriction of f to its range is the identity map. Note also that in this semigroup $f \leq g$ implies that the range of g includes the range of f, so minimal idempotents of X^X are the constant maps.*

Lemma 2.3 *If y is an idempotent and if I is a closed left-ideal, then the left-ideal Iy contains an idempotent x such that $x \leq y$.*

Proof. By Lemma 2.1, we can find an idempotent w in Iy. Choose $v \in I$ such that $w = vy$. Put $x = yw \ (= yvy)$. Then

$$x^2 = yvyyw = yvyw = yww = yw = x,$$

so x is an idempotent. The relation $x \leq y$ follows from

$$yx = yyw = yw = x$$

and

$$xy = yvyy = yvy = x.$$

$\square$

Corollary 2.4 *An idempotent is minimal if and only if it belongs to some minimal left-ideal.*

Proof. By Lemma 2.2, only the direct implication needs a proof. Consider a minimal idempotent y. Choose an arbitrary minimal left-ideal I of S. Then I is closed and Iy is also a minimal left ideal. By Lemma 2.3, the left-ideal Iy contains an idempotent x such that $x \leq y$. Since y is minimal we have that $x = y$ and therefore $y \in Iy$. $\square$

Corollary 2.5 *Any two-sided ideal of S contains all the minimal idempotents of S.*

Proof. Let J be a given two-sided ideal and let y be a minimal idempotent of S. By Corollary 2.4, there is a minimal left-ideal I such that $y \in I$. Note that $JI \subseteq I \cap J$, so $I \cap J$ is a nonempty left-ideal included in I. It follows that $I \cap J = I$ and therefore that $y \in I \subseteq J$. □

Let us reformulate this result in the form that is used in the rest of this chapter.

Corollary 2.6 *If y is an idempotent and J is a two-sided ideal of S then there is an idempotent $x \in J$ such that $x \leq y$.*

The following is yet another interesting property of any minimal idempotent of S.

Lemma 2.7 *If a is a minimal idempotent of S then Sa is a minimal left-ideal of S.*

Proof. By Lemma 2.4, there is a mimimal left ideal I of S such that $a \in I$. Then $Sa \subseteq I$, and since Sa is a left ideal of S, we must have the equality $Sa = I$. So, indeed, Sa is a minimal left ideal of S. □

Lemma 2.8 *If a is a minimal idempotent of S, then a is an identity of aSa and every element of aSa has a right inverse as well as a left inverse relative to a. In other words, aSa is a group with identity a.*

Proof. The fact that a is an identity of aSa is clear. Let $x = asa$ be a given member of aSa. Then $x \in Sa$, so Sx is a (closed) left ideal of Sa. By Lemma 2.7, $Sx = Sa$. It follows that $a \in Sx$, so we can find $t \in S$ such that $a = tx$. Let $y = ata$. Then $y \in aSa$, and

$$yx = atax = ataasa = atasa = atx = aa = a.$$

So, y is a left-inverse of x relative to a. To show that y is also a right-inverse of x in aSa relative to a, let z be left-inverse of y in aSa relative to a, i.e., $zy = a$. Multiplying the equation $yx = a$ by y from the right and using the fact that a is an identity of aSa, we get $yxy = y$. Multiplying this equation by z from the left, we get $zyxy = zy = a$. Since $zy = a$, this gives us $axy = a$. Since a is an identity of aSa this gives us the desired conclusion, $xy = a$. □

We have already pointed out that in this context right ideals of S enjoy fewer properties than left-ideals. However, they do have some useful properties.

Lemma 2.9 *For every minimal idempotent a of S, the set aS is a minimal right ideal of S.*

Proof. Suppose $\emptyset \neq J \subseteq aS$ and that $JS \subseteq J$. Pick $x \in J$. Then $x \in aS$, and so $xa \in aSa$. By Lemma 2.8, we can find $y \in aSa$, which is the right inverse of xa relative to a, i.e., such that $(xa)y = a$. It follows that

$$a = (xa)y = x(ay) \in JS \subseteq J.$$

Hence $a \in J$ and therefore $aS \subseteq JS \subseteq J$, as required. □

Lemma 2.10 *Every right ideal J of S includes a minimal right ideal.*

Proof. Let J be a given right ideal of S. Pick $b \in J$. We have just seen that S has a minimal right ideal, so let us fix one of them, R. We claim that $bR \subseteq J$ is a minimal right ideal of S and this will finish the proof. So let $Q \subseteq bR$ be a given right ideal of S. Let

$$P = \{x \in R \; : \; bx \in Q\}.$$

Then P is a nonempty subset of R. Consider a $p \in P$ and $y \in S$. Then $bp \in Q$, so $bpy \in Q$. Note that since $p \in R$, we have that $py \in RS \subseteq R$. It follows that $py \in P$. This shows that P is a right ideal of S, and therefore $P = R$. It follows that $bR = bP \subseteq Q$, as required. □

Lemma 2.11 *If x is a minimal idempotent and J is a right ideal of S, then there is an idempotent $y \in J$ such that $xy = x$.*

Proof. By Corollary 2.4 we can choose a minimal left-ideal I of S containing x. Moreover, we may assume that J is a minimal right ideal of S. Then IJ is a two-sided ideal of S, so by Corollary 2.5, $x \in IJ$. Find $u \in I$ and $v \in J$ such that $uv = x$. Since J is a minimal right ideal, $vJ = J$. So there is $w \in J$ such that $vw = v$. Then

$$xw = uvw = uv = x. \tag{2.1}$$

Let $y = wxw$. Then $y \in J$. Note that

$$y^2 = wxwwxw = wxwxw = wxxw = wxw = y, \tag{2.2}$$

so y is idempotent. Note also that

$$xy = xwxw = x^2 = x, \tag{2.3}$$

as required. □

Remark 2.12 Note that replacing y by yx, we obtain an idempotent in J that besides $xy = x$ has the additional property $yx = y$.

2.2 THE GALVIN-GLAZER THEOREM

A *partial semigroup* is a nonempty set S with a partial map $* : S^2 \to S$ that satisfies the associative law

$$(x * y) * z = x * (y * z), \tag{2.4}$$

i.e., whenever one side of the equation is defined, so is the other, and they are equal. A partial semigroup is *directed* if for every finite sequence $x_1, \dots, x_n$ of elements of S there exists $y \in S$ such that $y \neq x_i$ for all $i = 1, \dots, n$ and such that $x_1 * y, \dots, x_n * y$ are all defined.

Example 2.2.1 *For a fixed positive integer k, an example of a directed partial semigroup is the collection FIN_k of all maps*

$$p : \mathbb{N} \to \{0, \dots, k\}$$

that have finite supports $\text{supp}(p) = \{n : p(n) \neq 0\}$ *and that always achieve the maximal value* k. *We let the partial semigroup operation be the coordinate-wise addition*

$$(p+q)(n) = p(n) + q(n),$$

defined only in the case when p *and* q *have disjoint supports.*

Given a directed partial semigroup $(S, *)$, let γS be the space of all ultra-filters $\mathcal{U}$ on S such that

$$(\forall x \in S) \ \{y \in S \ : \ x * y \text{ is defined}\} \in \mathcal{U}.$$

We consider γS a nonempty (closed) subspace of the Čech-Stone compactification βS, i.e., we consider it a compact Hausdorff space with the topology generated by the sets of the form

$$\overline{A} = \{\mathcal{U} \in \gamma S \ : \ A \in \mathcal{U}\}, \ (A \subseteq S).$$

We extend the partial semigroup operation $*$ of S to a *total* operation $*$ on γS defined as

$$\mathcal{U} * \mathcal{V} = \{A \subseteq S : \{x \in S : \{y \in S : x * y \in A\} \in \mathcal{V}\} \in \mathcal{U}\},$$

or equivalently,

$$\mathcal{U} * \mathcal{V} = \{A \subseteq S : (\mathcal{U}x)(\mathcal{V}y) \ x * y \in A\}.$$

Lemma 2.13 $\mathcal{U} * \mathcal{V} \in \gamma S$ *whenever* $\mathcal{U}, \mathcal{V} \in \gamma S$.

Proof. Given $\mathcal{U}, \mathcal{V} \in \gamma S$, we need to check that $\mathcal{U} * \mathcal{V}$ is also an ultrafilter and member of γS. For this, it is convenient to have the notation

$$A/x = \{y \in S : \ x * y \in A\}$$

for $A \subseteq S$ and $x \in S$. Thus $A \in \mathcal{U} * \mathcal{V}$ iff $\{x \in S : A/x \in \mathcal{V}\} \in \mathcal{U}$. Note that $A \subseteq B$ implies $A/x \subseteq B/x$, giving easily the closure of $\mathcal{U} * \mathcal{V}$ under taking supersets. It is also clear that $\emptyset \notin \mathcal{U} * \mathcal{V}$. Note also that $(A \cap B)/x = (A/x) \cap (B/x)$ and $(S \setminus A)/x = S \setminus (A/x)$. This is useful in checking the remaining ultrafilter properties of $\mathcal{U} * \mathcal{V}$. To handle the closure under intersection note that

$$\{x \in S : \ (A \cap B)/x \in \mathcal{V}\} = \{x \in S : \ (A/x) \cap (B/x) \in \mathcal{V}\}$$

$$= \{x \in S : \ A/x \in \mathcal{V}\} \cap \{x \in S : \ B/x \in \mathcal{V}\}.$$

Suppose now that $A \notin \mathcal{U} * \mathcal{V}$. Then $\{x \in S : \ A/x \in \mathcal{V}\} \notin \mathcal{U}$, or equivalently, $\{x \in S : \ A/x \notin \mathcal{V}\} \in \mathcal{U}$, or equivalently $\{x \in S : \ (S \setminus A)/x \in \mathcal{V}\} \in \mathcal{U}$. Hence $S \setminus A \in \mathcal{U} * \mathcal{V}$.

To check that $\mathcal{U} * \mathcal{V} \in \gamma S$, note that this is equivalent to the following formula written using ultrafilter quantifiers

$$(\forall x)(\mathcal{U}y)(\mathcal{V}z) \ x * (y * z) \text{ is defined}.$$

This is clearly a consequence of our two assumptions,

$$(\forall x)(\mathcal{U}y) \ x * y \text{ is defined} \quad \text{and} \quad (\forall x)(\mathcal{V}y) \ x * y \text{ is defined}.$$

$\square$

Remark 2.14 Readers familiar with the ultrafilter quantifier have noticed that above proof can be shortened and perhaps made more clear if one uses the fact that ultrafilter quantifiers commute with all propositional connectives. This remark will apply to many proofs that follow, which are, however, written for readers who may lack this experience.

Lemma 2.15 $(\mathcal{U} * \mathcal{V}) * \mathcal{W} = \mathcal{U} * (\mathcal{V} * \mathcal{W})$.

Proof. To see this first note that

$$
\begin{aligned}
A \in \mathcal{U} * (\mathcal{V} * \mathcal{W}) \quad &\text{iff} \quad \{x \in S : \ A/x \in \mathcal{V} * \mathcal{W}\} \in \mathcal{U} \\
&\text{iff} \quad \{x \in S : \ \{y \in S : \ (A/x)/y \in \mathcal{W}\} \in \mathcal{V}\} \in \mathcal{U} \\
&\text{iff} \quad \{x \in S : \ \{y \in S : \ A/y \in \mathcal{W}\}/x \in \mathcal{V}\} \in \mathcal{U},
\end{aligned}
$$

since $\{y \in S : \ A/y \in \mathcal{W}\}/x = \{y \in S : \ (A/x)/y \in \mathcal{W}\}$. On the other hand, note that

$$
\begin{aligned}
A \in (\mathcal{U} * \mathcal{V}) * \mathcal{W} \quad &\text{iff} \quad \{x \in S : \ A/x \in \mathcal{W}\} \in \mathcal{U} * \mathcal{V} \\
&\text{iff} \quad \{y \in S : \ \{x \in S : A/x \in \mathcal{W}\}/y \in \mathcal{V}\} \in \mathcal{U}.
\end{aligned}
$$

$\square$

Lemma 2.16 *For every* $\mathcal{V} \in \gamma S$, *the map* $\mathcal{U} \mapsto \mathcal{U} * \mathcal{V}$ *is a continuous map from* γS *into* γS.

Proof. Fix $\mathcal{V} \in \gamma S$ and $A \subseteq S$. Let $B = \{x \in S : \ A/x \in \mathcal{V}\}$. It suffices to show that that the preimage of $\overline{A}$ under the map $\mathcal{U} \mapsto \mathcal{U} * \mathcal{V}$ is equal to $\overline{B}$. This amounts to showing that for an arbitrary $\mathcal{U} \in \gamma S$,

$$
A \in \mathcal{U} * \mathcal{V} \text{ if and only if } B = \{x \in S : \ A/x \in \mathcal{V}\} \in \mathcal{U}.
$$

$\square$

Corollary 2.17 *The space* $(\gamma S, *)$ *is a compact semigroup for every partial directed semigroup* $(S, *)$. $\square$

Remark 2.18 Note that another way to achieve the compactification when S is actually a full semigroup with identity is by taking the topological closure of the set of $\tau_s \in X^X$ $(s \in S)$, where $X = 2^S$ and $\tau_s(x)(t) = x(st)$.

Corollary 2.19 *For every directed partial semigroup* $(S, *)$ *that does not have idempotents itself or is left cancellative, there is a nonprincipal ultrafilter* $\mathcal{U}$ *in* S *such that* $\mathcal{U} * \mathcal{U} = \mathcal{U}$.

Proof. Note that if $\mathcal{U} \in \gamma S$ is an idempotent ultrafilter that is principal, i.e., contains some singleton $\{a\}$, then $a * a = a$. Note also that if $(S, *)$ is left cancellative, then $\gamma S \setminus S$ is a left ideal of γS, so it would contain an idempotent by Lemma 2.1. $\square$

Given a directed partial semigroup $(S, *)$, a (finite or infinite) sequence $X = \langle x_0, x_1, \dots, x_n, \dots \rangle$ of elements of S is *basic* if its elements are pairwise distinct and if

$$
x_{n_0} * x_{n_1} * \cdots * x_{n_k}
$$

is defined for every finite sequence $n_0 < n_1 < \cdots < n_k$ of indexes from the domain $|X|$ of X. For a given basic sequence $X = \langle x_n \rangle$, let

$$[X] = \{x_{n_0} * \cdots * x_{n_k} : k \in \mathbb{N},\ n_0 < n_1 < \cdots < n_k < |X|\}.$$

Note that if X is an infinite basic sequence of elements of S, then $([X], *)$ is also a directed partial semigroup.

Theorem 2.20 (Galvin-Glazer) *Suppose that $(S, *)$ is a partial semigroup that either has no idempotents or is left cancellative. Then for every finite coloring of $(S, *)$, there is an infinite basic sequence $X = \langle x_n \rangle_{n=0}^{\infty}$ of elements of S such that $[X]$ is monochromatic.*

Proof. Choose an ultrafilter $\mathcal{U} \in \gamma S \setminus S$ such that $\mathcal{U} * \mathcal{U} = \mathcal{U}$ and let $P_0 \in \mathcal{U}$ be a fixed monochromatic set relative to the given coloring of S. Then by the definition of $*$ on γS and the idempotence of $\mathcal{U}$,

$$(\mathcal{U}x)(\mathcal{U}y)\ \ x * y \in P_0,$$

so we can choose $x_0 \in P_0$ such that

$$P_1 = \{y \in P_0 : x_0 * y \in P_0\} \in \mathcal{U}.$$

Similarly,

$$(\mathcal{U}x)(\mathcal{U}y)\ \ x * y \in P_1,$$

so we can choose $x_1 \in P_1$ such that

$$P_2 = \{y \in P_1 : x_1 * y \in P_1\} \in \mathcal{U},$$

and so on. This procedure gives us an infinite basic sequence $X = \langle x_n \rangle_{n=0}^{\infty} \subseteq S$ with the following property.

Claim 2.20.1 $x_{n_0} * x_{n_1} * \cdots * x_{n_k} \in P_{n_0}$ *for every finite sequence $n_0 < n_1 < \cdots < n_k$ of nonnegative integers.*

Proof. Induction on k. The initial step $x_{n_0} \in P_{n_0}$ is given by our construction. To see the inductive step from k to $k + 1$, let $x = x_{n_1} * \cdots * x_{n_k}$. Then $x \in P_{n_1}$ by the inductive hypothesis. Since $n_1 \geq n_0 + 1$, we have that

$$x \in P_{n_1} \subseteq P_{n_0+1} = \{y \in P_{n_0} : x_{n_0} * y \in P_{n_0}\}.$$

It follows that $x_{n_0} * x \in P_{n_0}$, as required.   □

Note that this gives us the desired inclusion $[X] \subseteq P_0$.   □

Corollary 2.21 (Hindman) *For every finite coloring of $\mathbb{N}$ there is an infinite sequence $x_0 < x_1 < \cdots < x_n < \cdots$ of elements of $\mathbb{N}$ such that the set of all finite nonrepeating sums $x_{n_0} + \cdots + x_{n_k}$ is monochromatic.*   □

2.3 GOWERS'S THEOREM

Throughout this section, k will denote a positive integer. For $p : \mathbb{N} \to \{0, 1, \ldots, k\}$, let $\mathrm{supp}(p) = \{n : p(n) \neq 0\}$ and define

$$\mathrm{FIN}_k = \{p : \mathbb{N} \to \{0, 1, \ldots, k\} : \mathrm{supp}(p) \text{ is finite \& } k \in \mathrm{rang}(p)\}.$$

We consider FIN_k as a directed partial semigroup endowed with the operation $x + y$ of summing two disjointly supported elements. We extend this partial semigroup operation to the set

$$\mathrm{FIN}_{[1,k]} = \bigcup_{j=1}^{k} \mathrm{FIN}_j,$$

which will also be a partial semigroup of interest here. Thus, if $x \in \mathrm{FIN}_i$ and $y \in \mathrm{FIN}_j$ have disjoint supports, then their coordinatewise addition $x + y$ is a member of FIN_l, where $l = \max\{i, j\}$. Sometimes it will be convenient to add the identity to these semigroups, the mapping constantly equal to 0, though we do not change the notation for the semigroup when we make this addition. A *block sequence* is any finite or infinite sequence $B = \{b_n\}_{n=0}^{\leq \infty} \subseteq \mathrm{FIN}_k$ such that

$$\mathrm{supp}(b_i) < \mathrm{supp}(b_j) \text{ whenever } i < j.$$

Define $\quad T : \mathrm{FIN}_k \to \mathrm{FIN}_{k-1} \quad$ by

$$T(p)(n) = \max\{p(n) - 1, 0\}.$$

We call T a *tetris operation*. We shall soon see that there is a natural identification between FIN_k with an ε-net of the positive part of the sphere of the Banach space c_0 with $1 > \varepsilon > 0$ and k related by $\varepsilon(1 + \varepsilon)^{k-1} = 1$. In this identification the tetris operation T corresponds to scalar multiplication, and the following notion of a partial semigroup (or a combinatorial subspace) generated by a block sequence corresponds to talking about a linear subspace generated by the corresponding sequence of vectors in c_0.

For a given basic block sequence $B = \{b_n\}_{n=0}^{\leq \infty}$, we let the *partial subsemigroup* of FIN_k *generated by* B be the family of vectors of the forms

$$T^{(j_0)}(b_{n_0}) + \ldots + T^{(j_l)}(b_{n_l}),$$

where $n_0 < \ldots < n_l$ is a finite sequence from the domain of B and $j_0, \ldots, j_l$ is a sequence of elements of $\{0, 1, \ldots, k\}$ such that at least one of the $j_0, \ldots, j_l$ is 0.

The purpose of this section is to prove the following result.

Theorem 2.22 (Gowers) *For every finite coloring of* FIN_k *there is an infinite block sequence B of elements of* FIN_k *such that the partial subsemigroup generated by B is monochromatic.*

We say that an ultrafilter $\mathcal{U}$ on FIN_k is *cofinite* if

$$\{p \in \mathrm{FIN}_k : \mathrm{supp}(p) \cap \{0, \ldots, n\} = \emptyset\} \in \mathcal{U}$$

for all $n \in \mathbb{N}$. As in the proof of the Galvin-Glazer theorem, let $\gamma\mathrm{FIN}_k$ denote the family of all cofinite ultrafilters on FIN_k endowed with the topology generated by the basis

$$\overline{A} = \{\mathcal{U} \in \gamma\mathrm{FIN}_k : A \in \mathcal{U}\} \quad (A \subseteq \mathrm{FIN}_k),$$

which is the same as the one induced from the Čech-Stone compactification of the discrete FIN_k with the extension of the partial operation $+$:

$$A \in \mathcal{U} + \mathcal{V} \quad \text{iff} \quad (\mathcal{U}x)(\mathcal{V}y) \ x + y \in A.$$

This gives us a compact semigroup $(\gamma\mathrm{FIN}_k, +)$. Note also that for each $k > 1$, the set $\gamma\mathrm{FIN}_k$ is a two-sided ideal of any of the semigroups

$$\gamma\mathrm{FIN}_{[j,k]} = \bigcup_{i=j}^{k} \gamma\mathrm{FIN}_i$$

for $1 \leq j \leq k$, and in particular in the semigroup $\gamma\mathrm{FIN}_{[1,k]}$ which we work with from now on. We also need to extend the tetris operation on the space of ultrafilters as follows: $T : \gamma\mathrm{FIN}_k \to \gamma\mathrm{FIN}_{k-1}$,

$$T(\mathcal{U}) = \{A \subseteq \mathrm{FIN}_{k-1} : \{x \in \mathrm{FIN}_k : T(x) \in A\} \in \mathcal{U}\}$$

Note that for $k > 1$, $T(\mathcal{U})$ is indeed a cofinite ultrafilter on FIN_{k-1}.

Lemma 2.23 $T : \gamma\mathrm{FIN}_k \to \gamma\mathrm{FIN}_{k-1}$ *is a continuous onto homomorphism.*

Proof. To check that T preserves $+$, note that $T(\mathcal{U}+\mathcal{V})$ is a cofinite ultrafilter generated by

$$\{TA : \{x : \{y : x + y \in A\} \in \mathcal{V}\} \in \mathcal{U}\}$$

$$= \{B : \{x : \{y : T(x + y) \in B\} \in \mathcal{V}\} \in \mathcal{U}\}$$

$$= \{B : \{x : \{y : T(x) + T(y) \in B\} \in \mathcal{V}\} \in \mathcal{U}\}$$

$$= \{B : \{p : \{y : p + T(y) \in B\} \in \mathcal{V}\} \in T(\mathcal{U})\}$$

$$= \{B : \{p : \{q : p + q \in B\} \in T(\mathcal{V})\} \in T(\mathcal{U})\}.$$

It follows that $T(\mathcal{U} + \mathcal{V}) = T(\mathcal{U}) + T(\mathcal{V})$. $\qquad\qquad\square$

Lemma 2.24 *For every positive integer k, one can choose an idempotent $\mathcal{U}_k \in \gamma\mathrm{FIN}_k$ such that for all positive integers $i < j$:*

(1) $\mathcal{U}_i \geq \mathcal{U}_j$,

(2) $T^{(j-i)}(\mathcal{U}_j) = \mathcal{U}_i$.

Proof. The idempotents are chosen by induction on k. For $k = 1$ we let $\mathcal{U}_1$ be an arbitrary minimal idempotent of the semigroup γFIN_1. Suppose that $\mathcal{U}_j$ $(1 \leq j < k)$ have been selected satisfying (1) and (2). Let

$$S_k = \{\mathcal{X} \in \gamma\text{FIN}_k : T(\mathcal{X}) = \mathcal{U}_{k-1}\}.$$

By Lemma 2.23, S_k is a nonempty closed subset of γFIN_k and so is $S_k + \mathcal{U}_{k-1}$. Note that $S_k + \mathcal{U}_{k-1}$ is a subsemigroup of γFIN_k, since the sum $\mathcal{V} + \mathcal{U}_{k-1} + \mathcal{W} + \mathcal{U}_{k-1}$ of two members of $S_k + \mathcal{U}_{k-1}$ belongs to $S_k + \mathcal{U}_{k-1}$ by the equation

$$T(\mathcal{V} + \mathcal{U}_{k-1} + \mathcal{W}) = \mathcal{U}_{k-1} + \mathcal{U}_{k-2} + \mathcal{U}_{k-1} = \mathcal{U}_{k-1},$$

where in the case $k - 2$ the $\mathcal{U}_{k-2}$ is to be interpreted to be equal to the identity (say, the principal ultrafilter concentrating on the constant map 0) of all our semigroups. Pick an idempotent $\mathcal{W}$ in $S_k + \mathcal{U}_{k-1}$ and let $\mathcal{V} \in S_k$ be such that $\mathcal{W} = \mathcal{V} + \mathcal{U}_{k-1}$. Finally, let $\mathcal{U}_k = \mathcal{U}_{k-1} + \mathcal{V} + \mathcal{U}_{k-1}$. Then $\mathcal{U}_k \in FIN_k^*$ and $T(\mathcal{U}_k) = \mathcal{U}_{k-1}$. Note that

$$\mathcal{U}_k + \mathcal{U}_k = \mathcal{U}_{k-1} + \mathcal{V} + \mathcal{U}_{k-1} + \mathcal{U}_{k-1} + \mathcal{V} + \mathcal{U}_{k-1}$$

$$= \mathcal{U}_{k-1} + \mathcal{V} + \mathcal{U}_{k-1} + \mathcal{V} + \mathcal{U}_{k-1}$$

$$= \mathcal{U}_{k-1} + \mathcal{V} + \mathcal{U}_{k-1} = \mathcal{U}_k.$$

Thus $\mathcal{U}_k$ is an idempotent. Note that
$\mathcal{U}_k + \mathcal{U}_{k-1} = \mathcal{U}_{k-1} + \mathcal{U}_k = \mathcal{U}_k$,
which checks the inequality $\mathcal{U}_{k-1} \geq \mathcal{U}_k$. This finishes the inductive step as well as the proof of the lemma. $\qquad\square$

Proof of Theorem 2.22. Pick a piece P of the given finite partition of FIN_k such that $P \in \mathcal{U}_k$. Now we recursively build an infinite basic block sequence $x_0, x_1, \ldots$ of elements of FIN_k and for each $1 \leq l \leq k$ a decreasing sequence $A_0^l \supseteq A_1^l \supseteq \ldots$ of elements of $\mathcal{U}_l$ such that

(a) $A_0^k = P$,

(b) $x_n \in A_n^k$ and $T^{(k-l)}[A_n^k] = A_n^l$,

(c) $(\mathcal{U}_k x)[T^{(k-i)}(x_n) + T^{(k-j)}(x) \in A_n^{\max\{i,j\}}]$ for $1 \leq i, j \leq k$.

We start the recursion by letting $A_0^l = T^{(k-l)}(P)$, $(1 \leq l \leq k)$. By Lemma 2.24(2), $\mathcal{U}_k$-almost all $x_0 \in A_0^k$ satisfy (c) so there is a way to choose $x_0 \in A_0^k$ satisfying (a),(b) and (c). To see how to handle the inductive step, suppose $x_0, \ldots, x_{n-1}$ and $A_0^l \supseteq \ldots \supseteq A_{n-1}^l$ $(1 \leq l \leq k)$ have been constructed satisfying (a) − (c). For $1 \leq i, j \leq k$, and $m < n$, define

$$C_m^{ij} = \{x \in \text{FIN}_k : T^{(k-i)}(x_m) + T^{(k-j)}(x) \in A_m^{\max\{i,j\}}\}.$$

Set

$$A_n^k = A_{n-1}^k \cap \bigcap_{i,j \leq k,\ m < n} C_m^{ij}$$

$$A_n^l = T^{(k-l)}[A_n^k], \ (1 \le l < k).$$

Then $A_n^l \in \mathcal{U}_l$ $(1 \le l \le k)$. Note that by Lemma 2.24(2), $\mathcal{U}_k$-almost all $x_n \in A_n^k$ satisfy requirement (c), so we can choose x_n keeping the inductive hypothesis $(a) - (c)$ and moreover making sure that the support of x_n lies above the support of x_{n-1}. This finishes the inductive step.

Let us now show that the infinite basic block sequence $\{x_n\}_{n=0}^\infty$ that we have just constructed generates a subspace that is included in P. This will follow once we show by induction on p that

(d) $T^{(k-l_0)}(x_{n_0}) + \ldots + T^{(k-l_{p-1})}(x_{n_{p-1}}) + y \in A_{n_0}^{\max\{l_0,\ldots,l_p\}}$

for every choice of $n_0 < \ldots < n_p$, $l_0, \ldots, l_p \in \{1, \ldots k\}$, and $y \in A_{n_p}^{l_p}$. The case $p = 0$ follows from (c). To see the inductive step, let $p > 0$ and consider

$$y' = T^{(k-l_1)}(x_{n_1}) + \ldots + T^{(k-l_{p-1})}(X_{n_{p-1}}) + y.$$

By the inductive hypothesis we know that y' belongs to $A_{n_1}^{\max\{l_1,\ldots,l_p\}}$. Let $l = \max\{1,\ldots,l_p\}$. Pick $y^* \in A_{n_1}^k$ such that $y' = T^{(k-l)}(y^*)$. Then $y^* \in A_{n_0+1}^k$. Thus, in particular y^* belongs to the set $C_{n_0}^{l_0 l}$ as formed at the inductive step from n_0 to $n_0 + 1$ above. It follows that

$$T^{(k-l_0)}(x_{n_0}) + T^{(k-l)}(y^*) \in A_{n_0}^{\max\{l_0,l\}},$$

as required. This finishes the proof. $\qquad \square$

Corollary 2.25 (Hindman) *For every finite partition of the family* FIN *of all finite nonempty subsets of* $\mathbb{N}$*, there is an infinite block sequence* $B = (b_n)$ *of finite subsets of* $\mathbb{N}$ *such that the subsemigroup* $[B]$ *generated by* B*, i.e., the family of all unions of finite nonempty subfamilies of* B*, is monochromatic.*

Proof. This is just the case $k = 1$ of Theorem 2.22. $\qquad \square$

The relationship between FIN_k and the positive part PS_{c_0} of the sphere of c_0 can be explained as follows. Find a $0 < \delta < 1$ such that

$$\frac{1}{(1+\delta)^{k-1}} = \delta.$$

Let Δ_k be the collection of all finitely supported mappings

$$\xi : \mathbb{N} \to \left\{ 0, \frac{1}{(1+\delta)^{k-1}}, \frac{1}{(1+\delta)^{k-2}}, \ldots, \frac{1}{1+\delta}, 1 \right\}$$

such that $1 \in \mathrm{rang}(\xi)$. Note that Δ_k forms a δ-net in PS_{c_0} and that it is naturally isomorphic to FIN_k via the mapping

$$\Phi(x)(n) = \max \left\{ k - \left[\frac{\log x(n)}{\log (1+\delta)^{-1}} \right], 0 \right\}.$$

In this correspondence the tetris operation corresponds to a scalar multiplication. This establishes the following corollary of Theorem 2.22.

Corollary 2.26 *For every $0 < \delta < 1$ there is a δ-net Δ in PS_{c_0} with the property that for every finite partition of Δ there is an infinite-dimensional block subspace X of c_0 such that $S_X \cap \Delta$ is included in one of the pieces of the partition.*

In one of the following sections we shall present a corresponding result about δ-nets on the whole sphere of c_0.

2.4 A SEMIGROUP OF SUBSYMMETRIC ULTRAFILTERS

For a positive integer k, let $\mathrm{FIN}_k^{\pm}$ be the collection of all finitely supported functions

$$p : \mathbb{N} \to \{0, \pm 1, \dots, \pm k\}$$

that attain at least one of the values $\pm k$. The *tetris operation* $T : \mathrm{FIN}_k^{\pm} \to \mathrm{FIN}_{k-1}^{\pm}$ in this context is defined as follows:

$$T(p)(n) = \begin{cases} p(n) - 1 & \text{if } p(n) > 0, \\ 0 & \text{if } p(n) = 0, \\ p(n) + 1 & \text{if } p(n) < 0. \end{cases}$$

A *block sequence* of elements of $\mathrm{FIN}_k^{\pm}$ is defined as before. A *partial sub-semigroup of $\mathrm{FIN}_k^{\pm}$ generated by a basic block sequence* $B = \{b_n\}_{n=0}^{\infty}$ is the family of all functions of the form

$$\epsilon_0 T^{(j_0)}(b_{n_0}) + \epsilon_1 T^{(j_1)}(b_{n_1}) + \dots + \epsilon_l T^{(j_l)}(b_{n_l}),$$

where $\epsilon_i = \pm 1$ for $0 \le i \le l$, $n_0 < \dots < n_l$ is a finite sequence of elements of the index set of B, $j_0, \dots, j_l \in \{0, \dots, k-1\}$ and at least one of the $j_0, \dots, j_l$ is equal to 0.

We shall again consider only *cofinite* ultrafilters (or filters) on $\mathrm{FIN}_k^{\pm}$. We say that an ultrafilter $\mathcal{U}$ on $\mathrm{FIN}_k^{\pm}$ is *subsymmetric* if

$$-(A)_1 \in \mathcal{U} \quad \text{for all } A \in \mathcal{U}$$

Notation.

$$-B = \{-x : x \in B\} \quad \text{and} \quad (A)_\epsilon = \{q \in \mathrm{FIN}_k^{\pm} : \exists p \in A \ \|p - q\|_\infty \le \epsilon\}.$$

As in the case of FIN_k we define the operation $+$ on the space $\gamma\mathrm{FIN}_k^{\pm}$ of all cofinite ultrafilters on $\mathrm{FIN}_k^{\pm}$ and extend the tetris operation

$$T : \gamma\mathrm{FIN}_k^{\pm} \to \gamma\mathrm{FIN}_{k-1}^{\pm}.$$

This makes $(\gamma\mathrm{FIN}_k^{\pm}, +)$ $(k \ge 1)$ a compact semigroup and T a continuous homomorphism. Let $S_k^{\pm}$ be the collection of all cofinite subsymmetric ultra-filters on $\mathrm{FIN}_k^{\pm}$. The following is immediate from the definition and the way the topology of $\mathrm{FIN}_k^{\pm}$ is defined.

Lemma 2.27 $S_k^{\pm}$ *is a closed subsemigroup of* $\gamma \mathrm{FIN}_k^{\pm}$.

Lemma 2.28 $S_k^{\pm} \neq \emptyset$ *for all* $k \geq 1$.

Proof. Pick a cofinite ultrafilter $\mathcal{V}$ on $\mathrm{FIN}_k^{\pm}$ and set[1]

$$\begin{aligned}
\mathcal{U} = \quad & T^{(k-1)}(\mathcal{V}) - T^{(k-1)}(\mathcal{V}) + T^{(k-2)}(\mathcal{V}) - T^{(k-2)}(\mathcal{V}) + \\
& \dots + T(\mathcal{V}) - T(\mathcal{V}) + \mathcal{V} - \mathcal{V} + T(\mathcal{V}) - T(\mathcal{V}) + \\
& \dots + T^{(k-2)}(\mathcal{V}) - T^{(k-2)}(\mathcal{V}) + T^{(k-1)}(\mathcal{V}) - T^{(k-1)}(\mathcal{V}).
\end{aligned}$$

It is routine to check that $\mathcal{U}$ is a subsymmetric ultrafilter. $\qquad\square$

Lemma 2.29 $T(\mathcal{U}) \in S_{k-1}^{\pm}$ *for all* $\mathcal{U} \in S_k^{\pm}$.

Proof. This follows from the fact that $-T(x) = T(-x)$ and the fact that $T[(B)_1] \subseteq (T[B])_1$ for $B \in \mathrm{FIN}_k^{\pm}$. $\qquad\square$

Thus we have a sequence $S_j^{\pm}$ $(1 \leq j \leq k)$ of compact semigroups and homomorphisms $T^{(l-j)} : S_l^{\pm} \to S_j^{\pm}$ between them.

Lemma 2.30 *There is a sequence* $\mathcal{U}_j \in S_j^{\pm}$ $(1 \leq j \leq k)$ *such that for all* $1 \leq i \leq j \leq k$:

(1) $\mathcal{U}_i + \mathcal{U}_i = \mathcal{U}_i$,

(2) $\mathcal{U}_i + \mathcal{U}_j = \mathcal{U}_j + \mathcal{U}_i = \mathcal{U}_j$ $(i.e., \mathcal{U}_i \geq \mathcal{U}_j)$,

(3) $T^{(j-i)}(\mathcal{U}_j) = \mathcal{U}_i$.

Proof. Let $R_i^{\pm} = T^{(k-i)}[S_k^{\pm}]$ $(1 \leq i \leq k)$. Then $R_i^{\pm}$ is a nonempty compact subsemigroup of $\mathcal{S}_i^{\pm}$ and $T^{(i-j)} : R_j^{\pm} \to R_i^{\pm}$ is onto whenever $1 \leq i \leq j \leq k$. We shall pick $\mathcal{U}_j$ in $R_j^{\pm}$ $(1 \leq i \leq k)$ satisfying $(1)-(3)$. Let $\mathcal{U}_1$ be an arbitrary idempotent of $R_1^{\pm}$. Suppose $1 \leq j \leq k$ and $\mathcal{U}_i \in R_i^{\pm}$ $(1 \leq i < j)$ have been selected satisfying $(1)-(3)$. Let

$$P_j^{\pm} = \{x \in R_j^{\pm} : T(x) \in \mathcal{U}_{j-1}\}.$$

Then $P_j^{\pm}$ is a nonempty closed semigroup of $R_j^{\pm}$. As in the proof of Lemma 2.24, $P_j^{\pm} + \mathcal{U}_{j-1}$ is also a closed subsemigroup of $R_j^{\pm}$, so we can pick an idempotent $\mathcal{W}$ that belongs to it. Pick $\mathcal{V} \in P_j^{\pm}$ such that $\mathcal{W} = \mathcal{V} + \mathcal{U}_{j-1}$. Let

$$\mathcal{U}_j = \mathcal{U}_{j-1} + \mathcal{V} + \mathcal{U}_{j-1}.$$

As before, one shows that $\mathcal{U}_j \in R_j^{\pm}$ continue to satisfy $(1)-(3)$. $\qquad\square$

Theorem 2.31 (Gowers) *For every finite partition of* $\mathrm{FIN}_k^{\pm}$, *there is a piece* P *of the partition such that* $(P)_1$ *contains a partial subsemigroup of* $\mathrm{FIN}_k^{\pm}$ *generated by an infinite basic block sequence.*

---

[1]The subtraction $-\mathcal{W}$ of an ultrafilter $\mathcal{W}$ here means the image of $\mathcal{W}$ the under the reflection map $x \mapsto -x$, i.e., $\mathcal{W} = \{-A : A \in \mathcal{W}\}$.

Proof. We shall use the sequence $\mathcal{U}_j$ $(1 \le j \le k)$ of ultrafilters given by Lemma 2.30. Let P be a piece of the partition such that $P \in \mathcal{U}_k$. Since $\mathcal{U}_k$ is subsymmetric, we know that $-(P)_1 \in \mathcal{U}_k$. So $(P)_1 \cap -(P)_1$ is a *symmetric element* of $\mathcal{U}_k$. As before, we recursively define a basic block sequence $\{x_n\}_{n=0}^{\infty}$, and for each l such that $1 \le l \le k$, a decreasing sequence $\{A_n^l\}_{n=0}^{\infty}$ of sets such that

(1) $A_0^k = (P)_1 \cap -(P)_1$, $A_n^l = T^{(k-l)}[A_n^k]$,

(2) $A_n^l = -A_n^l \in \mathcal{U}_l$,

(3) $\pm x_n \in A_n^k$ and $\pm T^{(k-l)}(x_n) \in A_n^l$,

(4) $(\mathcal{U}_l y)[\pm T^{(k-j)}(x_n) \pm y \in A_n^{\max\{j,l\}}]$ for $1 \le j, l \le k$.

The fact that for $\mathcal{U}_k$-almost all choices of $x_0 \in A_0^k$ satisfy (3) and (4) follows from the basic relationships between the ultrafilters $\mathcal{U}_i$ given in Lemma 2.30. At the inductive step at some $n > 1$, for $1 \le l \le k$, let A_n^l be the intersection of all sets of the form

$$\{y \in A_{n-1}^l : \pm T^{(k-j)}(x_n) \pm y \in A_{n-1}^{\max\{j,l\}}\}, \ (1 \le j \le k).$$

By the inductive hypothesis, A_n^l is a symmetric member of $\mathcal{U}_l$ for all $1 \le l \le k$. Again, the fact that $\mathcal{U}_k$-almost all choices of $x_n \in A_n^k$ satisfy (3) and (4) follows from the basic relationships between the ultrafilters $\mathcal{U}_l$ given in Lemma 2.30.

The proof of Theorem 2.31 is complete once we show, by induction on p, that

$(5)_p$ $\quad \epsilon_0 T^{(k-l_0)}(x_{n_0}) + \ldots + \epsilon_p T^{(k-l_{p-1})}(x_{n_{p-1}}) + \epsilon_p y \in A_{n_0}^{\max\{l_0,\ldots,l_p\}}$

for all choices of $n_0 < \ldots < n_p$ in $\mathbb{N}$, $l_0, \ldots l_p$ in $\{1, \ldots, k\}$, $\epsilon_0, \ldots, \epsilon_p \in \{\pm 1\}$, and $y \in A_{n_p}^{l_p}$. The case $p = 1$ reduces to (4) above, so let us assume $p > 1$ and that $(5)_{p-1}$ is true. This in particular means that

$$y_1 = \epsilon_1 T^{(k-l_1)}(X_{n_1}) + \ldots + \epsilon_{p-1} T^{(k-l_{p-1})}(X_{n_{p-1}}) + \epsilon_p y \in A_{n_1}^{\max\{l_1,\ldots,l_p\}}.$$

Let $l = \max\{l_1, \ldots, l_p\}$. Then $A_{n_1}^l \subseteq A_{n_0+1}^l$ so at the inductive step from n_0 to $n_0 + 1$, we have made sure that from $y_1 \in A_{n_1}^l \subseteq A_{n_0+1}^l$ we can conclude that

$$\epsilon_0 T^{(k-l_0)}(X_{n_0}) + y_1 \in A_{n_0}^{\max\{l_0,l_1\}},$$

and this is exactly the conclusion of $(5)_p$. $\qquad \square$

Pick $0 < \delta < 1$ such that $(1 + \delta)^{(1-k)} = \delta$ and let $\Delta_k^{\pm}$ be the collection of all finitely supported maps

$$p : \mathbb{N} \to \{0, \pm(1+\delta)^{1-k}, \pm(1+\delta)^{2-k}, \ldots, \pm(1+\delta)^{-1}, \pm 1\}$$

that attain at least one of the values ± 1. Note that $\Delta_k^{\pm}$ is a δ-net on the sphere S_{c_0} and that the distance between distinct members of $\Delta_k^{\pm}$ is at least δ^2. Define $\varphi : \mathbb{R} \to \mathbb{R} \cup \{+\infty\}$ by

$$\varphi(w) = \frac{\log|w|}{\log(1+\delta)^{-1}}$$

with the convention that $\log \ 0 = +\infty$. Note that $\varphi(\pm(1 + \delta)^{-l}) = \pm l$ for every positive integer l.

Define $\Phi : S_{c_0} \to \text{FIN}_k^{\pm}$ by

$$\Phi(x)(n) = \text{sign}(x(n)) \cdot \max\{k - \lfloor \varphi(x(n)) \rfloor, 0\}.$$

Note that

$$\Phi(-x) = -\Phi(x) \text{ and } \Phi(x + y) = \Phi(x) + \Phi(y)$$

for every $x, y \in S_{c_0}$ such that $\text{supp}(x) \cap \text{supp}(y) = \emptyset$. Let $\Psi = \Phi \upharpoonright \Delta_k^{\pm}$. Note that Ψ is a bijection and that $\text{supp}(p) = \text{supp}(\Psi(p))$ for all $p \in \Delta_K^{\pm}$. The following is also easy to check.

Lemma 2.32 *For every* $x \in \Delta_k^{\pm}$ *and* $0 \leq \lambda \leq 1$,

$$\phi(\lambda \cdot x) = T^{(j)}(\psi(x))$$

for $j = \min\{k, \lfloor \varphi(\lambda) \rfloor\}$.

This leads us to the following geometrical interpretation of Theorem 2.31.

Corollary 2.33 *For every finite partition of* $\Delta_k^{\pm}$, *there is an infinite dimensional block subspace* X *of* c_0 *and there is some piece* P *of the partition such that* $S_X \subseteq (P)_\delta$.

Corollary 2.34 *For every Lipschitz map* $f : S_{c_0} \to \mathbb{R}$ *and* $\epsilon > 0$, *there is an infinite-dimensional block subspace* X *of* c_0 *such that* $\text{osc}(f, S_X) \leq \epsilon$.

Proof. Let K be the Lipschitz constant of f. Find a sufficiently large integer $k \geq 1$ such that if $(1 + \delta)^{(1-k)} = \delta$ then $\delta \cdot K \leq \frac{\epsilon}{2}$. Let $\{A\}$ be a finite partition of the range of f into sets of diameter $\leq \frac{\epsilon}{2}$. Now apply Corollary 2.33 to the partition $\{f^{-1}(A) \cap \Delta_k^{\pm}\}$. □

2.5 THE HALES-JEWETT THEOREM

Let $L = \bigcup_{n=0}^{\infty} L_n$ be a given alphabet decomposed into an increasing chain of finite subsets L_n and v be a variable distinct from all the symbols from L. We let W_L (or simply W) denote the set of all *words* over L and let W_{Lv} be the set of all *variable-words* over L, i.e., all finite strings of elements of $L \cup \{v\}$ in which v occurs at least once. If $s = s[v] \in W(v)$ and $a \in L \cup \{v\}$ then by $s[a]$ we denote the element of W or W_{Lv} depending on whether $a \neq v$ or not, obtained by replacing *every* occurrence of v in s by a.

For a (finite or infinite) sequence $X = \langle x_0, x_1, \ldots \rangle$ of elements of W_{Lv}, we denote by $[X]_L$, respectively by $[X]_{Lv}$, the *partial subsemigroup* of W_L, respectively of W_{Lv}, *generated by* X defined as follows:

$$[X]_L = \{x_{n_0}[\lambda_0]^\frown \ldots ^\frown x_{n_k}[\lambda_k] \in W_L : n_0 < \ldots < n_k, \ \lambda_i \in L_{n_i} \ (i \leq k)\}$$

$$[X]_{Lv} = \{x_{n_0}[\lambda_0]^\frown \ldots ^\frown x_{n_k}[\lambda_k] \in W_{Lv} : n_0 < \ldots < n_k,$$

$$\lambda_i \in L_{n_i} \cup \{v\} \ (i \le k)\}.$$

Theorem 2.35 (Infinite Hales-Jewett Theorem) *For every finite coloring of $W_L \cup W_{Lv}$, there is an infinite sequence $X = (x_n)$ of elements of W_{Lv} such that the partial subsemigroups $[X]_L$ and $[X]_{Lv}$ are both monochromatic.*

Proof. We use Glazer's idea and extend the word semigroup $S = W_L \cup W_{Lv}$ to its compactification $(\beta S, ^\frown)$. We shall actually work only with the closed subsemigroup $S^* = \beta S \setminus S$, consisting of nonprincipal ultrafilters on S. Note that

$$S_L^* = \{\mathcal{U} \in S^* : W_L \in \mathcal{U}\}$$

is a closed subsemigroup of S^* and that

$$S_{Lv}^* = \{\mathcal{U} \in S^* : W_{Lv} \in \mathcal{U}\}$$

is a two-sided ideal of S^*. By Lemma 2.3, we can choose a minimal idempotent $\mathcal{W}$ in S_L^*. Applying again Lemma 2.3, we can find a minimal idempotent $\mathcal{V} \le \mathcal{W}$ belonging to the two-sided ideal S_{Lv}^*. Each letter $\lambda \in L$ determines the substitution map $x \mapsto x[\lambda]$ from $W_{Lv} \cup W_L$ into W_L, which is clearly the identity on W_L and which extends to a continuous homomorphism $\mathcal{U} \mapsto \mathcal{U}[\lambda]$ from $S_{Lv}^* \cup S_L^*$ into S_L^*, which is the identity on S_L^*.

Claim 2.35.1 $\mathcal{V}[\lambda] = \mathcal{W}$ *for all* $\lambda \in L$.

Proof. Since $\mathcal{U} \mapsto \mathcal{U}[\lambda]$ is a homomorphism, $\mathcal{V}[\lambda]$ is an idempotent of S_L^* and $\mathcal{V}[\lambda] \le \mathcal{W}[\lambda] = \mathcal{W}$. Since $\mathcal{W}$ is minimal in S_L^*, we have that $\mathcal{V}[\lambda] = \mathcal{W}$. $\square$

Let P_v be the color of the given coloring that belongs to $\mathcal{V}$ and let P_W be the color which belongs to $\mathcal{W}$. By recursion on n, we build an infinite sequence $X = (x_k)$ of variable-words and two infinite decreasing sequences $\{P_W^n\}$ and $\{P_v^n\}$ of elements of $\mathcal{W}$ and $\mathcal{V}$, respectively, such that for all n

$(a)_n$ $x_n \in P_v^n$,
$(b)_n$ $\forall \lambda \in L_n \ \forall x \in P_v^n \ x[\lambda] \in P_W^n$,
$(c)_n$ $(\mathcal{V}y)(\forall \lambda \in L_n \cup \{v\}) \ x_n[\lambda]^\frown y \in P_v^n$,
$(d)_n$ $(\mathcal{W}t) \ x_n \ ^\frown t \in P_v^n$.

We start by letting $P_W^0 = P_W \cap W_L$ and

$$P_v^0 = \{x \in P_v \cap W_{Lv} : \forall \lambda \in L_0 \ x[\lambda] \in P_W^0\}.$$

By Claim 2.35.1 and the fact that $P_W^0 \in \mathcal{W}$, the set P_v^0, being a finite intersection of members of $\mathcal{V}$, belongs to $\mathcal{V}$. Rewriting the fact

$$(\forall \lambda \in L_0 \cup \{v\}) \ P_v^0 \in \mathcal{V}[\lambda]^\frown \mathcal{V}$$

using the ultrafilter quantifier, we get

$$(\forall x)(\forall y)(\forall \lambda \in L_0 \cup \{v\}) \ x[\lambda]^\frown y \in P_v^0.$$

Similarly, reading the fact $P_v^0 \in \mathcal{V} = \mathcal{V}^\frown \mathcal{W}$, we get

$$(\forall x)(\mathcal{W}t) \ x \ ^\frown t \in P_v^0.$$

It follows that we can choose $x_0 \in P_v^0$ such that $(a)_0 - (d)_0$ are satisfied.

The inductive step from n to $n+1$ is done similarly. By $(d)_n$ the set

$$P_W^{n+1} = \{t \in P_W^n : x_n \ ^\frown t \in P_v^n\}$$

belongs to $\mathcal{W}$. By $(c)_n$, the set

$$Q_v^n = \{y \in P_v^n : (\forall \lambda \in L_n \cup \{v\})x_n[\lambda]^\frown y \in P_v^n\}$$

belongs to $\mathcal{V}$. So, as before it follows from Claim 2.35.1 that the set

$$P_v^{n+1} = \{x \in Q_v^n : (\forall \lambda \in L_{n+1})x[\lambda] \in P_W^{n+1}\}$$

belongs to $\mathcal{V}$. As before, we argue that $\mathcal{V}$-almost all choices of x_{n+1} from P_v^{n+1} satisfy $(c)_{n+1}$ and $(d)_{n+1}$.

Claim 2.35.2

(1) $x_{n_0}[\lambda_0]^\frown \ldots ^\frown x_{n_{k-1}}[\lambda_{k-1}]^\frown y \in P_v^{n_0}$ *for every* $k > 0$, $n_0 < \ldots < n_k$, $\lambda_i \in L_{n_i} \cup \{v\} \ (i < k)$ *and* $y \in P_v^{n_k}$.

(2) $x_{n_0} \ ^\frown x_{n_1}[\lambda_1]^\frown \ldots ^\frown x_{n_k}[\lambda_k] \in P_v^{n_0}$ *for every* $k \geq 0$, $n_0 < \ldots < n_k$, $\lambda_i \in L_{n_i} \ (0 < i \leq k)$.

(3) $x_{n_0}[\lambda_0]^\frown \ldots ^\frown x_{n_k}[\lambda_k] \in P_W^{n_0}$ *for every* $k \geq 0$, $n_0 < \ldots < n_k$, $\lambda_i \in L_{n_i} \ (i \leq k)$.

Proof. The proof is by induction on k. The case $k = 1$ of (1) follows from the way we have made the recursive step from n_0 to $n_0+1 : \ P_v^{n_1} \subseteq P_v^{n_0+1} \subseteq Q_v^{n_0}$. So, let us suppose (1) at k and prove it for $k+1$. So let $n_0 < \ldots < n_{k+1}$, $\lambda_i \in L_{n_i} \cup \{v\} \ (i < k)$ and $y \in P_v^{n_{k+1}}$ be given. Let

$$y' = x_{n_1}[\lambda_1]^\frown \ldots ^\frown x_{n_k}[\lambda_k]^\frown y.$$

Then $y' \in P_v^{n_1}$ by the inductive hypothesis. By the way that we made the recursive step from n_0 to $n_0 + 1$, we have that $y' \in P_V^{n_1} \subseteq P_v^{n_0+1} \subseteq Q_v^{n_0}$. It follows that $y = x_{n_0}[\lambda_0] \ ^\frown y' \in P_v^{n_0}$. Note that (3) follows from (1) with $y = x_{n_k}$ and $(b)_{n_0}$. It remains to prove (2). So let $k \geq 0$, $n_0 < \ldots < n_k$ and $\lambda_i \in L_{n_i} \ (0 < i \leq k)$ be given. Let

$$t' = x_{n_1}[\lambda_1] \ ^\frown \ldots ^\frown x_{n_k}[\lambda_k].$$

By (3) we know that $t' \in P_W^{n_1}$. According to the recursion step from n_0 to $n_0 + 1$ we have that

$$t' \in P_W^{n_1} \subseteq P_W^{n_0+1} = \{t \in P_W^{n_0} : \ x_{n_0} \ ^\frown t \in P_v^{n_0}\},$$

hence $x_{n_0} \ ^\frown t' \in P_v^{n_0}$, which is the conclusion of (2). $\qquad\square$

We are now in a situation to finish the proof of the infinitary Hales-Jewett theorem by showing that if $X = (x_n)_{n=0}^\infty$ is the sequence just produced, then $[X]_L \subseteq P_W$ and $[X]_{Lv} \subseteq P_v$. The first conclusion follows from Claim 2.35.2(3). To show the second, consider an expression of the form

$$x = x_{n_0}[\lambda_0] \frown \ldots \frown x_{n_k}[\lambda_k],$$

where $n_0 < \ldots < n_k$ and $\lambda_i \in L_{n_i} \cup \{v\}$ $(i < k)$ are such that at least one of the λ_i is equal to v. Let $l = \max\{i \leq k : \lambda_i = v\}$. By Claim 2.35.2(2),

$$y = x_{n_l}[\lambda_l] \frown x_{n_{l+1}}[\lambda_{l+1}] \frown \ldots \frown x_{n_k}[\lambda_k] \in P_v^{n_l}.$$

It follows that

$$x = x_{n_0}[\lambda_0] \frown \ldots \frown x_{n_{l-1}}[\lambda_{l-1}] \frown y.$$

is an expression whose terms satisfy the hypothesis of Claim 2.35.2(1). Therefore, we can conclude that $x \in P_v^{n_0} \subseteq P_v$. This completes the proof. □

Remark 2.36 Note that the above proof shows that whenever we have an idempotent $\mathcal{U}$ concentrating on variable-words and a set P of nonvariable words such that $(\forall \lambda \in L)(\mathcal{U}x)\, x[\lambda] \in P$, then there is a procedure that gives us an infinite sequence $X = (x_n)$ of variable-words such that $[X]_L \subseteq P$.

The proof of Theorem 2.35 admits many variations. To state one, working still with L, W_L and W_{Lv} as above, we say that an $x \in W_{Lv}$ is a *left variable word* if the first letter of x is v.

Theorem 2.37 (Infinite Hales-Jewett Theorem for Left Variable Words) *If the alphabet L is finite, then for every finite coloring of W_L there is an infinite sequence $X = (x_n)_{n=0}^\infty$ of left variable-words and a variable-free word w_0 such that the translate $w_0 \frown [X]_L$ of the partial subsemigroup of W_L generated by X is monochromatic.*

Proof. As in the proof of Theorem 2.35, we work with the semigroup

$$(S = W_L \cup W_{Lv}, \frown)$$

and its extension $(S^*, \frown)$ where $S^* = \beta S \setminus S$. As before we take a minimal idempotent $\mathcal{W}$ in the subsemigroup $\{\mathcal{X} \in S^* : W_L \in \mathcal{X}\}$ and an idempotent $\mathcal{V} \leq \mathcal{W}$ minimal in S^*. Since $\{\mathcal{X} \in S^* : W_{Lv} \in \mathcal{X}\}$ is a two-sided ideal of S^*, the ultrafilter $\mathcal{V}$ must belong to it, or in other words, $\mathcal{V}$ concentrates on variable-words. Let $v \frown W_{Lv}$ denote the right-ideal of S consisting of all left-sided variable-words over L. Then $J = \{\mathcal{X} \in S^* : v \frown W_{Lv} \in \mathcal{X}\}$ is a right-ideal of S^*. So, by Lemma 2.11 and Remark 2.12 there is idempotent $\mathcal{U} \in J$ such that

$$\mathcal{V} \frown \mathcal{U} = \mathcal{V} \text{ and } \mathcal{U} \frown \mathcal{V} = \mathcal{U}.$$

Using this and the Claim 2.35.1 from the proof of Theorem 2.35, we get that

$$(\forall \lambda \in L) \ \mathcal{W} \frown \mathcal{U}[\lambda] = \mathcal{W} \text{ and } \mathcal{U}[\lambda] \frown \mathcal{W} = \mathcal{U}[\lambda].$$

As before, for a subset Q of W_L and $w \in W_L$, let

$$Q/w = \{t \in W_L : w \frown t \in Q\}.$$

Note that the equations $W \frown W = W$ and $W \frown \mathcal{U}[\lambda] = W$ for $\lambda \in L$ mean that if $Q \in \mathcal{W}$, then the set of all $w \in Q$ such that

$$Q/w \in \mathcal{W} \text{ and } Q/w \in \mathcal{U}[\lambda] \text{ for all } \lambda \in L$$

belongs to $\mathcal{W}$. Let ∂Q denote this subset of Q.

Let $P \in \mathcal{W}$ be a set that is monochromatic relative to the given coloring. Then ∂P belongs to $\mathcal{W}$ and so in particular is nonempty, which means that we can pick $w_0 \in P$ such that P/w_0 belongs to $\mathcal{W}$ and all the ultrafilters $\mathcal{U}[\lambda]$ for $(\lambda \in L)$. It follows, in particular, that

$$(\mathcal{U}x)(\forall \lambda \in L) \; x[\lambda] \in P/w_0.$$

Using this and the equations $\mathcal{U}[\lambda] \frown W = \mathcal{U}[\lambda]$, for $\lambda \in L$, we can find a left variable-word x_0 such that for all $\lambda \in L$ there is a set $P_\lambda \in \mathcal{W}$ such that $x_0[\lambda] \in P/w_0$ and $x_0[\lambda] \frown w \in P/w_0$ for all $w \in P_\lambda$. Let Q_0 be the intersection of P and the finitely many sets P_λ $(\lambda \in L)$. Then $Q_0 \in \mathcal{W}$ and therefore $\partial Q_0 \in \mathcal{W}$ as well. So in particular, we can pick $w_1 \in Q_0$ such that

$$(\forall \lambda \in L)[Q_0/w_0 \frown x_0[\lambda] \frown w_1 \in \mathcal{W} \text{ and } (\forall \mu \in L) \; Q_0/w_0 \frown x_0[\lambda] \frown w_1 \in \mathcal{U}[\mu]].$$

Let P_1 be the intersection of Q_0 and the finitely many sets

$$Q_0/w_0 \frown x_0[\lambda] \frown w_1 (\lambda \in L).$$

Working as above, we can find a left variable-word x_1 and for each $\lambda \in L$ a set $P_{1\lambda} \in \mathcal{W}$ such that $x_1[\lambda] \in P_1$ and $x_1[\lambda] \frown w \in P_1$ for all $w \in P_{1\lambda}$. Let Q_1 be the intersection of P_1 and the finitely many sets $P_{1\lambda}$ $(\lambda \in L)$, and so on. Proceeding in this way, we construct a decreasing sequence $P_1 \supseteq P_2 \supseteq \ldots$ of subsets of P and a sequence

$$w_0, x_0, w_1, x_1, \ldots, w_n, x_n, \ldots$$

of variable-free words w_n and left variable words x_n such that

$$w_0 \frown x_{n_0}[\lambda_0] \frown w_{n_0+1} \frown \ldots \frown x_{n_k}[\lambda_k] \frown w_{n_k+1} \in P_{n_0} \subseteq P$$

for all $n_0 < \ldots < n_k$ and $\lambda_i \in L$ for $0 \le i \le k$. It follows that if we let $y_n = x_n \frown w_{n+1}$ for $n = 0, 1, \ldots$, then we get a sequence of left variable-words together with the variable-free word w_0, satisfies the conclusion of the theorem. $\qquad \square$

Remark 2.38 Note that Theorem 2.37 is no longer true if we allow infinite alphabets L even if we restrict the substitutions $x_n[\lambda]$ in the nth variable-word x_n to letters λ belonging to the nth piece L_n of a fixed increasing decomposition $L = \bigcup_{n=0}^{\infty} L_n$ of L into its finite subalphabets. Fixing an enumeration of L, for a given w in W_L, let M_w be the set of all integers k with the property that kth letter of L is in the kth position in w. Then if we color w from W_L by the parity of the cardinality of the set M_w, then no combinatorial subspace of the form $w_0 \frown [X]_L$ could be monochromatic.

We finish this section with the original Hales-Jewett theorem which is clearly an immediate consequence of Theorem 2.35.

Theorem 2.39 (Finite Hales-Jewett Theorem) *For every finite alphabet L and a positive integer k, there corresponds a positive integer n such that for every k-coloring of the set $W_L(n)$ of all L-words of length n there is a variable-word x of length n such that the set $\{x[\lambda] : \lambda \in L\}$ is monochromatic.*

2.6 PARTIAL SEMIGROUP OF LOCATED WORDS

We start again with an alphabet $L = \bigcup_{n=0}^{\infty} L_n$ expressed as an increasing union of finite alphabets L_n. Let $v \notin L$ be a fixed variable. A *located word* over L is a function from a finite nonempty subset of $\mathbb{N}$ into L. Let FIN_L be the collection of all located words over L. A *located variable word* over L is a finite partial function from $\mathbb{N}$ into $L \cup \{v\}$ that takes the value v at least once. Let FIN_{Lv} be the collection of all located variable-words. A sequence (finite or infinite) $X = (x_n)$ of elements of FIN_L or FIN_{Lv} is a *block sequence* if

$$\mathrm{dom}(x_n) < \mathrm{dom}(x_m) \text{ whenever } n < m.$$

For a given block sequence $X = (x_n)$ of located variable-words, we define the corresponding subspaces of FIN_L and FIN_{Lv} as before,

$$[X]_L = \{x_{n_0}[\lambda_0] \cup \ldots \cup x_{n_k}[\lambda_k] \in \mathrm{FIN}_L : n_0 < \ldots < n_k,$$

$$\lambda_i \in L_{n_i} \ (i \le k)\},$$

$$[X]_{Lv} = \{x_{n_0}[\lambda_0] \cup \ldots \cup x_{n_k}[\lambda_k] \in \mathrm{FIN}_{Lv} : n_0 < \ldots < n_k,$$

$$\lambda_i \in L_{n_i} \cup \{v\} \ (i \le k)\}.$$

(For $x \in \mathrm{FIN}_{Lv}$ and $\lambda \in L \cup \{v\}$, we denote by $x[\lambda]$ the function with the same domain as x that agrees with x on all places where x does not take the value v and that is equal to λ at any place where x is equal to v.) The proof of the infinitary Hales-Jewett theorem easily adapts to a proof of the following result.

Theorem 2.40 (Infinite Hales-Jewett Theorem for Located Words) *For every finite coloring of the set $\mathrm{FIN}_L \cup \mathrm{FIN}_{Lv}$, there is an infinite block sequence $X = (x_n)$ of variable located words such that $[X]_L$ and $[X]_{Lv}$ are both monochromatic.*

Proof. Let S be the collection of all *cofinite* ultrafilters on $\mathrm{FIN}_L \cup \mathrm{FIN}_{Lv}$, i.e., ultrafilters that contain each of the sets of the form

$$\{x \in \mathrm{FIN}_L \cup \mathrm{FIN}_{Lv} : \mathrm{dom}(x) \cap \{0, \ldots, n\} = \emptyset\}$$

for $n \in \mathbb{N}$. For $\mathcal{U}, \mathcal{V} \in S$, let $\mathcal{U} * \mathcal{V}$ be the collection of all subsets A of $\mathrm{FIN}_L \cup \mathrm{FIN}_{Lv}$ such that

$$\{x \in \mathrm{FIN}_L \cup \mathrm{FIN}_{Lv} : \{y \in \mathrm{FIN}_L \cup \mathrm{FIN}_{Lv} : x < y \ \& \ x \cup y \in A\} \in \mathcal{V}\} \in \mathcal{U}.$$

Then as in the case of the Galvin-Glazer theorem, one has that $*$ is an associative operation on S and that $\mathcal{U} \mapsto \mathcal{U} * \mathcal{V}$ is continuous for all $\mathcal{V} \in S$. Thus $(S, *)$ is a topological semigroup, so we can apply the theory of minimal idempotents.

Let S_L be the closed subsemigroup of S consisting of all cofinite ultra-filters that concentrate on FIN_L and let S_{Lv} be the two sided ideal of S consisting of all cofinite ultrafilters that concentrate on FIN_{Lv}. Choose a minimal idempotent $\mathcal{W}$ of S_L. Using Lemma 2.3 again, we can find a min-imal idempotent $\mathcal{V}$ of S such that $\mathcal{V} \le \mathcal{W}$ and $\mathcal{V} \in S_{Lv}$. For $\lambda \in L \cup \{v\}$, the map $x \mapsto x[\lambda]$ from $\mathrm{FIN}_L \cup \mathrm{FIN}_{Lv}$ into itself extends to a continuous homomorphism $\mathcal{U} \mapsto \mathcal{U}[\lambda]$ from S into S. Note that this map is the identity on S_L if $\lambda \in L$, and the identity on S_{Lv} if $\lambda = v$.

Claim 2.40.1 $\mathcal{V}[\lambda] = \mathcal{W}$ *for all* $\lambda \in L$.

Proof. Since $\mathcal{U} \mapsto \mathcal{U}[\lambda]$ is a homomorphism from S into S_L, the ultrafilter $\mathcal{V}[\lambda]$ is an idempotent of S_L. Moreover $\mathcal{V} \le \mathcal{W}$ implies $\mathcal{V}[\lambda] \le \mathcal{W} = \mathcal{W}$. Since $\mathcal{W}$ is minimal in S_L, we have that $\mathcal{V}[\lambda] = \mathcal{W}$. $\qquad\square$

Let P_v be the color of the given coloring of $\mathrm{FIN}_L \cup \mathrm{FIN}_{Lv}$ that belongs to $\mathcal{V}$ and let P_W be the color that belongs to $\mathcal{W}$. As in the proof of the infinite Hales-Jewett theorem, starting from P_v and P_W we build the sequences $\{P_v^n\}$ and $\{P_W^n\}$ of members of $\mathcal{V}$ and $\mathcal{W}$, respectively, and the infinite block-sequence $X = (x_n)$ of variable located words, such that $[X]_{Lv} \subseteq P_v$ and $[X]_L \subseteq P_W$. $\qquad\square$

Theorem 2.41 (Hindman) *For every finite coloring of the set* FIN *of all nonempty finite subsets of* $\mathbb{N}$, *there is an infinite block sequence* $X = (x_n)$ *of members of* FIN *such that the set* $[X]$ *of all finite unions of members of* X *is monochromatic.*

Proof. This is just the case $L = \emptyset$ of the previous result. $\qquad\square$

NOTES TO CHAPTER TWO

The theory of compact left-topological semigroups exposed above is an old subject of topological dynamics (see, e.g., Ellis [28], Furstenberg-Katznelson [32], Hindman-Strauss [49]). Glazer's proof of Hindman's theorem (see, e.g., Comfort [17]) gives the added feature to classical theory of enveloping semi-groups that even partial semigroups lead to compact left-topological semi-groups whose idempotents have meaningful Ramsey theoretic interpretations back in the original partial semigroups. The extent of the applicability of this observation can perhaps be most easily seen in the proofs of the two theorems of Gowers (Theorems 2.22 and 2.31 above), appearing originally in his paper [37] and giving the positive solution to the distortion problem for the Banach space c_0. This could be seen equally well in the proof of the Bergelson-Blass-Hindman theorem (Theorem 2.40 above) which appears originally in [8] and is also inspired by Glazer's proof of the well-known theorem of Hindman [47]. The finite version of the Hales-Jewett theorem appears in the original paper [43] of Hales and Jewett. The first infinite versions of the Hales-Jewett theorem were proved by Carlson-Simpson [16] and Carlson [13] although the proofs of these extensions presented above are more closely related to the proofs appearing in [32], [49], and [48]. Needless to say, they are all inspired by Glazer's proof of Hindman's theorem, but of course they have some added features such as the use of more than one idempotent or the use of the ordering $\leq$ between idempotents.

Chapter Three

Trees and Products

3.1 VERSIONS OF THE HALPERN-LÄUCHLI THEOREM

In this section by a *tree* we mean a rooted finitely branching tree of height ω with no terminal nodes. Given a tree T, and $n \in \omega$, let $T(n)$ denote the nth level of T. A *subtree* of T is a subset of T with an induced tree-ordering. Note that in general for a subtree S of T the nth level $S(n)$ may not be a *level set*, i.e., included in some level $T(m)$ of T although we typically work with subsets S of T for which all levels are level subsets of T. One such subtree is the subtree

$$T(A) = \bigcup_{n \in A} T(n)$$

for some infinite set $A \subseteq \omega$. Another such subtree is a so-called *strong subtree* of T, i.e., a subtree S of T for which we can find an *infinite* set $A \subseteq \omega$ of levels such that

(1) $S \subseteq T(A)$ and $S \cap T(n) \neq \emptyset$ for all $n \in A$,

(2) if $m < n$ are two successive elements of A and if s is a node belonging to $S \cap T(m)$, then every immediate successor of s in T has exactly one extension in $S \cap T(n)$.

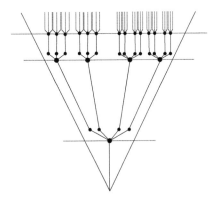

Figure 3.1 A strong subtree.

Any tree T has its natural topology τ_0 generated by sets of the form $T^s = \{t \in T : s \leq t\}$ for $s \in T$. This topology has its standard notions like *nowhere dense, somewhere dense* and *dense*, but in order to avoid confusions with similar notions for trees that we introduce below, we shall keep the reference to τ_0 whenever we use this topological notions. Thus, we say that a subset X of T is *nowhere τ_0-dense* if for every $t \in T$, there is a $u \geq t$ such that $x \notin X$ for all $x \geq u$. On the other hand, a subset X of T is τ_0-*dense* if for every $t \in T$, there is an $x \in X$ such that $t \leq x$. If there is an $s \in T$ such that for every $t \in T$ with $t \geq s$ there is $x \in X$ such that $t \leq x$, then we say that X is *somewhere τ_0-dense*. Thus, X is somewhere τ_0-dense if it is dense in a subtree of T of the form

$$T[s] = \{x \in T : x \leq s \text{ or } s \leq x\}.$$

The Halpern-Läuchli theorem is about finite analogs of these standard notions. A subset X of T is k-*dense* if it dominates every node of T of height k. For $x \in T$ and $k \in \omega$ we say that a subset X of T is k-x-*dense* if X dominates every node of $T[x] \cap T(k)$. If $X \subseteq T$ is k-x-dense for some $x \in T$ and some $k > \text{level}(x)$, or equivalently for $k = \text{level}(x) + 1$, then X is said to be *somewhere dense*.

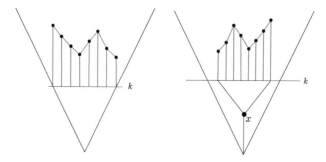

Figure 3.2 A k-dense set and a k-x-dense set.

Suppose now we are given a (finite) sequence T_i ($i < d$) of trees. We consider their product $\prod_{i<d} T_i$ an ordered set with the coordinatewise ordering. A k-*dense matrix* of $\prod_{i<d} T_i$ is a subproduct of the form $\prod_{i<d} X_i$, where X_i is a k-dense subset of T_i for all i. A k-*dense subset* of $\prod_{i<d} T_i$ is a subset P of $\prod_{i<d} T_i$ that dominates (in the cartesian ordering) every element of $\prod_{i<d} T_i(k)$. Thus, every k-dense matrix is a k-dense subset but not vice versa. For $\vec{x} \in \prod_{i<d} T_i$ and $k \in \omega$, a k-$\vec{x}$-*dense matrix* is a product of the form $\prod_{i<d} X_i$, where X_i is k-x_i-dense for each $i < d$. A product $\prod_{i<d} X_i$ is called a *somewhere dense matrix* if $\prod_{i<d} X_i$ is k-$\vec{x}$-dense for some $k \in \omega$ and $\vec{x} \in \prod_{i<d} T_i$ such that

$$k > \max_{i<d}(\text{level}(x_i)).$$

In the next section we shall actually prove the following version of the well-known partition theorem of Halpern and Läuchli.

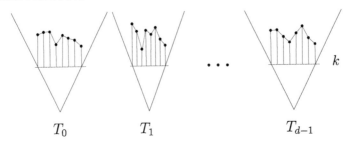

Figure 3.3 A k-dense matrix.

Theorem 3.1 (Halpern-Läuchli) *For every finite sequence T_i $(i < d)$ of trees and every finite partition of their product $\prod_{i<d} T_i$, one of the pieces of the partition must contain a somewhere dense matrix.*

As we shall consider several equivalent formulations of this result, let us call this version, the *somewhere dense matrix version* and denote it by SDHL_d to emphasize the dimension d as well. To state another formulation of this result we need a higher-dimensional version of the notion of strong subtree. Thus, we say that a sequence S_i $(i < d)$ is a *sequence of strong subtrees* of T_i $(i < d)$ if there is a *single* infinite set $A \subseteq \omega$ such that for all $i < d$, S_i is a strong subtree of T_i with A as the witness, i.e., satisfying conditions (1) and (2) above.

Theorem 3.2 (Strong Subtree Version of the Halpern-Läuchli Theorem) *For every finite partition of a finite product $\prod_{i<d} T_i$ of trees, there is one piece P of the partition and there is a sequence S_i $(i < d)$ of strong sub-trees of T_i $(i < d)$ as witnessed by the same infinite set $A \subseteq \omega$ such that $\bigcup_{n \in A} \prod_{i<d} S_i(n) \subseteq P$.*

Proof. Note that by applying the compactness principle, Theorem 3.1 gives us the following finite version of SDHL_d: For every integer $l \geq 1$, there is an integer n such that every partition

$$c : \prod_{i<d} (T_i \restriction n) \to l = \{0, 1, \ldots, l-1\}$$

is constant on a somewhere dense matrix $\prod_{i<d} X_i$ such that $X_i \subseteq T_i \restriction n$ for $i < d$.

Let $c : \prod_{i<d} T_i \to l$ be a given partition and let n be an integer which satisfies the finite form of SDHL_d for this l. Choose

$$h_i : T_i \restriction n \to T_i(n) \ (i < d)$$

such that $h_i(t) \geq_{T_i} t$ for all $i < d$ and $t \in T_i \restriction n$. Define

$$\bar{c} : \prod_{i<d} (T_i \restriction n) \to l$$

by

$$\bar{c}(t_0, \ldots, t_{d-1}) = c(h_0(t_0), \ldots, h_{d-1}(t_{d-1})).$$

By finite $\mathrm{SDHL}_d(l)$ for the trees $T_i \upharpoonright n$ $(i < d)$, the coloring $\bar{c}$ is constant on some somewhere dense matrix $\prod_{i<d} X_i$ such that $X_i \subseteq T_i \upharpoonright n$ for all $i < d$. Let

$$\bar{X}_i = h_i[X_i] \ (i < d).$$

Then $\bar{X}_i$ $(i < d)$ is a sequence of *level subsets* of T_i $(i < d)$ forming a somewhere dense matrix $\prod_{i<d} \bar{X}_i$ of $\prod_{i<d} T_i$ on which c is constant. It follows that without loss of generality we may assume that $\mathrm{SDHL}_d(l)$ gives us a monochromatic matrix of *level subsets*. Given this observation, we now claim that for a given $c : \prod_{i<d} T_i \to l$, there is a level sequence $(x_0, \ldots, x_{d-1}) \in \prod_{i<d} T_i$ such that for all k there is a sequence X_i $(i < d)$ of level subsets of T_i $(i < d)$ such that X_i is k-dense in $T_i[x_i]$ for all $i < d$ and such that c is constant on $\prod_{i<d} X_i$. From this, the conclusion of the strong subtree version of the Halpern-Läuchli theorem follows easily. Toward a contradiction, suppose that this strengthening of $\mathrm{SDHL}_d(l)$ is false. So for each $\vec{x} \in \prod_{i<d} T_i$ we can associate the maximal integer $k(\vec{x})$ for which we can find a c-homogeneous $k(\vec{x})$-$\vec{x}$-dense sequence of level subsets of T_i $(i < d)$. So we can build a strictly increasing sequence $\{n_j\}_{j=0}^{\infty}$ of positive integers such that for every j,

$$n_{j+1} > k(\vec{x}) \quad \text{for all} \ \ \vec{x} \in \bigcup_{n \le n_j} \prod_{i<d} T_i(n).$$

For $i < d$, set

$$T_i^* = \bigcup_{j=0}^{\infty} T_i(n_j).$$

Applying $\mathrm{SDHL}_d(l)$ to the restriction of c to $\prod_{i<d} T_i^*$, we find a somewhere dense level matrix $\prod_{i<d} X_i^*$ of $\prod_{i<d} T_i^*$ on which c is constant. This means that there exists $j < j^+ < \omega$ and $\vec{x} \in \prod_{i<d} T_i(n_j)$ such that for all $i < d$,

$$X_i^* \ \text{dominates} \ T_i[x_i] \cap T_i(n_{j^+}).$$

It follows that $\prod_{i<d} X_i^*$ is an n_{j^+}-$\vec{x}$-dense c-monochromatic level matrix, contradicting the definition of $k(\vec{x})$ and the fact that

$$n_{j^+} \ge n_{j+1} > k(\vec{x}).$$

This finishes the proof. $\square$

Remark 3.3 Note that the proof just given shows that in either of the two versions of the Halpern-Läuchli theorem we may restrict ourselves to partitions with just two pieces without losing any generality. This leads us to an asymmetric version of this result. To state this version we need a definition.

Definition 3.4 *A subset P of some finite product $\prod_{i<d} T_i$ of trees is* highly dense *if for every k there is some n such that $P \cap M$ is a k-dense subset of $\prod_{i<d} T_i$ for every n-dense matrix $M = \prod_{i<d} X_i$ of $\prod_{i<d} T_i$.*

Note that a subset P of $\prod_{i<d} T_i$ whose complement does not include a somewhere dense matrix is a highly dense set since for a given integer k the value $n = k + 1$ works. The following is a restatement of this observation.

Lemma 3.5 *A given subset P of $\prod_{i<d} T_i$ is either highly dense in the full product, or there is $\vec{x} \in \prod_{i<d} T_i$ such that its complement P^c contains a k-$\vec{x}$-dense matrix for every positive integer k.*

The following is yet another formulation of the Halpern-Läuchli theorem.

Theorem 3.6 (Highly Dense Set Version of the Halpern-Läuchli Theorem) *Every highly dense subset of some finite product $\prod_{i<d} T_i$ of trees contains a k-dense matrix for every positive integer k.*

Corollary 3.7 (Asymmetric Version of the Halpern-Läuchli Theorem) *For every coloring $\prod_{i<d} T_i = K_0 \cup K_1$ of some finite product $\prod_{i<d} T_i$ of finitely branching trees either the first color K_0 contains a k-dense matrix for every integer k, or there is some $\vec{x} \in \prod_{i<d} T_i$ such that the second color K_1 contains a k-$\vec{x}$-dense matrix for every integer k.*

Remark 3.8 (1) It is clear that the version of the Halpern-Läuchli theorem appearing in Theorem 3.6, call it HDHL, has the somewhere dense matrix version SDHL as immediate corollary. As we see later, a simple argument gives HDHL as a corollary of SDHL. It turns out, however, that it is more convenient to prove HDHL$_d$ by induction on d rather than SDHL$_d$.
(2) Note that by replacing T_i by $T_i \setminus \bigcup_{0<n<k} T_i(n)$ $(i < d)$, HDHL$_d$ is equivalent to the statement that any highly dense subset of $\prod_{i<d} T_i$ contains a 1-dense matrix. Note also that by replacing T_i with $T_i(A)$ for an appropriately chosen set of levels, HDHL$_d$ is equivalent to the statement that a subset P of some product $\prod_{i<d} T_i$ of trees is either disjoint from a somewhere dense matrix or it includes a 1-dense matrix (and therefore includes a k-dense matrix for all k). This suggests the following finitary formulation of the Halpern-Läuchli theorem.

Theorem 3.9 (Finite Halpern-Läuchli Theorem) *For every finite product $\prod_{i<d} T_i$ of trees and for every integer k, there is an integer l such that for every l-dense matrix M of $\prod_{i<d} T_i$ and every coloring $c : M \to 2$, either there is a somewhere dense matrix $M_1 \subseteq M$ such that $c[M_1] = \{1\}$, or there is a k-dense matrix $M_0 \subseteq M$ such that $c[M_0] = \{0\}$.*

We finish this section with a metric reformulation of the Halpern-Läuchli theorem.

Definition 3.10 *For a given $\varepsilon > 0$, a subset D of some metric space (X, ρ) is ε-dense if for every $x \in X$, there is a $y \in D$ such that $\rho(x, y) < \varepsilon$. A subset D of some product $\prod_{i<d}(X_i, \rho_i)$ of metric spaces is an ε-dense matrix if $D = \prod_{i<d} D_i$, where D_i is ε-dense in X_i for $i < d$.*

Recall also that a metric space (X, ρ) is *totally bounded* if for every $\varepsilon > 0$ there is a finite ε-dense subset of X.

Definition 3.11 *A subset D of some finite product $\prod_{i<d}(X_i, \rho_i)$ of metric spaces is* large *if it contains an ε-dense matrix for every $\varepsilon > 0$. We say that D is* somewhere large *if there is a sequence $U_i \subseteq X_i$ ($i < d$) of nonempty open sets such that the intersection of D with $\prod_{i<d} U_i$ is large in that subproduct.*

Theorem 3.12 (Metric Halpern-Läuchli Theorem) *Let $D = \bigcup_{l=0}^{k} C_l$ be a finite coloring of a large subset D of some finite product $\prod_{i<d}(X_i, \rho_i)$ of totally bounded metric spaces. Then one of the colors C_l must be somewhere large.*

Proof. Since the metric spaces are totally bounded, our assumption that D is large in the product $\prod_{i<d}(X_i, \rho_i)$ permits us to construct a sequence T_i ($i < d$) of finitely branching rooted trees such that

(1) $\bigcup_{n=0}^{\infty} \prod_{i<d} T_i(n) \subseteq D$,

(2) for every $i < d$ and $t \in T_i$ there is a $\delta > 0$ such that for every $\varepsilon > 0$ there is an integer m such that for all $n \geq m$, the set
$$\{x \in T_i(n) : t \leq x\} \cap B_\delta(t)$$
is ε-dense in the ball $B_\delta(t) = \{x \in X_i : \rho_i(t, x) < \delta\}$.

By (1) the coloring of D transfers to a coloring of $\prod_{i<d} T_i$ in a natural way. More precisely, we have a coloring of level-sequences $\vec{x} \in \prod_{i<d} T_i$. One extends the coloring to the rest of the product by taking the projection of $\vec{x}$ on $\min_{i<d}\text{level}(x_i)$. It is clear that the conclusion of the Halpern-Läuchli theorem applied to this coloring via the condition (2) gives us the conclusion of the theorem. $\square$

It should now be clear that we can similarly reformulate the highly dense set version of the Halpern-Läuchli theorem. The following notion corresponds to the notion of a highly dense subset of some product of trees introduced above in Definition 3.4.

Definition 3.13 *Fix a large subset D of some finite product $\prod_{i<d}(X_i, \rho_i)$ of metric spaces. We shall say that a subset P of D is* highly dense *in D if for every $\varepsilon > 0$ there is a $\delta > 0$ such that for every δ-dense matrix $C = \prod_{i<d} C_i \subseteq D$, the intersection $P \cap \prod_{i<d} C_i$ is ε-dense in the product metric space $(Y, \rho) = \prod_{i<d}(X_i, \rho_i)$.*

By embedding a product $\prod_{i<d} T_i$ of finitely branching trees inside D as in the proof of Theorem 3.12 one gets the following result out of the highly dense set version of the Halpern-Läuchli theorem.

Theorem 3.14 (Highly Dense Metric Halpern-Läuchli Theorem) *A highly dense subset of a large subset of some finite product of totally bounded metric spaces is itself a large subset of that product.*

3.2 A PROOF OF THE HALPERN-LÄUCHLI THEOREM

In this section we give a proof of HDHL_d. As indicated above the proof will be done by induction on d. The case $d = 1$ is immediate, so we concentrate on the induction step from d to $d+1$. So, let P be a given subset of some product $\prod_{i \leq d} T_i$ of trees such that its complement does not include any somewhere dense matrix. We need to show that P includes a 1-dense matrix. For $y \in T_d$, we need to consider its section,

$$P_y = \{\vec{x} \in \prod_{i<d} T_i : \vec{x}^\frown y \in P\}.$$

Applying the inductive assumption HDHL_d in the strong-subtree form and using a simple fusion procedure, we find strong subtrees $S_i \subseteq T_i$ $(i < d)$ as witnessed by the same infinite subset $A = \{n_k\}_{k=0}^\infty$ of ω enumerated increasingly such that $n_0 = 0$, i.e., $\text{root}(S_i) = \text{root}(T_i)$ for all $i < d$, and such that for all k, $y \in T_d(k)$ and $\vec{x} \in \bigcup_{l>k} \prod_{i<d} S_i(l)$, $(\subseteq \bigcup_{l>k} \prod_{i<d} T_i(n_l))$,

$$\vec{x}^\frown y \in P \text{ iff } (\vec{x} \restriction n_{k+1})^\frown y \in P.$$

So, to simplify the notation we assume that our initial trees T_i $(i \leq d)$ and the set $P \subseteq \prod_{i \leq d} T_i$ already have this property, or in other words,

(*) For every $y \in T_d$ and every level-vector $\vec{x} \in \prod_{i<d} T_i$ with $\text{level}(\vec{x}) > \text{level}(y) = k$, $\vec{x} \in P_y$ iff $\vec{x} \restriction k + 1 \in P_y$.

To take the full advantage of this property and the same time avoid repetitions, in what follows, matrices and vectors of $\prod_{i<d} T_i$ are implicitly assumed to be level-matrices and level-vectors, respectively. Considering T_d a topological space with the topology τ_0 generated by the basic open sets $\{y : x \leq y\}$ $(x \in T_d)$, we have the following fact.

Lemma 3.15 *For every k and every somewhere τ_0-dense $X \subseteq T_d$ there exists an integer n and somewhere τ_0-dense $Y \subseteq X$ such that for every $y \in Y$ and every n-dense matrix M of $\prod_{i<d} T_i$, the intersection $M \cap P_y$ is a k-dense subset of $\prod_{i<d} T_i$.*

Proof. Suppose the conclusion of the lemma fails for some k and X. Then for every n the set

$$X_n = \{y \in X : \text{there is } n\text{-dense matrix } M \text{ such that} P_y \cap M \text{ is not } k\text{-dense}\}$$

has a nowhere τ_0-dense complement in X. Choose a minimal subset $D \subseteq \prod_{i<d} T_i(k)$ with the property that for some integer n_0 and somewhere τ_0-dense $Y \subseteq X$ the following holds:

(1) $(\forall n \geq n_0)(\forall y \in Y \cap X_n)(\exists \vec{x} \in D)(\exists n\text{-}\vec{x}\text{-dense matrix } M)\ M \cap P_y = \emptyset.$

Then for every $\vec{x} \in D$ and every integer n, the set

$$Y_n(\vec{x}) = \{y \in Y \cap X_n : (\exists n\text{-}\vec{x}\text{-dense matrix } M)\ M \cap P_y = \emptyset\}$$

is τ_0-dense in Y, since otherwise, removing $\vec{x}$ from D, we obtain a smaller set satisfying (1).

Fix Y and D satisfying (1) and an arbitrary $\vec{x} = (x_0, \ldots, x_{d-1}) \in D$. Pick $t \in T_d$ of some height $l \geq k$ such that Y is τ_0-dense in $T_d[t]$. Let $t_0, \ldots, t_p$ be a list of all immediate successors of t in T_d. Then

(2) $(\forall q \leq p)(\forall n)(\exists y \in Y \cap X_n \cap T^{t_q})(\exists n\text{-}\vec{x}\text{-dense matrix } M)\ M \cap P_y = \emptyset.$

Apply (2) for $q = 0$ and get $y_0 \geq t_0$ in $Y \cap X_{l+1}$ such that for some $\vec{x}\text{-}(l+1)$-dense matrix M_0 of $\prod_{i<d} T_i$, the intersection $M_0 \cap P_{y_0}$ is empty. Let n_1 be an integer that is bigger than the level set of the matrix M_0 and the height of y_0. Apply (2) for $q = 1$ and get $y_1 \geq t_1$ in $Y \cap X_{n_1}$ such that for some $\vec{x}\text{-}n_1$-dense matrix M_1 of $\prod_{i<d} T_i$, we have that $M_1 \cap P_{y_1} = \emptyset$, and so on. Proceeding in this way, we obtain $n_p > n_{p-1} > \ldots > n_0 = l+1$, $y_p \geq t_p$ in $Y \cap X_{n_p}$, and some $\vec{x}\text{-}n_p$-dense matrix M_p such that $M_p \cap P_{y_p} = \emptyset$. Note that by the property $(*)$ and the choices of $n_p > \ldots > n_0$ we have that

(3) $(\forall q \leq p)\ M_p \cap P_{y_q} = \emptyset.$

It follows that $M = M_p \times \{y_0, \ldots, y_p\}$ is an $(l+1)\text{-}(\vec{x}^\frown t)$-dense matrix of $\prod_{i<d} T_i$ that avoids P. Since $l+1$ is bigger than the height of any of the nodes of $\vec{x}^\frown t$, this contradicts our initial assumption about P. This proves the lemma. $\qquad \square$

Lemma 3.16 *For every $t \in T_d$ and $k \in \omega$, there is $y \geq t$ in T_d and a k-dense matrix M of $\prod_{i<d} T_i$ such that the level of M lies above the level of y and such that $M \subseteq P_y$.*

Proof. By Lemma 3.15, starting from $n_0 = k$ and $X_0 = \{y \in T_d : y \geq t\}$, we build a strictly increasing sequence $\{n_p\}_{p=0}^\infty$ of integers and a decreasing sequence $\{X_p\}_{p=0}^\infty$ of somewhere τ_0-dense-sets such that

(4) For every $p \in \omega$, every $y \in X_{p+1}$ and every n_{p+1}-dense matrix M, the intersection $M \cap P_y$ is an n_p-dense subset of $\prod_{i<d} T_i$.

Applying now the finite version of the Halpern-Läuchli theorem for the sequence of trees

$$T_i^* = \{\text{root}(T_i)\} \cup \bigcup_{p=0}^\infty T_i(n_p) \quad (i < d),$$

we find an integer $l > 1$ with the property that for every l-dense matrix M of $\prod_{i<d} T_i^*$ and every $c : M \to 2$, either there is a somewhere dense matrix $M^1 \subseteq M$ of the product $\prod_{i<d} T_i^*$ on which c is constantly 1, or a

1-dense matrix $M^0 \subseteq M$ of $\prod_{i<d} T_i^*$ on which c is constantly 0. Choose $y \in X_l$ and choose $l^* \geq l$ such that $n_{l^*} > \text{level}(y)$. Chose an n_l-dense matrix $M = \prod_{i<d} E_i \subseteq \prod_{i<d} T_i(n_{l^*})$ such that the projections from E_i onto $T_i(n_l)$ are all one-to-one. Define $c : M \to 2$ by

$$c(\vec{x}) = 0 \ \text{ iff } \ \vec{x} \in P_y.$$

Using (4), we conclude that c cannot be constantly 1 on any somewhere dense submatrix of M relative to $\prod_{i<d} T_i^*$. So there is a 1-dense submatrix M^0 of M relative to $\prod_{i<d} T_i^*$ on which c is constantly 0, i.e., such that $M_0 \subseteq P_y$. Now note that being 1-dense in $\prod_{i<d} T_i^*$ implies being k-dense in $\prod_{i<d} T_i$. This completes the proof. $\hfill\square$

We are now ready to finish the proof of HDHL_{d+1}. Let $t_0, \ldots, t_p$ be a list of all immediate successors of the root of T_d. Apply Lemma 3.16 to $t = t_0$ and $k = 1$ we get $y_0 \geq t_0$ and a 1-dense matrix M_0 of $\prod_{i<d} T_i$ such that $M_0 \subseteq P_{y_0}$ and $\text{levelset}(M_0) > \text{level}(y_0)$. Let l_0 be the level of the matrix M_0 and choose $u_1 \geq t_1$ at level l_0. Apply now Lemma 3.16 to $k = l_0$ and $t = u_1$ and obtain an l_0-dense matrix M_1 of $\prod_{i<d} T_i$ and $y_1 \geq u_1$ such that $M_1 \subseteq P_{y_1}$ and $\text{levelset}(M_1) > \text{level}(y_1)$, and so on. Proceeding in this way, we arrive at $l_{p-1}(=\text{the level of } M_{p-1})$-dense matrix M_p and $y_p \geq u_p \geq t_p$ such that $M_p \subseteq P_{y_p}$. Note that by $(*)$,

(5) $M_p \subseteq P_{y_q}$ for all $q \leq p$.

It follows that $M = M_p \times \{y_0, \ldots, y_p\}$ is a 1-dense matrix of $\prod_{i<d} T_i$ such that $M \subseteq P$. This completes the inductive step and therefore the proof of the highly dense set version of the Halpern-Läuchli theorem. $\hfill\square$

3.3 PRODUCTS OF FINITE SETS

The purpose of this section is to present a basic Ramsey theoretic result about a special kind of finitely branching tree of height ω, the trees of the form

$$T(\vec{H}) = \bigcup_k \prod_{i<k} H_i$$

for some infinite sequence $\vec{H} = (H_i)_{i=0}^{\infty}$ of nonempty finite sets of integers. We call such trees *product trees*. A product tree $T(\vec{H})$ is also called an $\vec{m}$-*tree* for some infinite sequence $\vec{m} = (m_i)_{i=0}^{\infty}$ of positive integers provided that $m_i = |H_i|$ for all i. A product tree $T(\vec{J})$ is a *subtree* of a product tree $T(\vec{H})$ if $J_i \subseteq H_i$ for all i. In this section we prove a particular strengthening of the following basic Ramsey theoretic result about product trees.

Theorem 3.17 (Product Tree Ramsey Theorem) *For every infinite sequence $\vec{m}$ of positive integers there is an infinite sequence $\vec{n}$ of positive integers such that for every 2-coloring c of an $\vec{n}$-tree $T(\vec{H})$ there exists an $\vec{m}$-subtree $T(\vec{J})$ of $T(\vec{H})$ such that c is constant on infinitely many levels of $T(\vec{J})$.*

We shall use the symbol $\vec{n} \rightarrow \vec{m}$ to denote conclusion of this theorem and $\vec{n} \nrightarrow \vec{m}$ to denote its negation. Identifying an infinite subset M of $\mathbb{N}$ with its increasing enumeration $M = (m_i)_{i=0}^{\infty}$, we extend the arrow-notation $N \rightarrow M$ to pairs N and M of infinite subsets of $\mathbb{N}$. Let $\mathbb{N}^{[\infty]}$ denote the collection of all infinite subsets of $\mathbb{N}$. The Product Tree Ramsey Theorem follows from the following result, where for an infinite subset $M = (m_i)_{i=0}^{\infty}$ of $\mathbb{N}$ enumerated increasingly, we let

$$M_e = (m_{2i})_{i=0}^{\infty} \text{ and } M_o = (m_{2i+1})_{i=0}^{\infty}.$$

Lemma 3.18 *There is an infinite subset N of $\mathbb{N}$ such that $M_o \rightarrow M_e$ for all infinite subsets M of N.*

The most natural way to try to prove this lemma is to apply the infinite-dimensional Ramsey theorem to the coloring

$$\mathbb{N}^{[\infty]} = \{M : M_o \rightarrow M_e\} \cup \{M : M_o \nrightarrow M_e\}$$

by eliminating the possibility that the second color contains an infinite-dimensional cube $N^{[\infty]}$ for some infinite subset N of $\mathbb{N}$. In Section 1.4 above, we found an optimal condition on such colorings that guarantees the existence of a monochromatic cube $N^{[\infty]}$, a condition that in particular allows all colorings of certain descriptive complexity. Observant readers will notice that here we need exactly the infinite-dimensional Ramsey theorem for colors belonging to the second level of the projective hierarchy, a result that is not exactly provable without use of some additional axioms of set theory so we have to find a different route to prove Lemma 3.18.

Before starting the proof we need the following definition and lemma.

Definition 3.19 *Define[1] $S : \mathbb{N}_{+}^{<\infty} \rightarrow \mathbb{N}_{+}$, by letting $S(\emptyset) = 0$, $S(m_0) = 2m_0 - 1$, and*

$$S(m_0, \ldots, m_{k+1}) = 2(m_{k+1} - 1) \prod_{i=0}^{k} \left(\frac{S(m_0, \ldots, m_i)}{m_i} \right) + 1.$$

Lemma 3.20 *For every finite sequence $m_0, \ldots, m_k$ of positive integers and every coloring*

$$c : \prod_{i=0}^{k} S(m_0, \ldots, m_i) \rightarrow \{0, 1\},$$

there exist $H_i \subseteq S(m_0, \ldots, m_i)$,[2] $|H_i| = m_i$ for $i \leq k$ such that the subproduct $\prod_{i=1}^{k} H_i$ is monochromatic.

Proof. The proof is by induction on k. It clearly holds for $k = 0$. So let us assume the conclusion if true for $m_0, \ldots, m_k$ and prove it for $m_0, \ldots, m_k, m_{k+1}$.

---

[1]Recall that by $\mathbb{N}_{+}$, we denote the set of all *positive* integers.
[2]Recall that we identify the integer $n = S(m_0, \ldots, m_{k+1})$ with the set $\{0, 1, \ldots, n-1\}$ of nonnegative integers smaller than n.

Let $c : \prod_{i=0}^{k+1} S(m_0, \ldots, m_i) \to 2$ be a given coloring. Then for every $0 \leq x < n$ we can consider the induced coloring

$$c_x : \prod_{i=0}^{k} S(m_0, \ldots, m_i) \to 2$$

defined by $c_x(x_0, \ldots, x_k) = c(x_0, \ldots, x_k, x)$. By the induction hypothesis for each $0 \leq x < n$, we can fix sets $H_i^x \subseteq S(m_0, \ldots, m_i)$, $|H_i^x| = m_i$ for $i \leq k$ such that c_x is constant on their product $\prod_{i=1}^{k} H_i^x$. Let ε_x be the constant value of c_x on that product. From the definition of $n = S(m_0, \ldots, m_k, m_{k+1})$, we infer that there must be $H_{k+1} \subseteq n$, $\varepsilon \in \{0, 1\}$, and $H_i \subseteq S(m_0, \ldots, m_i)$, $|H_i| = m_i$ for $i \leq k$ such that $\varepsilon_x = \varepsilon$ and $H_i = H_i^x$ for all $i \leq k$ and $x \in H_{k+1}$. Then c is constant on $\prod_{i=1}^{k+1} H_i$, as required. $\qquad\square$

Now, we are ready to start with the proof of Lemma 3.18. Suppose that the conclusion of Lemma 3.18 is false. Thus, for every infinite subset N of $\mathbb{N}$, there is an infinite subset M of N such that $M_o \not\to M_e$. Let $\mathcal{R}$ be the collection of all pairs (N, M, c), where $M = (m_i)_{i=0}^{\infty} \subseteq N$ are infinite subsets of $\mathbb{N}$ and $c : \mathbb{N}^{<\infty} \to 2$ is a mapping whose restriction to the product tree

$$T(M_o) = \bigcup_{k=0}^{\infty} \prod_{i=0}^{k} m_{2i+1}$$

is a witness of $M_o \not\to M_e$. Clearly, $\mathcal{R}$ is a coanalytic subset of the product-space $\mathbb{N}^{[\infty]} \times \mathbb{N}^{[\infty]} \times 2^{(\mathbb{N}^{<\infty})}$ with the property that for every $N \in \mathbb{N}^{[\infty]}$, there exists $(M, c) \in \mathbb{N}^{[\infty]} \times 2^{(\mathbb{N}^{<\infty})}$ such that $\mathcal{R}(N, M, c)$ holds. By the Ramsey Uniformization Theorem 1.59 there is an infinite subset P of $\mathbb{N}$ and a continuous map

$$F : P^{[\infty]} \to \mathbb{N}^{[\infty]} \times 2^{(\mathbb{N}^{<\infty})}$$

such that $\mathcal{R}(N, F(N))$ for all $N \in P^{[\infty]}$. Let $F_0 : P^{[\infty]} \to \mathbb{N}^{[\infty]}$ and $F_1 : P^{[\infty]} \to 2^{(\mathbb{N}^{<\infty})}$ be the compositions of F with the first and second projections of the product space $\mathbb{N}^{[\infty]} \times 2^{(\mathbb{N}^{<\infty})}$, respectively. Thus $F(N) = (F_0(N), F_1(N))$ for every $N \in P^{[\infty]}$. Let $\mathcal{X}$ be the range of F_0. Then $\mathcal{X}$ is an analytic subset of $P^{[\infty]}$ with the property that for all $N \in P^{[\infty]}$ there is an $M \in \mathcal{X}$ such that $M \subseteq N$. Applying Silver's theorem we conclude that there is an $N \in P^{[\infty]}$ such that $N^{[\infty]} \subseteq \mathcal{X}$. Working directly from the given mapping F, or applying the Ramsey-Uniformization theorem again and shrinking N if necessary, we may assume we have a continuous map $M \mapsto c_M$ from $N^{[\infty]}$ into $2^{(\mathbb{N}^{<\infty})}$ such that the restriction of c_M to the product tree $T(M_o) = \bigcup_{k=0}^{\infty} \prod_{i=0}^{k} m_{2i+1}$ is a witness of $M_o \not\to M_e$. From now on, we fix such an infinite subset N of $\mathbb{N}$ and a continuous map $M \mapsto c_M$ from $N^{[\infty]}$ into $2^{(\mathbb{N}^{<\infty})}$ with this property and work towards a contradiction that will finish the proof of Lemma 3.18.

Given an infinite set $M = (m_i)_{i=0}^{\infty}$ of N, let $M^* = (m_{\sigma(i)})_{i=0}^{\infty}$ denote its subset determined by the strictly increasing map $\sigma : \mathbb{N} \to \mathbb{N}$ defined by

$\sigma(0) = 0$, $\sigma(2k+2) = \sigma(2k+1)+1$, and

$$\sigma(2k+1) = 1 + 2k + \sum_{i=1}^{k} 2m_{\sigma(2i)}.$$

Thus, M^* is obtained by taking the first element m_0 of M and if $m \in M$ is chosen to occupy the index $2k$ in the increasing enumeration of M^*, or in other words $m = m_{\sigma(2k)}$, then between m and the next element of M^* there are exactly $2m$ elements of M. We let $H_k^0(M)$ be the set formed by the first m members of M after m and $H_k^1(M)$ be the set formed by the next m members of M. Note that the corresponding product trees $T(\vec{H}_k^0(M))$ and $T(\vec{H}_k^1(M))$ are both subtrees of the product tree $T(M_o^*)$ and that both of them are $M_e^* = (m_{\sigma(2i)})_{i=0}^\infty$-trees. It follows that c_{M^*} cannot be constant on infinitely many levels of either of these two product subtrees. So for $\varepsilon \in \{0, 1\}$, we can define

$$k_\varepsilon(M) = \max \left\{ k : c_{M^*} \upharpoonright \prod_{i=0}^{k} H_i^\varepsilon(M) \text{ is constant} \right\}.$$

This allows us to consider the coloring

$$N^{[\infty]} = \{M : k_0(M) < k_1(M)\} \cup \{M : k_0(M) \geq k_1(M)\},$$

which is clearly a Borel coloring, given the fact that our mapping $M \mapsto c_M$ is continuous. Applying the Galvin-Prikry theorem, we will reach the desired contradiction if we show that neither of the two colors can contain a cube of the form $M^{[\infty]}$, where M is an infinite subset of N. Since the two colors are rather symmetric, it suffices to show the following.

Claim 3.20.1 *Every infinite subset $X = (x_i)_{i=0}^\infty$ of N contains an infinite subset M such that $k_0(M) > k_1(M)$.*

Proof. Let $Y = (x_\tau(i))_{i=0}^\infty$, where $\tau : \mathbb{N} \to \mathbb{N}$ is the strictly increasing map defined by $\tau(0) = 0$, $\tau(2k+2) = \tau(2k+1)+1$, and

$$\tau(2k+1) = 1 + 2k + x_{\tau(2k)} + S(x_{\tau(0)}, x_{\tau(2)}, \ldots, x_{\tau(2k)}),$$

where $S : \mathbb{N}^{<\infty} \to \mathbb{N}$ is the map given above in Definition 3.19. Then for each k, the open interval between $x_{\tau(2k)}$ and $x_{\tau(2k+1)}$ contains exactly $x_{\tau(2k)} + S(x_{\tau(0)}, x_{\tau(2)}, \ldots, x_{\tau(2k)})$ many elements of X, so let J_k denote the $S(x_{\tau(0)}, x_{\tau(2)}, \ldots, x_{\tau(2k)})$ elements, and let H_k^1 denote the last $x_{\tau(2k)}$ elements of X in this interval. This gives us an infinite sequence $\vec{H}^1 = (H_i^1)_{i=0}^\infty$ such that the corresponding product tree $T(\vec{H}^1)$ is a subtree of the product tree $T(Y_o)$, so by the choice of c_Y, there is a maximal integer k such that c_Y is constant on $\prod_{i=0}^{k} H_i^1$. By the property of S stated in Lemma 3.20 there exist sets $H_i^0 \subseteq J_i$ for $0 \leq i \leq k+1$ such that

(1) $|H_i^0| = x_{\tau(2i)}$, and

(2) c_Y is constant on the product $\prod_{i=0}^{k+1} H_i^0$.

For $i > k + 1$, choose $H_i^0 \subseteq J_i$, $|H_i^0| = x_{\tau(2i)}$ arbitrarily. Let

$$M = Y \cup \bigcup_{i=0}^{\infty} (H_i^0 \cup H_i^1).$$

Then M is an infinite subset of X, $M^* = Y$, and $H_i^\varepsilon(M) = H_i^\varepsilon$ for all $i \in \mathbb{N}$ and $\varepsilon \in \{0, 1\}$. It follows that

$$k_0(M) = k < k + 1 \leq k_1(M).$$

Thus, we have found an infinite subset of X of the second color. Clearly, a symmetric argument will give us an infinite subset of X of the first color. So we have reached the desired contradiction that finishes the proof of the claim and therefore the proof of Lemma 3.18. □

We now state a theorem that is stronger than Theorem 3.17 and which we use in Chapter Nine below.

Theorem 3.21 *There is an $R : \mathbb{N}_+^{<\infty} \to \mathbb{N}_+$ such that for every infinite sequence $(m_i)_{i=0}^{\infty}$ of positive integers and for every coloring*

$$c : \bigcup_{k=0}^{\infty} \prod_{i=0}^{k} R(m_0, \ldots, m_i) \to 2,$$

there exist $H_i \subseteq R(m_0, \ldots, m_i)$, $|H_i| = m_i$, for $i \in \mathbb{N}$, such that c is constant on the product

$$\prod_{i=0}^{k} H_i$$

for infinitely many k.

Proof. Pick an infinite subset $N = (n_i)_{i=0}^{\infty}$ of $\mathbb{N}_+$ enumerated increasingly and satisfying Lemma 3.18. Set

$$R(m_0, \ldots, m_k) = n_{2(\sum_{i=0}^{k} m_i) + 1}.$$

Then for every infinite sequence $(m_i)_{i=0}^{\infty}$ of positive integers, if we let

$$P = \{n_{2(\sum_{i=0}^{k} m_i) + \varepsilon} : k \in \mathbb{N}, \varepsilon \in \{0, 1\}\},$$

we get an infinite subset of N such that

$$P_o = (R(m_0, \ldots, m_i))_{i=0}^{\infty},$$

while the sequence P_e pointwise dominates our given sequence $(m_i)_{i=0}^{\infty}$. It follows that $P_o \to P_e$ gives us the desired conclusion of Theorem 3.21. □

We finish this section with the following natural question regarding the Product Tree Ramsey Theorem.

Question 3.22 *Is there a (primitive) recursive sequence $(n_i)_{i=0}^{\infty}$ of positive integers such that for every coloring*

$$c : \bigcup_{k=0}^{\infty} \prod_{i=0}^{k} \{0, 1, \ldots, n_i\} \to \{0, 1\}$$

there exists sets $H_i \subseteq \{0, 1, \ldots, n_i\}$, $|H_i| = 2$, $(i = 0, 1, 2, \ldots)$ such that the restriction

$$c \upharpoonright \prod_{i=0}^{k} H_i$$

is constant for infinitely many k?

NOTES TO CHAPTER THREE

The Halpern-Läuchli theorem was originally proved by Halpern and Läuchli [44] as a lemma that was needed for a construction of a model of set theory in which the Boolean Prime Ideal Theorem (BPI) is true but not the full Axiom of Choice (AC). This consistency result itself appears in the paper of Halpern-Levy [45]. It turns out that the Halpern-Läuchli theorem can in fact be deduced from this consistency result, provided one states it in the form that every countable transitive model of set theory has a forcing extension satisfying BPI but failing to satisfy AC in the particular way described by Halpern-Levy [45]. This information may have been motivation for a proof of the Halpern-Läuchli theorem due to L. Harrington (unpublished), which uses analysis of the forcing relation of the poset $\mathcal{C}_I$ of all finite partial functions from I into $\{0, 1\}$ for a sufficiently large index set I that depends on the dimension d. Harrington's proof appears to us as the conceptually simplest proof of this deep result, but there are several other proofs that, unlike Harrington's, all use induction on the the dimension d and which vary in complexity depending on which form of the theorem they are proving. The first known reformulation of the Halpern-Läuchli theorem in terms of metric spaces appears in the paper of Argyros-Felouzis-Kanellopoulos [2] and goes in a slightly different direction than the one given above in Theorem 3.12. The Product Tree Ramsey Theorems 3.17 and 3.21 appear in the paper of DiPrisco-Llopis-Todorcevic [22], but the proof there uses an additional set theoretic assumption that we know now is unnecessary. It should be mentioned that the original motivation for this result was again in finding a consistency proof, this time the consistency proof showing that the infinite-dimensional form of the Ramsey theorem is not a consequence of its "polarized" (product) version. It turns out that Theorem 3.17 can be deduced from the fact that a generic extension of $L(\mathbb{R})$ (the Gödel constructible universe over the reals) satisfies the infinite-dimensional polarized Ramsey theorem that we give in Chapter Eight of this book. So this is another instance where a consistency question is essentially equivalent to a Ramsey-theoretic problem.

Chapter Four

Abstract Ramsey Theory

4.1 ABSTRACT BAIRE PROPERTY

An infinite-dimensional Ramsey theoretic result is usually given under some restriction on the colorings. It turns out that an appropriate variation of the classical topological notion of sets with the Baire property leads us to a restriction that seems optimal. The purpose of this section is to present a variation that is used in the rest of the book.

Definition 4.1 *Let X be a given set and let $\mathcal{P}$ be a collection of nonempty subsets of X that we call basic sets. For $P \in \mathcal{P}$, let*

$$\mathcal{P} \upharpoonright P = \{Q \in \mathcal{P} : Q \subseteq P\}.$$

We say that a subset Y of X is $\mathcal{P}$-Baire if for every $P \in \mathcal{P}$ there is a $Q \in \mathcal{P} \upharpoonright P$ such that $Q \subseteq Y$ or $Q \cap Y = \emptyset$. If for every $P \in \mathcal{P}$ we can find $Q \in \mathcal{P} \upharpoonright P$ such that $Q \cap Y = \emptyset$, we say that Y is $\mathcal{P}$-meager.

Clearly, $\mathcal{P}$-Baire sets form a field of subsets of X, and the collection of $\mathcal{P}$-meager subsets of X forms an *ideal* of subsets of X, i.e., it is closed under taking subsets and finite unions.

Lemma 4.2 *If every member of $\mathcal{P}$ is $\mathcal{P}$-Baire and if the ideal of $\mathcal{P}$-meager sets is a σ-ideal, then the collection of all $\mathcal{P}$-Baire subsets of X is a σ-field.*

Proof. Let (Y_n) be a given sequence of $\mathcal{P}$-Baire sets, let $Y = \bigcup_{n=0}^{\infty} Y_n$ and let P be a given member of $\mathcal{P}$. If there is n such that $Y_n \cap P$ is not $\mathcal{P}$-meager i.e., it includes a basic set Q, we would be done. So we are left with the case that $Y_n \cap P$ is $\mathcal{P}$-meager for all n.
By the assumption of the lemma,

$$Y \cap P = \bigcup_{n=0}^{\infty} (Y_n \cap P)$$

is also $\mathcal{P}$-meager. So there is a basic set $Q \subseteq P$ such that $Q \cap Y = \emptyset$ as required. $\qquad\square$

Example 4.1.1   *Let X be a topological space in which no nonempty open set is meager and let $\mathcal{P}$ be the collection of all nonempty subsets of X of the form $G \setminus \bigcup_{k=0}^{\infty} N_k$ where $G \subseteq X$ is open and all the N_k are nowhere dense* *in X. Then a subset Y of X is $\mathcal{P}$-Baire if and only if Y has the property of Baire in X, i.e., Y can be written as*

$$Y = (U \setminus M) \cup (M \setminus U)$$

for some open set $U \subseteq X$ and a meager subset M of X, i.e., a subset that can be covered by countably many nowhere dense subsets of X. It is also clear that a subset Y of X is $\mathcal{P}$-meager if and only if it is meager in X. Clearly, the ideal of meager subsets of X is σ-additive so we have that the property of Baire subsets of X form a σ-field.

Example 4.1.2 *Let $X = [0,1]$ and let $\mathcal{P}$ be the collection of all compact subsets of $[0,1]$ of positive Lebesgue measure. Then a subset Y of $[0,1]$ is $\mathcal{P}$-Baire if and only if it is Lebesgue-measurable. Moreover, a subset Y of $[0,1]$ is $\mathcal{P}$-meager if and only if it has Lebesgue measure-zero. So in this case, we also have that $\mathcal{P}$-Baire subsets form a σ-field.*

Definition 4.3 *A $\mathcal{P}$-envelope of a subset Y of X is any $\mathcal{P}$-Baire set $\Phi(Y)$ that includes Y and has the property that every $\mathcal{P}$-Baire subset of the difference $\Phi(Y) \setminus Y$ must be $\mathcal{P}$-meager.*

Lemma 4.4 *Let X be a topological space and let $\mathcal{P}$ be the collection of basic subsets given above in Example 4.1.1. Then every subset of X admits an F_σ-envelope relative to $\mathcal{P}$.*

Proof. For a subset Y of X, let $D(Y)$ be the set of all $x \in X$ such that $Y \cap U$ is not meager for any open subset U of X containing x. Then by the Banach Category Theorem, $Y \setminus D(Y)$ is a meager set, so there is a meager F_σ-set $M(Y)$ including $Y \setminus D(Y)$. Let

$$\Phi(Y) = D(Y) \cup M(Y).$$

Then $\Phi(Y)$ is an F_σ-superset of Y and every property of Baire subset of $\Phi(Y) \setminus Y$ is meager. □

Similarly, one has the following result about the family of basic sets given in Example 4.1.2.

Lemma 4.5 *Let $X = [0,1]$ and $\mathcal{P}$ be the family of the compact subsets of $[0,1]$ of positive measure. Then every $Y \subseteq X$ has a $\mathcal{P}$-envelope. In particular, we can take $\Phi(Y)$ to be a G_δ-superset of Y whose measure is equal to the outer measure of Y.*

Definition 4.6 *Recall that a Souslin operation $\mathcal{A}$ is an operation that turns a family Y_s ($s \in \mathbb{N}^{<\infty}$) of sets indexed by finite sequences of nonnegative integers, a Souslin scheme, into the set*

$$\mathcal{A}(Y_s : s \in \mathbb{N}^{<\infty}) = \bigcup_{x \in \mathbb{N}^\infty} \bigcap_{n \in \mathbb{N}} Y_{x \restriction n}.$$

Theorem 4.7 (Marczewski) *Suppose $\mathcal{P}$ is a family of nonempty subsets of some set X such that every member of $\mathcal{P}$ is $\mathcal{P}$-Baire and such that the ideal of $\mathcal{P}$-meager sets is σ-additive. Suppose further that every subset of X admits a $\mathcal{P}$-envelope. Then the field of $\mathcal{P}$-Baire subsets of X is closed under the Souslin operation.*

Proof. Let Y_s ($s \in \mathbb{N}^{<\infty}$) be a given Souslin scheme of subsets of X that are $\mathcal{P}$-Baire. Let $Y = \mathcal{A}(Y_s : s \in \mathbb{N}^{<\infty})$ be the result of the Souslin operation and let $P \in \mathcal{P}$ be a given basic set. We need to find $Q \subseteq P$ in $\mathcal{P}$ such that $Q \subseteq Y$ or $Q \cap Y = \emptyset$.
Clearly, we may assume that $Y_t \subseteq Y_s$ whenever $s \sqsubseteq t$. For $s \in \mathbb{N}^{<\infty}$, set

$$Y_s^* = \bigcup_{x \sqsupseteq s} \bigcap_n Y_{x \restriction n},$$

and pick a $\mathcal{P}$-envelope $\Phi(Y_s^*)$ of Y_s^* such that

$$Y_s^* \subseteq \Phi(Y_s^*) \subseteq Y_s.$$

Let

$$M_s = \Phi(Y_s^*) \setminus \bigcup_n \Phi(Y_{s^\frown n}^*).$$

Then M_s is a $\mathcal{P}$-Baire set that is included in the difference $\Phi(Y_s^*) \setminus Y_s^*$, since clearly

$$Y_s^* = \bigcup_n Y_{s^\frown n}^*.$$

It follows that M_s is a $\mathcal{P}$-meager set for all $s \in \mathbb{N}^{<\infty}$. So we can find $Q_0 \subseteq P$ in $\mathcal{P}$ such that

$$Q_0 \cap \left(\bigcup_{s \in \mathbb{N}^{<\infty}} M_s \right) = \emptyset.$$

Since $Y_\emptyset^* = Y$, the proof is finished once we show that

$$Q_0 \cap \Phi(Y_\emptyset^*) = Q_0 \cap Y_\emptyset^*,$$

since by our assumption the set $Q_0 \cap \Phi(Y_\emptyset^*)$ is a $\mathcal{P}$-Baire set and so we can find $Q \subseteq Q_0$ that is either contained in or is disjoint from this set. To check the equality, pick an $x \in Q_0 \cap \Phi(Y_\emptyset^*)$. Since $Q_0 \cap M_\emptyset = \emptyset$, there is n_0 such that $x \in \Phi(Y_{\langle n_0 \rangle}^*)$. Since $Q_0 \cap M_{\langle n_0 \rangle} = \emptyset$, there is n_1 such that $x \in \Phi(Y_{\langle n_0\ n_1 \rangle}^*)$, and so on. Continuing this way we obtain an infinite sequence $a = (n_k) \in \mathbb{N}^\infty$ such that $x \in \Phi(Y_{a \restriction k}^*)$ for all k. Since $\Phi(Y_s^*) \subseteq Y_s$ for all s, we conclude that

$$x \in \bigcap_k \Phi(Y_{a \restriction k}^*) \subseteq \bigcap_k Y_{a \restriction k} \subseteq Y = Y_\emptyset.$$

This finishes the proof.   □

Corollary 4.8 (Nikodym) *The field of property of Baire subsets of any topological space is closed under the Souslin operation.*

Let us now give one sufficient condition for the existence of $\mathcal{P}$-envelopes.

Definition 4.9 *A subset $\mathcal{D}$ of $\mathcal{P}$ is* predense *in $\mathcal{P}$ if for every $P \in \mathcal{P}$ there exist $D \in \mathcal{D}$ and $Q \in \mathcal{P}$ such that $Q \subseteq P \cap D$. A subset $\mathcal{D} \in \mathcal{P}$ is* dense *in $\mathcal{P}$ if for every $P \in \mathcal{P}$, there exists a $D \in \mathcal{D}$ such that $D \subseteq P$. A subset O of $\mathcal{P}$ is* open *if $P \subseteq Q$, $P \in \mathcal{P}$, and $Q \in O$ imply $P \in O$.*

Note that every predense-set that is open is in fact dense.

Definition 4.10 *We say that $\mathcal{P}$ has the* disjoint-refinement *property if for every dense-open $\mathcal{D} \subseteq \mathcal{P}$ there is an $\mathcal{A} \subseteq \mathcal{D}$ that is predense in $\mathcal{P}$ and that consists of pairwise disjoint basic sets.*

Example 4.1.3 *Let $X = [0, 1]$ and let $\mathcal{P}$ be the collection of all compact subsets of $[0, 1]$ of positive Lebesgue measure. Then $\mathcal{P}$ has the disjoint refinement property.*

Theorem 4.11 *If $\mathcal{P}$ has the disjoint-refinement property then every subset of X has a $\mathcal{P}$-envelope.*

Proof. Let Y be a given subset of X and consider the following set

$$\mathcal{D} = \{P \in \mathcal{P} : P \cap Y = \emptyset \vee (\forall Q \in \mathcal{P})(Q \subseteq P \rightarrow Q \cap Y \neq \emptyset)\}.$$

Then $\mathcal{D}$ is a dense open subset of $\mathcal{P}$. Choose a predense-set $\mathcal{A} \subseteq \mathcal{D}$ consisting of pairwise disjoint basic sets. Let

$$\Phi(Y) = X \setminus \bigcup \{P \in \mathcal{A} : P \cap Y = \emptyset\}.$$

We need to check that every $\mathcal{P}$-Baire subset M of $\Phi(Y) \setminus Y$ must be $\mathcal{P}$-meager. Otherwise, using the fact that $\Phi(Y)$ is $\mathcal{P}$-Baire, we can find $P \in \mathcal{P}$ such that $P \subseteq M$. Since $\mathcal{A}$ is predense in $\mathcal{P}$, there are $Q \in \mathcal{A}$ and $R \in \mathcal{P}$ such that

$$R \subseteq P \cap Q.$$

Since $Q \in \mathcal{D}$ and we have found $R \in \mathcal{P}$ such that $R \subseteq Q$ and $R \cap Y = \emptyset$, it must be that $Q \cap Y = \emptyset$. It follows that $Q \cap \Phi(Y) = \emptyset$, a contradiction. This finishes the proof. $\square$

Corollary 4.12 *Suppose that every $P \in \mathcal{P}$ is $\mathcal{P}$-Baire and that the ideal of $\mathcal{P}$-meager sets is σ-additive. If $\mathcal{P}$ has the disjoint-refinement property then the σ-field of $\mathcal{P}$-Baire sets is closed under the Souslin operation.*

Corollary 4.13 (Luzin-Sierpinski) *Lebesgue measurability is invariant under the Souslin operation.*

It turns out that the disjoint-refinement property is shared by many families of basic sets that are considered in this book. To establish that a given family $\mathcal{P}$ of basic sets has the disjoint-refinement property, one usually uses some sort of noneffective method, like the well-ordering principle or Zorn's lemma. As one might have expected, these uses can be avoided in all cases considered here by applying some local version of the disjoint refinement lemma that involves only countably many sets. It is still of interest to know under which conditions every subset Y of X does have a $\mathcal{P}$-envelope. So let us give an example of one such condition.

Definition 4.14 *A family $\mathcal{P}$ of basic subsets of some set X is* regular *if for every $P \in \mathcal{P}$ and $\mathcal{A} \subseteq \mathcal{P}$ consisting of pairwise disjoint basic sets, either P has a non-P-meager intersection with one member of $\mathcal{A}$, or else there is a $Q \in \mathcal{P}$ such that $Q \subseteq P$ and $Q \cap R = \emptyset$ for all $R \in \mathcal{A}$.*

Example 4.1.4 *Let X be a topological space in which no nonempty open subset is meager. Let $\mathcal{P}$ be the collection of all subsets of X of the form $G \setminus M$, where G is a nonempty open subset of X and M is a meager subset of X. Then $\mathcal{P}$ is regular.*

Theorem 4.15 *Every regular family $\mathcal{P}$ of basic subsets of some set X has the disjoint-refinement property.*

Proof. We may assume that $\mathcal{P}$ is infinite. Let $\leq_w$ be a well-ordering of $\mathcal{P}$ of minimal possible order type. Let $\mathcal{D} \subseteq \mathcal{P}$ be dense-open. Recursively on $\leq_w$, for each $P \in \mathcal{P}$, we choose $Q_P \subseteq P$ in $\mathcal{D} \cup \{\emptyset\}$ as follows. If there is a $P' <_w P$ such that $Q_{P'} \cap P$ is not $\mathcal{P}$-meager, we let $Q_P = \emptyset$; otherwise, we choose $Q_P \in \mathcal{D}$ such that $Q_P \subseteq P$ and such that

$$Q_{P'} \cap Q_P = \emptyset \text{ for all } P' <_w P.$$

To show that the choice of Q_P is possible, one uses the regularity of $\mathcal{P}$ and the inductive assumption that $\{Q_{P'} : P' <_w P\}$ consists of pairwise disjoint sets. This finishes the proof. $\square$

Corollary 4.16 *Suppose $\mathcal{P}$ is a regular family of basic subsets of some set X, that every member of $\mathcal{P}$ is $\mathcal{P}$-Baire, and that the ideal of $\mathcal{P}$-meager sets is σ-additive. Then the σ-field of $\mathcal{P}$-Baire subsets of X is closed under the Souslin operation.*

We finish this section with a standard example of a regular family of basic sets.

Example 4.1.5 *Let $X = \mathbb{R}$ and let $\mathcal{P}$ be the family of all perfect subsets of $\mathbb{R}$, i.e., nonempty compact subsets of $\mathbb{R}$ that do not have isolated points.*

The field of subsets of $\mathbb{R}$ that are $\mathcal{P}$-Baire is known in the literature as the Marczewski field. The ideal of $\mathcal{P}$-meager sets of reals is σ-additive and it is known in the literature as the Marczewski ideal or the s_0-ideal.

Lemma 4.17 *The family $\mathcal{P}$ of perfect subsets of $\mathbb{R}$ has the disjoint refinement property.*

Proof. To see that $\mathcal{P}$ has the disjoint refinement property via a recursive construction of length continuum, it suffices to show for every $P \in \mathcal{P}$ and every family $\mathcal{A}$ of disjoint perfect subsets of $\mathbb{R}$ of size smaller than the continuum, there is a perfect set $Q \subseteq P$ that is either disjoint from every set from $\mathcal{A}$, or there is $Q \in \mathcal{A}$ such that $Q \cap P$ contains a perfect set of reals. If there is a $Q \in \mathcal{A}$ such that $Q \cap P$ is uncountable, the intersection contains a perfect set as required. If $Q \cap P$ is countable for all $Q \in \mathcal{A}$, the set $(\bigcup \mathcal{A}) \cap P$ has size less than continuum. Splitting $\mathcal{P}$ into a family of size continuum disjoint perfect subsets yields that one of the perfect sets does not intersect $\bigcup \mathcal{A}$, as required. □

Corollary 4.18 *The Marczewski field on $\mathbb{R}$ generated by the family $\mathcal{P}$ of all perfect subsets of $\mathbb{R}$ is closed under the Souslin operation.*

4.2 THE ABSTRACT RAMSEY THEOREM

In this section we introduce the basic notion of this book, the notion of *Ramsey space*, a structure of the form

$$(\mathcal{R}, \mathcal{S}, \leq, \leq^o, r, s),$$

satisfying certain conditions. We shall think of $\mathcal{R}$, $\mathcal{S}$ as families of infinite sequences of objects and

$$r : \mathcal{R} \times \omega \to \mathcal{AR} , \ s : \mathcal{S} \times \omega \to \mathcal{AS}$$

as functions giving us finite approximations

$$r_n(A) = r(A, n) \text{ and } s_n(X) = s(X, n)$$

to these objects. We let $\mathcal{AR}_n$ be the range of r_n and $\mathcal{AS}_n$ the range of s_n, and we assume that

$$\mathcal{AR} = \bigcup_{n \in \mathbb{N}} \mathcal{AR}_n \text{ and } \mathcal{AS} = \bigcup_{n \in \mathbb{N}} \mathcal{AS}_n.$$

The relation $\leq$ is a reflexive and transitive relation on $\mathcal{S}$ while $\leq^o$ is a subset of $\mathcal{R} \times \mathcal{S}$, i.e., a relation between elements of $\mathcal{R}$ and elements of $\mathcal{S}$ such that

$$A \leq^o X \leq Y \text{ implies } A \leq^o Y.$$

We introduce the following axiom about these objects which tells us that $\mathcal{R}$ and $\mathcal{S}$ are collections of infinite sequences of objects and that, on the other hand, $\mathcal{AR}$ and $\mathcal{AS}$ are collections of finite sequences that approximate them.

A.1. (Sequencing) For any choice of $(\mathcal{P}, p) \in \{(\mathcal{R}, r), (\mathcal{S}, s)\}$ the following conditions are satisfied:

(1) $p_0(P) = p_0(Q)$ for all $P, Q \in \mathcal{P}$,

(2) $P \neq Q$ implies that $p_n(P) \neq p_n(Q)$ for some n,

(3) $p_n(P) = p_m(Q)$ implies $n = m$ and $p_k(P) = p_k(Q)$ for all $k \leq m$.

On the basis of this axiom, we can make identifications of objects from $\mathcal{R}$ and $\mathcal{S}$ with the infinite sequences $(r_n(A))$, $(s_n(X))$ of objects. Similarly, $a \in \mathcal{AR}$ and $x \in \mathcal{AS}$ are identified with finite sequences

$$(r_k(A))_{k<n} \text{ and } (s_k(X))_{k<m},$$

where n and m are unique integers such that $a = r_n(A)$ for some $A \in \mathcal{R}$ and $x = s_m(X)$ for some $X \in \mathcal{S}$. We call n the *length* of a and m the *length* of x and use the notation length(a) and length(x), or $|a|$ and $|x|$, respectively. This identification of $\mathcal{AR}$ and $\mathcal{AS}$ gives us also the relation $\sqsubseteq$ of *end-extension* defined naturally on these sets of finite approximations.

Notation. In what follows we shall reserve the letters $A, B, C, \dots$ for members of $\mathcal{R}$, the letters $X, Y, Z, \dots$ for members of $\mathcal{S}$, the letters $a, b, c, \dots$ for members of $\mathcal{AR}$, and the letters $x, y, z, \dots$ for members of $\mathcal{AS}$. We let $a \sqsubseteq b$ mean that there exist $m \leq n$ and $A \in \mathcal{R}$ such that $a = r_m(A)$ and $b = r_n(A)$. Similarly, we let $x \sqsubseteq y$ mean that there exist $m \leq n$ and $X \in \mathcal{S}$ such that $x = s_m(X)$ and $y = s_n(X)$.

We shall continue our description of $(\mathcal{R}, \mathcal{S}, \leq, \leq^o, r, s)$ having in mind the following prototype example.

The Hales-Jewett space. We fix a countable alphabet $L = \bigcup_{n=0}^{\infty} L_n$ and its decomposition into an increasing union of finite alphabets L_n. We also fix a variable $v \notin L$ and consider the corresponding semigroups W_L and W_{Lv} of *words* over L and *variable-words* over L, respectively.

A finite or infinite sequence $X = (x_n)$ of elements of $W_L \cup W_{Lv}$ is said to be *rapidly increasing* if

$$|x_n| > \sum_{i=0}^{n-1} |x_i| \text{ for all } n.$$

Let $W_L^{[\infty]}$ and $W_{Lv}^{[\infty]}$ denote the collection of all infinite rapidly increasing sequences of elements of W_L and W_{Lv}, respectively. Of course one similarly considers the families $W_L^{[<\infty]}$ and $W_{Lv}^{[<\infty]}$ the finite rapidly increasing sequences to serve as families of finite approximations to $W_L^{[\infty]}$ and $W_{Lv}^{[\infty]}$, respectively, using the usual restriction maps

$$r_n : W_L^{[\infty]} \to W_L^{[n]} \text{ and } s_n : W_{Lv}^{[\infty]} \to W_{Lv}^{[n]}$$

over the nonnegative integers n. Recall the notion of the combinatorial subspaces

$$[X]_L \subseteq W_L \text{ and } [X]_{Lv} \subseteq W_{Lv}$$

generated by an $X = (x_n) \in W_{Lv}^{[\infty]}$ given above in Section 2.5. Note that the fact that X is rapidly increasing gives us that every $w \in [X]_L \cup [X]_{Lv}$ has uniquely defined support $\operatorname{supp}_X(w)$ the set $n_0 < \ldots < n_k$ of indexes such that

$$w = x_{n_0}[\lambda_0]^\frown \ldots ^\frown x_{n_k}[\lambda_k]$$

for some choices of $\lambda_i \in L_{n_i}$ or $\lambda_i \in L_{n_i} \cup \{v\}$ $(i \leq k)$ depending on whether w is simply a word over L or it has a variable v as one of its entries. We shall say that $Y = (y_n) \in W_{Lv}^{[\infty]}$ is a *block subsequence* of $X = (x_n) \in W_{Lv}^{[\infty]}$ if $y_n \in [X]_{Lv}$ for all n, and

$$\operatorname{supp}_X(y_n) < \operatorname{supp}_X(y_m) \text{ whenever } n < m.$$

We use the notation $Y \leq X$ to denote this relation between X and Y. Similarly we define $W = (w_n) \leq^o X = (x_n)$ for $W \in W_L^{[\infty]}$ and $X \in W_{Lv}^{[\infty]}$ if $w_n \in [X]_L$ for all n and if

$$\operatorname{supp}_X(w_n) < \operatorname{supp}_X(w_m) \text{ whenever } n < m.$$

Of course, all these notions also make perfect sense for the families $W_L^{[<\infty]}$ and $W_{Lv}^{[<\infty]}$ of finite rapidly increasing sequences. In other words, the relations $\leq$ and $\leq^o$ *admit finitizations* $\leq_{\text{fin}}$ and $\leq_{\text{fin}}^o$, respectively:

$$y \leq_{\text{fin}} x \text{ iff } y \leq x \text{ and } y \not\leq x{\upharpoonright} l \text{ for all } l < \operatorname{length}(x)$$

$$a \leq_{\text{fin}}^o x \text{ iff } a \leq^o x \text{ and } y \not\leq^o x{\upharpoonright} l \text{ for all } l < \operatorname{length}(x).$$

This leads us to the following condition on the space $(\mathcal{R}, \mathcal{S}, \leq, \leq^o, r, s)$ we are trying to describe.

A.2. (Finitization) There is a relation $\leq_{\text{fin}}^o \subseteq \mathcal{AR} \times \mathcal{AS}$ and there is a transitive and reflexive relation $\leq_{\text{fin}} \subseteq \mathcal{AS} \times \mathcal{AS}$ such that

(1) $\{a : a \leq_{\text{fin}}^o x\}$ and $\{y : y \leq_{\text{fin}} x\}$ are finite for all $x \in \mathcal{AS}$,

(2) $X \leq Y$ iff $\forall n \, \exists m \, s_n(X) \leq_{\text{fin}} s_m(Y)$,

(3) $A \leq^o X$ iff $\forall n \, \exists m \, r_n(A) \leq^o_{\text{fin}} s_m(X)$,

(4) $\forall a \in \mathcal{AR} \; \forall x, y \in \mathcal{AS} \; [a \leq^o_{\text{fin}} x \leq_{\text{fin}} y \rightarrow a \leq^o_{\text{fin}} y]$,

(5) $\forall a, b \in \mathcal{AR} \; \forall x \in \mathcal{AS} \; [a \sqsubseteq b \; \& \; b \leq^o_{\text{fin}} x \rightarrow \exists y \sqsubseteq x \; a \leq^o_{\text{fin}} y]$.

The basic sets. For $a \in \mathcal{AR}$, $x \in \mathcal{AS}$, $m \in \mathbb{N}$, and $Y \in \mathcal{S}$, set

$$[a, Y] = \{A \in \mathcal{R} : A \leq^o Y \; \& \; \exists n \, r_n(A) = a\},$$

$$[x, Y] = \{X \in \mathcal{S} : X \leq Y \; \& \; \exists n \, s_n(X) = x\},$$

$$[m, Y] = [s_m(Y), Y].$$

For $a \in \mathcal{AR}$, let

$$[a] = \{A \in \mathcal{R} : r_{|a|}(A) = a\}.$$

The sets of the form $[a]$ ($a \in \mathcal{AR}$) form a neighborhood base of the metrizable topology of $\mathcal{R}$. Note the following immediate consequence of the two axioms A.1 and A.2, where by metric on $\mathcal{R}$ we mean the first difference metric obtained by its identification with a subset of $\mathcal{AR}^{\mathbb{N}}$ as discussed above.

Lemma 4.19 *Suppose* $(\mathcal{R}, \mathcal{S}, \leq, \leq^o, r, s)$, *satisfies A.1 and A.2. Then every basic set* $[a, X]$ *is closed relative to the metrizable topology of* $\mathcal{R}$.

Notation. For $a \in \mathcal{AR}$ and $Y \in \mathcal{S}$,

$$\text{depth}_Y(a) = \begin{cases} \min\{k : a \leq^o_{\text{fin}} s_k(Y)\} & \text{if } (\exists k) \, a \leq^o_{\text{fin}} s_k(Y), \\ \infty & \text{otherwise.} \end{cases}$$

The notion of depth leads us to the following natural requirement on our space $(\mathcal{R}, \mathcal{S}, \leq, \leq^o, r, s)$:

A.3. (Amalgamation)

(1) $\forall a \in \mathcal{AR} \; \forall Y \in \mathcal{S}$
$$[d = \text{depth}_Y(a) < \infty \rightarrow \forall X \in [d, Y] \, ([a, X] \neq \emptyset)],$$

(2) $\forall a \in \mathcal{AR} \; \forall X, Y \in \mathcal{S}$
$$[X \leq Y \; \& \; [a, X] \neq \emptyset \rightarrow \exists Y' \in [\text{depth}_Y(a), Y] \, ([a, Y'] \subseteq [a, X])].$$

Note that in A.3(2) we have $[a, Y'] \neq \emptyset$ by A.3(1). We shall frequently use the following immediate fact.

Lemma 4.20 *Suppose* $(\mathcal{R}, \mathcal{S}, \leq, \leq^o, r, s)$ *satisfies A.1, A.2 and A.3. If* $a \sqsubseteq b$ *and if* $[b, Y] \neq \emptyset$ *then* $[a, Y] \neq \emptyset$ *and* $\mathrm{depth}_Y(a) \leq \mathrm{depth}_Y(b) < \infty$.

The Hales-Jewett space (cont.) Let us check that

$$(W_L^{[\infty]}, W_{Lv}^{[\infty]}, \leq, \leq^o, r, s)$$

has the amalgamation property. Suppose $a \in \mathcal{AR}$ and $Y = (y_n)$ satisfy the hypothesis of A.3(1). Note that $d = \mathrm{depth}_Y(a)$ is the minimal integer ≥ 0 such that a is a block sequence in $[(y_n)_{n=0}^{d-1}]_L$. Consider an $X = (x_n)$ from $[d, Y]$. Then X is an infinite block subsequence of Y such that $x_n = y_n$ for all $n < d$. It follows that a is a finite block sequence in $[(x_n)_{n=0}^{d-1}]_L$ and d is the minimal integer with this property. Pick an arbitrary letter $\lambda_0 \in L_0$ and for $i \geq 0$, set

$$w_{l+i} = x_{d+i}[\lambda_0],$$

where $l = \mathrm{length}(a)$. This defines an element $A = a^\frown (w_i)_{i \geq l}$ of $\mathcal{R}$ such that $A \in [a, X]$. This shows that $[a, X] \neq \emptyset$ for all $X \in [d, Y]$.

To check A.3(2), let $a \in W_L^{[<\infty]}$, $X, Y \in W_{Lv}^{[<\infty]}$ be such that $X \leq Y$ and $[a, X] \neq \emptyset$. Then a is a finite block subsequence of both X and Y. Hence, the top term of a, call it w_{l-1} where $l = \mathrm{length}(a)$, has two representations

$$w_{l-1} = x_{n_0}[\lambda_0]^\frown \ldots ^\frown x_{n_k}[\lambda_k], \text{ and}$$

$$w_{l-1} = y_{m_0}[\mu_0]^\frown \ldots ^\frown x_{m_p}[\mu_p].$$

Since $X = (x_n)$ and $Y = (y_n)$ are assumed to be rapidly increasing, both representations are unique. From $X \leq Y$ we know that each x_{n_i} has a unique representation as

$$x_{n_i} = y_{m_0^i}[\mu_0^i]^\frown \ldots ^\frown y_{m_{p_i}^i}[\mu_{p_i}^i].$$

It follows that $I_i = \{m_0^i < \ldots < m_{p_i}^i\}$ $(i \leq k)$ must be convex subsets of $\{m_0 < \ldots < m_p\}$ and they must cover the set. In particular, $m_{p_k}^k = m_p$. Let $Y' = (y_i')$ be determined as follows:

$$y_i' = y_i \text{ for } i \leq m_p,$$

$$y_{m_p+i}' = x_{n_k+i} \text{ for } i > 0.$$

Note that $\mathrm{depth}_Y(a) = m_p + 1$, so $Y' \in [\mathrm{depth}_Y(a), Y]$. Note also that any $A \in [a, Y']$ must start like a, i.e., have the $(l-1)$st term equal to w_{l-1} and the l-tail A/l of A must be a block subsequence of the tail $Y'/m_p + 1$, which by definition is a subsequence of the tail $X/n_k + 1$. This shows that $[a, Y'] \subseteq [a, X]$.

The following is the final but most important condition that we put on our space $(\mathcal{R}, \mathcal{S}, \leq, \leq^o, r, s)$.

A.4. (Pigeon hole) Suppose $a \in \mathcal{AR}$ has length l and $\mathcal{O}$ is a subset of $\mathcal{AR}_{l+1}$. Then for every $Y \in \mathcal{S}$ such that $[a, Y] \neq \emptyset$, there exists $X \in [\mathrm{depth}_Y(a), Y]$ such that $r_{l+1}[a, X] \subseteq \mathcal{O}$ or $r_{l+1}[a, X] \subseteq \mathcal{O}^c$.

The Hales-Jewett space (cont.) Let us check that the space

$$(W_L^{[\infty]}, W_{Lv}^{[\infty]}, \leq, \leq^o, r, s)$$

satisfies A.4. Let $Y = y_n$ be a given member of $W_{Lv}^{[\infty]}$, let a be a finite block sequence of elements of $[Y]_L$ of some length l, and let $\mathcal{O}$ be a given subset of $\mathcal{AR}_{l+1}$. Let $d = \mathrm{depth}_Y(a)$. Let S be the collection of all *cofinite ultrafilters* on the partial semigroup

$$[Y/d]_L \cup [Y/d]_{Lv}$$

i.e., ultrafilters $\mathcal{U}$ on $[Y/d]_L \cup [Y/d]_{Lv}$ with the property that

$$[Y/k]_L \cup [Y/k]_{Lv} \in \mathcal{U} \text{ for all } k \geq d,$$

or in other words, ultrafilters $\mathcal{U}$ such that $\{w : \mathrm{supp}_Y(w) > k\} \in \mathcal{U}$ for all k. Note that S is a closed subsemigroup of the semigroup S^* considered in Section 2.5 above. Let S_L be the closed subsemigroup of S consisting of ultrafilters concentrating on $[Y/d]_L$ and let S_{Lv} be the two-sided ideal consisting of ultrafilters from S concentrating on $[Y/d]_{Lv}$. As before we pick idempotents $\mathcal{W} \in S_L$ and $\mathcal{V} \in S_{Lv}$ such that $\mathcal{V} \leq \mathcal{W}$ and $\mathcal{V}[\lambda] = \mathcal{W}$ for all $\lambda \in L$. Let $\mathcal{O}_0$ be the collection of all $w \in W_L$ such that $a^\frown w \in \mathcal{O}$ and let $\mathcal{O}_1 = W_L \setminus \mathcal{O}_0$. Then there is $P_W \in \{\mathcal{O}_0, \mathcal{O}_1\}$ such that $P_W \in \mathcal{W}$. Repeating the argument from the proof of the infinite Hales-Jewett theorem given above in Section 2.5, and using the fact that $\mathcal{W}$ and $\mathcal{V}$ are idempotent ultrafilters that are cofinite relative to Y/d, it is clear that we can construct an infinite rapidly increasing sequence $X' = (x'_n)$ of members of $[Y/d]_{Lv}$ such that $[X']_L \subseteq P_W$ and such that

$$|x'_0| > \sum_{n=0}^{d-1} |y_n|.$$

Define $X = (x_n) \in W_{Lv}^{[\infty]}$ by the following two requirements

$$x_n = y_n \text{ for } n < d$$

$$x_{d+i} = x'_i \text{ for } i \geq 0.$$

Then $X \in [d, Y]$ and $r_{l+1}[a, X] \subseteq \mathcal{O}$ or $r_{l+1}[a, X] \subseteq \mathcal{O}^c$, depending whether $P_W = \mathcal{O}_0$ or $\mathcal{O}_1$. This establishes A.4 for the Hales-Jewett space, which can be restated in the following form which is slightly stronger that Theorem 2.35 above.

Theorem 4.21 (Infinite Hales-Jewett Theorem) *For every finite coloring of $W_L \cup W_{Lv}$ and every $Y \in W_{Lv}^{[\infty]}$, there is $X \leq Y$ in $W_{Lv}^{[\infty]}$ such that $[X]_L$ and $[X]_{Lv}$ are both monochromatic.*

The following definition gives us the main notions to study in this section.

Definition 4.22 *A set $\mathcal{X} \subseteq \mathcal{R}$ is S-Ramsey if for every nonempty basic set $[a, Y]$ there is $X \in [\mathrm{depth}_Y(a), Y]$ such that $[a, X] \subseteq \mathcal{X}$ or $[a, X] \subseteq \mathcal{X}^c$.*
If for every $[a, Y] \neq \emptyset$ we can find $X \in [\mathrm{depth}_Y(a), Y]$ such that $[a, X] \cap \mathcal{X} = \emptyset$ we call $\mathcal{X}$ an S-Ramsey null set of $\mathcal{R}$.

Any infinite-dimensional Ramsey result says in some way or other that the family of Ramsey subsets of a given Ramsey space is in some sense *large*. The following seemingly less restrictive condition on a subset of $\mathcal{R}$ will give us the language to express how rich is the field of $\mathcal{S}$-Ramsey subsets of $\mathcal{R}$.

Definition 4.23 *A set* $\mathcal{X} \subseteq \mathcal{R}$ *is* $\mathcal{S}$-*Baire if for every nonempty basic set* $[a, X]$ *there exist a* $\sqsubseteq b \in \mathcal{AR}$ *and* $Y \leq X$ *such that* $[b, Y] \neq \emptyset$ *and* $[b, Y] \subseteq \mathcal{X}$ *or* $[b, Y] \subseteq \mathcal{X}^c$. *If for every* $[a, X] \neq \emptyset$ *we can find a* $\sqsubseteq b \in \mathcal{AR}$ *and* $Y \leq X$ *such that* $[b, Y] \neq \emptyset$ *and* $[b, Y] \cap \mathcal{X} = \emptyset$, *then* $\mathcal{X}$ *is said to be* $\mathcal{S}$-*meager.*

Recall that we identify $\mathcal{R}$ with a subset of the infinite power $\mathcal{AR}^{\mathbb{N}}$. The power $\mathcal{AR}^{\mathbb{N}}$ has its natural first-difference metric, so our identification induces a metric on $\mathcal{R}$. The following simple fact shows that our notion of $\mathcal{S}$-Baire agrees well with this topology on $\mathcal{R}$.

Lemma 4.24 *Suppose* $(\mathcal{R}, \mathcal{S}, \leq, \leq^o, r, s)$, *satisfies A.1 and A.2. Then every metrically open subset of* $\mathcal{R}$ *is* $\mathcal{S}$-*Baire.*

Proof. Consider a metrically open subset $\mathcal{O}$ of $\mathcal{R}$ and a basic set $[a, X]$ and assume $[a, X] \not\subseteq \mathcal{O}^c$. Choose $A \in [a, X] \cap \mathcal{O}$. Choose $l \geq |a|$ such that if $b = r_l(A)$ then $A \in [b] \subseteq \mathcal{O}$. Then $[b, X]$ is a nonempty basic subset of $[a, X]$ included in $\mathcal{O}$, which is what we wanted to find. $\square$

Thus, in analogy, $\mathcal{S}$-Baire sets correspond to $\mathcal{S}$-Ramsey sets and $\mathcal{S}$-meager sets to $\mathcal{S}$-Ramsey null sets. As the definitions suggest one should expect $\mathcal{S}$-meager sets to form a σ-ideal of subsets of $\mathcal{R}$. It turns out that it is much easier to show that $\mathcal{S}$-Ramsey null subsets of $\mathcal{R}$ form a σ-ideal under a suitable assumption on $\mathcal{S}$. Considering $\mathcal{AS}$ as a discrete space, the infinite power $\mathcal{AS}^{\mathbb{N}}$ gets its Tychonov product topology, a completely metrizable topology. We have already remarked that we consider $\mathcal{S}$ a subset of $\mathcal{AS}^{\mathbb{N}}$ via the identification $X \to (s_n(X))$. Thus it is natural to call $\mathcal{S}$ *closed* if in this identification it corresponds to a closed subset of $\mathcal{AS}^{\mathbb{N}}$. Whenever $\mathcal{S}$ is closed, the following procedure will be quite useful in building members of $\mathcal{S}$ with desired properties.

Definition 4.25 *A sequence* $([n_k, Y_k])$ *of basic subsets of* $\mathcal{S}$ *is a* fusion se-quence *if it is infinite and if*

(1) (n_k) *is a nondecreasing sequence of integers converging to* ∞,

(2) $Y_{k+1} \in [n_k, Y_k]$ *for all* k.

The limit *of the fusion sequence (if it exists) is the unique element* Y_{∞} *of* $\mathcal{S}$ *such that* $r_{n_k}(Y_{\infty}) = r_{n_k}(Y_k)$ *for all* k.

Note that when $\mathcal{S}$ is closed and when $(\mathcal{R}, \mathcal{S}, \leq, \leq^o, r, s)$ satisfies A.1 and A.2, the limit Y_{∞} of any fusion sequence exists and $Y_{\infty} \in [n_k, Y_k]$ for all k.

Lemma 4.26 *Suppose* $(\mathcal{R}, \mathcal{S}, \leq, \leq^o, r, s)$ *satisfies A.1 to A.3 and also that* $\mathcal{S}$ *is closed. Then* $\mathcal{S}$-*Ramsey null subsets of* $\mathcal{R}$ *form a* σ-*ideal.*

Proof. Suppose $\mathcal{X} = \bigcup_{k=0}^{\infty} \mathcal{X}_k$, where each $\mathcal{X}_k$ is $\mathcal{S}$-Ramsey null. Since the union of two $\mathcal{S}$-Ramsey null sets is a $\mathcal{S}$-Ramsey null set, we may assume that $\mathcal{X}_n \subseteq \mathcal{X}_m$ whenever $n \leq m$. Let $[a, X]$ be a given nonempty basic subset of $\mathcal{R}$. Let $d = \mathrm{depth}_X(a)$. We build a fusion sequence $([n_k, X_k])$ such that

(1) $X_0 = X$,

(2) $n_k = d + k$,

(3) $[b, X_{k+1}] \cap \mathcal{X}_k = \emptyset$ for all $b \in \mathcal{AR}$ such that $\mathrm{depth}_{X_k}(b) = n_k$.

To see that this can be done assume $[n_k, X_k]$ has been defined. By $A.2$ the set

$$\{b \in \mathcal{AR} : \mathrm{depth}_{X_k}(b) = n_k\}$$

is finite so we can list it as $b_0, \dots, b_{p-1}$ for some integer p. Using the assumption that the set $\mathcal{X}_k$ is $\mathcal{S}$-Ramsey null, we can recursively on i, choose Y_i ($0 \leq i \leq p$) in $\mathcal{S}$ such that

(4) $Y_0 = X_k$,

(5) $Y_{i+1} \in [n_k, Y_i]$,

(6) $[b_i, Y_{i+1}] \cap \mathcal{X}_k = \emptyset$.

Let $X_{k+1} = Y_p$. It is clear that $(1) - (3)$ remain satisfied. Finally, let $X_\infty = \lim_k X_k$ be the limit of the fusion sequence $([n_k, X_k])$ that exists by our assumption that $\mathcal{S}$ is closed. Moreover $X_\infty \in [n_k, X_k]$ for all k. We claim that $[a, X_\infty]$ is the desired basic subset of $\mathcal{R}$ i.e., that $[a, X_\infty] \cap \mathcal{X} = \emptyset$. Suppose not and pick A in the intersection. Pick a minimal k such that $A \in \mathcal{X}_k$. By $A.2(3)$ and $A.2(1)$, we can find n such that

$$\infty > \mathrm{depth}_{X_\infty}(r_n(A)) \geq n_k.$$

Let $b = r_n(A)$. Then $\mathrm{depth}_{X_k}(b) = n_l$ for some $l \geq k$. From (3), we conclude that $[b, X_{l+1}] \cap \mathcal{X}_l = \emptyset$, and therefore $[b, X_{l+1}] \cap \mathcal{X}_k = \emptyset$ since $\mathcal{X}_k \subseteq \mathcal{X}_l$. But $X_\infty \in [n_l, X_l]$ and therefore $[b, X_\infty]$, which is a subset of $[b, X_{l+1}]$, does not intersect $\mathcal{X}_k$, a contradiction, since clearly $A \in [b, X_\infty] \cap \mathcal{X}_k$. This completes the proof. □

We finish this section with the statement of the Abstract Ramsey Theorem, whose proof is given in the following section.

Theorem 4.27 (Abstract Ramsey Theorem) *Suppose we are given a structure $(\mathcal{R}, \mathcal{S}, \leq, \leq^o, r, s)$ that satisfies axioms A.1 to A.4 and that $\mathcal{S}$ is closed. Then the field of $\mathcal{S}$-Ramsey subsets of $\mathcal{R}$ is closed under the Souslin operation and it coincides with the field of $\mathcal{S}$-Baire subsets of $\mathcal{R}$. Moreover, the ideals of $\mathcal{S}$-Ramsey null subsets of $\mathcal{R}$ and $\mathcal{S}$-meager subsets of $\mathcal{R}$ are σ-ideals and they also coincide.*

To state a useful consequence of this result, we need a definition.

Definition 4.28 *Let T be a given topological space. The field of Souslin-measurable subsets of T is the minimal field of subsets of T that contains all open subsets and that is closed under the Souslin operation. Thus, in particular, every Borel or analytic subset of T is Souslin-measurable.*

By Lemma 4.24, we have the following immediate consequence of the Abstract Ramsey Theorem, which gives us some idea about the richness of the field of $\mathcal{S}$-Ramsey sets and therefore an idea about the potential applicability and power behind the Abstract Ramsey Theorem.

Corollary 4.29 *Suppose we are given a structure $(\mathcal{R}, \mathcal{S}, \leq, \leq^o, r, s)$ that satisfies axioms A.1 to A.4 and that $\mathcal{S}$ is closed. Then for every finite coloring of $\mathcal{R}$ that is Souslin-measurable relative to the metric topology of $\mathcal{R}$, there is X in $\mathcal{S}$ such that the basic subset $[\emptyset, X]$ of $\mathcal{R}$ is monochromatic.*

4.3 COMBINATORIAL FORCING

The purpose of this section is to give the proof of the Abstract Ramsey Theorem. We fix for a while a subset $\mathfrak{X}$ of $\mathcal{R}$. So throughout this section we suppose that we are given a structure $(\mathcal{R}, \mathcal{S}, \leq, \leq^o, r, s)$ satisfying axioms A.1 to A.4 and that $\mathcal{S}$ is closed. We reserve the letters X, Y, Z to denote members of $\mathcal{S}$; A, B, C to denote members of $\mathcal{R}$; a, b, c to denote members of $\mathcal{AR}$, and x, y, z to denote members of $\mathcal{AS}$.

Definition 4.30 (combinatorial forcing)   *We say that Y accepts a if $[a, Y] \subseteq \mathfrak{X}$. We say that Y rejects a if $[a, Y] \neq \emptyset$ and there is no $X \in [\text{depth}_Y(a), Y]$ accepting a. We say that Y decides a if Y accepts a or if Y rejects a.*

The following summarizes the immediate properties of these notions.

Lemma 4.31   (1) *Y accepts any a such that $[a, Y] = \emptyset$.*

(2) *If Y accepts a and $X \leq Y$, then X accepts a.*

(3) *If Y rejects a, $X \leq Y$ and $[a, X] \neq \emptyset$, then X rejects a.*

(4) *If Y decides a and $X \leq Y$, then X decides a.*

(5) *If $\text{depth}_Y(a) < \infty$, then there is $X \in [\text{depth}_Y(a), Y]$, which decides a.*

(6) *If $\text{depth}_Y(a) < \infty$ and Y decides a, then every $X \in [\text{depth}_Y(a), Y]$ decides a in the same way Y does.*

(7) *If Y accepts a, then Y accepts every member of $r_{|a|+1}[a, Y]$.*

Proof. To check (3), let $a, X \leq Y$ be given and assume that $[a, X] \neq \emptyset$, yet X does not reject a. So there is $X' \in [\text{depth}_X(a), X]$ which accepts a. By $A.3(1)$, $[a, X'] \neq \emptyset$ and $X' \leq Y$. By $A.3(2)$ there is $Y' \in [\text{depth}_Y(a), Y]$ such that $[a, Y'] \subseteq [a, X'] \subseteq \mathfrak{X}$. Then Y' accepts a, and therefore, Y does not reject a.

To check (7), assume that Y accepts a and let $l = |a| (= \text{length}(a))$. Consider $b \in r_{l+1}[a, Y]$. Then $[b, Y] \subseteq [a, Y]$, so Y also accepts b. $\square$

Lemma 4.32 *Given that Y rejects some $a \in \mathcal{AR}$, then there is no X in $[\text{depth}_Y(a), Y]$ that accepts every member of $r_{|a|+1}[a, X]$.*

Proof. Assume such an X exists. Consider an $A \in [a, X]$. Then
$$A \in [r_{|a|+1}(A), X] \subseteq \mathfrak{X}.$$
Since A was an arbitrary member of $[a, X]$ this shows that $[a, X] \subseteq \mathfrak{X}$, and so X accepts a. This contradicts the hypothesis that Y rejects a. $\square$

Lemma 4.33 *For every $n \geq 0$ and $Y \in \mathcal{S}$ there is $X \in [n, Y]$ that decides every $b \in \mathcal{AR}$ such that $\text{depth}_X(b) \geq n$.*

Proof. We construct a fusion sequence $([n_k, Y_k])_k$ such that

(1) $Y_0 = Y$,

(2) $n_k = n + k$,

(3) Y_{k+1} decides every $b \in \mathcal{AR}$ such that $\text{depth}_{Y_k}(b) = n_k$.

Suppose Y_k has been defined. Let $b_1, \ldots, b_p$ be an enumeration of
$$\{b \in \mathcal{AR} : \text{depth}_{Y_k}(b) = n_k\} \tag{4.1}$$
We define X_i $(0 \leq i \leq p)$ by recursion on i so that

(4) $X_0 = Y_k$,

(5) $X_{i+1} \in [n_k, X_i]$,

(6) X_i decides b_i.

Suppose X_i has been defined. By $A.3(1)$, $[b_{i+1}, X_i] \neq \emptyset$. Note also that
$$\text{depth}_{X_i}(b_{i+1}) = \text{depth}_{Y_k}(b_{i+1}) = n_k.$$
By Lemma 4.31 (5) there exists $X_{i+1} \in [n_k, X_i]$ which decides b_{i+1}. This defines X_i $(0 \leq i \leq p)$.

Let $Y_{k+1} = X_p$. Then $Y_{k+1} \in [n_k, Y_k]$, and by lemma 4.31 (4), Y_{k+1} decides every $b \in \mathcal{AR}$ such that
$$\text{depth}_{Y_{k+1}}(b) = \text{depth}_{Y_k}(b) = n_k. \tag{4.2}$$
Let $X = \lim_k Y_k$. We claim that X satisfies the conclusion of the Lemma. Consider a $b \in \mathcal{AR}$ such that $\text{depth}_X(b) \geq n$. Then $\text{depth}_X(b) = n_k$ for some $k \geq 0$. Since $X \in [n_k, Y_k]$ we have that
$$\text{depth}_{Y_k}(b) = \text{depth}_X(b) = n_k. \tag{4.3}$$
So b appears in the list used in constructing Y_{k+1}. Hence, Y_{k+1} decides b. Since $X \leq Y_{k+1}$ by Lemma 4.31 (4), X also decides b. $\square$

Lemma 4.34 *If Y rejects some $a \in \mathcal{AR}$ of length l, then there exists an $X \in [\mathrm{depth}_Y(a), Y]$ such that X rejects every member of $r_{l+1}[a, X]$.*

Proof. Put $n = \mathrm{depth}_Y(a) \geq 0$. Use Lemma 4.33 to obtain $Z \in [n, Y]$, which decides every b with $\mathrm{depth}_Z(b) \geq n$. By $A.3(1)$, $[a, Z] \neq \emptyset$ and therefore Z rejects a (see Lemma 4.31(3)). Let

$$\mathcal{O} = \{b \in \mathcal{AR}_{l+1} : Z \text{ accepts } b\}. \tag{4.4}$$

By $A.4$, there is $X \in [n, Z]$ such that either

(a) $r_{l+1}[a, X] \subseteq \mathcal{O}$, or

(b) $r_{l+1}[a, X] \subseteq \mathcal{O}^c$.

By $A.3(1)$, we have that $[a, X] \neq \emptyset$. So, by Lemma 4.31 (3), X rejects a because Z does. By Lemma 4.32, alternative (a) is not possible, so we have that the second alternative $r_{l+1}[a, X] \subseteq \mathcal{O}^c$ must hold. Consider $b \in r_{l+1}[a, X]$. Then $[b, Z] \neq \emptyset$, or else b would belong to $\mathcal{O}$ (by Lemma 4.31(1)). By Lemma 4.20, $\mathrm{depth}_Z(b) \geq \mathrm{depth}_Z(a) = n$, so Z decides b. It must reject it or else we would have that $b \in \mathcal{O}$. By Lemma 4.31(3) and the facts that $X \leq Z$ and $[b, X] \neq \emptyset$, we conclude that X rejects b. $\qquad\square$

Lemma 4.35 *Suppose $[a, Y] \neq \emptyset$ and that Y decides every $b \in \mathcal{AR}$ such that $\mathrm{depth}_Y(b) \geq \mathrm{depth}_Y(a)$. If Y rejects a, then there is $X \in [\mathrm{depth}_Y(a), Y]$, which rejects every $b \in \mathcal{AR}$ end-extending a such that $[b, X] \neq \emptyset$.*

Proof. We construct by recursion a fusion sequence $([n_k, Y_k])_k$ so that

(1) $Y_0 = Y$ and $n_0 = \mathrm{depth}_Y(a)$,

(2) $n_k = n_0 + k$,

(3) if $\mathrm{depth}_{Y_k}(b) = n_k$ and Y_k rejects b, then Y_{k+1} rejects every member of $r_{l+1}[b, Y_{k+1}]$, where $l = \mathrm{length}(b)$.

Assume Y_k has been defined and list the elements of the finite set

$$\{b \in \mathcal{AR} : a \sqsubseteq b, \mathrm{depth}_{Y_k}(b) = n_k \text{ and } Y_k \text{ rejects } b\} \tag{4.5}$$

as $b_0, \ldots, b_{p-1}$. Define now recursively on i a sequence X_i $(0 \leq i \leq p)$ so that

(4) $X_0 = Y_k$,

(5) $X_{i+1} \in [n_k, X_i]$,

(6) X_{i+1} rejects every member of $r_{l+1}[b_i, X_{i+1}]$, where $l = length(b)$.

Assume X_i has been defined. By (4) and (5), $\mathrm{depth}_{X_i}(b_i) = n_k$, and moreover by $A.3(1)$, $[b_i, X_i] \neq \emptyset$. By Lemma 4.31 (3), we conclude that X_i rejects b_i. By Lemma 4.34, we obtain $X_{i+1} \in [n_k, X_i]$ which rejects every member of $r_{|b_i|+1}[b_i, X_{i+1}]$. Having obtained X_p, set $Y_{k+1} = X_p$. By $A.3(1)$ and Lemma 4.31(3), for all $i < p$, Y_{k+1} rejects every member of $r_{|b_i|+1}[b_i, Y_{k+1}]$.

Having completed the recursion on k, let $X = \lim_k Y_k$. We show that X rejects every b such that $[b, X] \neq \emptyset$ and such that b end-extends a, by induction on $|b|$. The case $b = a$ has been taken care of by the assumption. Suppose that $|b| = j + 1 > |a|$ and $[b, X] \neq \emptyset$. Pick $C \in [b, X]$ and let $c = r_j(C)$. Then c end-extends a and therefore

$$\text{depth}_X(c) \geq \text{depth}_X(a) = \text{depth}_Y(a) = n_0. \tag{4.6}$$

Hence there is a $k \geq 0$ such that $n_k = \text{depth}_X(c)$. Since $s_{n_k}(X) = s_{n_k}(Y_k)$, we conclude that $\text{depth}_{Y_k}(c) = n_k$. By the induction hypothesis, X rejects c. Since $\text{depth}_Y(c) \geq \text{depth}_Y(a)$, by the hypothesis of the lemma, Y decides c. By Lemma 4.31(4), Y_k decides c. But Y_k must reject c, since $X \in [n_k, Y_k]$ rejects it. By (3), Y_{k+1} rejects every member of $r_{j+1}[c, Y_{k+1}]$. Note that $b \in r_{j+1}[c, Y_{k+1}]$, so Y_{k+1} rejects b, and therefore X rejects b. This finishes the proof. $\square$

This finishes the sequence of basic Lemmas about combinatorial forcing relative to a fixed subset $\mathfrak{X}$ of $\mathcal{R}$. From now on, we apply these lemmas for various $\mathfrak{X} \subseteq \mathcal{R}$, and therefore we use the terms "$\mathfrak{X}$-accepts", "$\mathfrak{X}$-rejects" and "$\mathfrak{X}$-decides."

Lemma 4.36 *The field of $\mathcal{S}$-Ramsey subsets of $\mathcal{R}$ coincides with the field of $\mathcal{S}$-Baire subsets of $\mathcal{R}$.*

Proof. It is clear that every $\mathcal{S}$-Ramsey set is $\mathcal{S}$-Baire so the content of this lemma is in the converse. Let $\mathfrak{X} \subseteq \mathcal{R}$ be a given $\mathcal{S}$-Baire set and let $[a, Y] \neq \emptyset$ be a given basic set. Let

$$n = \text{depth}_Y(a). \tag{4.7}$$

By Lemma 4.33 there is an $X \in [n, Y]$ that $\mathfrak{X}$-decides as well as $\mathfrak{X}^c$-decides every $b \in \mathcal{AR}$ such that $\text{depth}_X(b) \geq n$.

If X either $\mathfrak{X}$-accepts a or $\mathfrak{X}^c$-accepts a, we are done. Assume then, for a proof by a contradiction, that X both $\mathfrak{X}$-rejects a and $\mathfrak{X}^c$-rejects a. Applying Lemma 4.35 twice, we obtain $Z \in [n, X]$, which both $\mathfrak{X}$-rejects and $\mathfrak{X}^c$-rejects every b end-extending a such that $[b, Z] \neq \emptyset$.

Since $\mathfrak{X}$ is $\mathcal{S}$-Baire, there are b end-extending a and $Z_0 \leq Z$ such that $[b, Z_0] \neq \emptyset$, and either

(a) $[b, Z_0] \subseteq \mathfrak{X}$, or

(b) $[b, Z_0] \subseteq \mathfrak{X}^c$.

By A.3(2) there is a $Z'_0 \in [\text{depth}_Z(b), Z]$ such that $\emptyset \neq [b, Z'_0] \subseteq [b, Z_0]$. If alternative (a) holds, then Z'_0 is $\mathfrak{X}$-accepting b, which contradicts the fact that Z $\mathfrak{X}$-rejects b (see Lemma 4.31(3)). If alternative (b) holds then Z'_0 is $\mathfrak{X}^c$-accepting b, which contradicts the fact that Z $\mathfrak{X}^c$-rejects b. This contradiction finishes the proof of this lemma. $\square$

Lemma 4.37 *The field of $\mathcal{S}$-Ramsey subsets of $\mathcal{R}$ is a σ-field.*

Proof. Let $\mathfrak{X} = \bigcup_{k=0}^{\infty} \mathfrak{X}_k$, where each $\mathfrak{X}_k$ is $\mathcal{S}$-Ramsey. Let $[a, Y] \neq \emptyset$ be given. Let $n_0 = \mathrm{depth}_Y(a)$. By Lemma 4.33, there is an $X \in [n, Y]$ that $\mathfrak{X}$-decides every b such that $\mathrm{depth}_X(b) \geq n_0$. If X $\mathfrak{X}$-accepts a, we are done. Assume that X $\mathfrak{X}$-rejects a. Applying Lemma 4.35, we obtain $Y_0 \in [n_0, X]$, which $\mathfrak{X}$-rejects every $b \in \mathcal{AR}$ end-extending a such that $[b, Y_0] \neq \emptyset$.

Starting with $[n_0, Y_0]$, we build a fusion sequence $([n_k, Y_k])_k$ such that

(1) $n_k = n_0 + k$,

(2) for every $b \in \mathcal{AR}$ such that $\mathrm{depth}_{Y_k}(b) = n_k$ and for all $n \leq k$ either $[b, Y_{k+1}] \subseteq \mathfrak{X}_n$ or $[b, Y_{k+1}] \subseteq \mathfrak{X}_n^c$.

Assume Y_k has been defined. Let $b_0, \ldots, b_{p-1}$ be a list of all
$$\{b \in \mathcal{AR} : \mathrm{depth}_{Y_k} = n_k\}. \tag{4.8}$$
Build X_i $(0 \leq i \leq p)$ such that

(3) $X_0 = Y_k$,

(4) $X_{i+1} \in [n_k, X_i]$,

(5) for all $n \leq n_k$, either $[b_i, X_{i+1}] \subseteq \mathfrak{X}_n$ or $[b_i, X_{i+1}] \subseteq \mathfrak{X}_n^c$.

Clearly, there is no problem in producing such a sequence X_i $(0 \leq i \leq p)$ since $\mathfrak{X}_n$ $(n \leq k)$ are all $\mathcal{S}$-Ramsey. Let $Y_{k+1} = X_p$. Then Y_{k+1} satisfies (2). Let $Y_\infty = \lim_k Y_k$. It suffices to show that $[a, Y_\infty] \subseteq \mathfrak{X}^c$. Take $A \in [a, Y_\infty]$ and $n \in \mathbb{N}$. Then by A.2 there exists $l \geq \mathrm{length}(a)$ such that $\mathrm{depth}_{Y_\infty}(r_l(A)) \geq n$. Let $b = r_l(A)$. Find k such that $n_k = \mathrm{depth}_{Y_\infty}(b)$. Since
$$r_{n_k}(Y_\infty) = r_{n_k}(Y_k), \tag{4.9}$$
we conclude that $\mathrm{depth}_{Y_k}(b) = n_k$. Applying (2), we conclude that either

(a) $[b, Y_{k+1}] \subseteq \mathfrak{X}_n$ or

(b) $[b, Y_{k+1}] \subseteq \mathfrak{X}_n^c$.

We claim that the alternative (b) must hold. This follows from our assumption that Y_0 $\mathfrak{X}$-rejects any $b \in \mathcal{AR}$ that end-extends a and has the property that $[b, Y_0] \neq \emptyset$.

It follows that $A \in [b, Y_{k+1}] \subseteq \mathfrak{X}_n^c$. Since n was chosen arbitrarily, we conclude that $A \in \mathfrak{X}^c$, as claimed. □

Lemma 4.38 *The ideals of $\mathcal{S}$-meager and $\mathcal{S}$-Ramsey null subsets of $\mathcal{R}$ coincide, and therefore they are both σ-ideals.*

Proof. Let $\mathfrak{X}$ be a given $\mathcal{S}$-meager set and let $[a, Y] \neq \emptyset$ be a given basic set. By Lemma 4.36, there is an $X \in [\mathrm{depth}_Y(a), Y]$ such that $[a, X] \subseteq \mathfrak{X}$ or $[a, X] \subseteq \mathfrak{X}^c$. The first alternative is impossible, since applying the assumption that $\mathfrak{X}$ is $\mathcal{S}$-meager to the nonempty basic set $[a, X]$, we would get $Z \leq X$ and $b \in \mathcal{AR}$ end-extending a such that $[b, Z] \neq \emptyset$ and $[b, Z] \cap \mathfrak{X} = \emptyset$. However, note that $[a, X]$ includes $[b, Z]$. □

The following lemma is the crucial step toward the proof of the Abstract Ramsey Theorem.

Lemma 4.39 *The field of S-Ramsey subsets of $\mathcal{R}$ is closed under the Souslin operation.*

Proof. Let $\mathfrak{X}_s$ ($s \in \mathbb{N}^{[<\infty]}$) be a given Souslin scheme of S-Ramsey subsets of $\mathcal{R}$ indexed by the family $\mathbb{N}^{[<\infty]}$ of all finite subsets of $\mathbb{N}$. Consider the corresponding result of the Souslin operation

$$\mathfrak{X} = \bigcup_{A \in \mathbb{N}^{[\infty]}} \bigcap_{n \in \mathbb{N}} \mathfrak{X}_{r_n(A)}, \tag{4.10}$$

where $\mathbb{N}^{[\infty]}$ denotes the family of all infinite subsets of $\mathbb{N}$ and where $r_n(A)$ is the finite set formed by taking the first n elements of A according to its increasing enumeration. Thus, if $A = (n_i)_{i=0}^{\infty}$ is a member of $\mathbb{N}^{[\infty]}$ given with its increasing enumeration and if $l \in \mathbb{N}$ then $r_l(A) = \{n_0, \ldots, n_{l-1}\}$. For $s \in \mathbb{N}^{[<\infty]}$, let

$$\widetilde{\mathfrak{X}}_s = \bigcup_{A \in \mathbb{N}^{[\infty]}, \, A \sqsupseteq s} \bigcap_{n \geq |s|} \mathfrak{X}_{r_n(A)}. \tag{4.11}$$

Note that $\widetilde{\mathfrak{X}}_s \subseteq \mathfrak{X}_s$ and that $\widetilde{\mathfrak{X}}_\emptyset = \mathfrak{X}$. To show that $\mathfrak{X}$ is S-Ramsey we start from a given basic set $[a, Y] \neq \emptyset$. By recursion on k, we define a fusion sequence $([n_k, Y_k])_k$ such that

(1) $n_0 = \operatorname{depth}_Y(a)$, $Y_0 = Y$,

(2) $n_k = n_0 + k$,

(3) for every $s \subseteq \{0, \ldots, n_k\}$ and every $b \in \mathcal{AR}$ such that $\operatorname{depth}_{Y_k}(b) = n_k$, Y_{k+1} $(\widetilde{\mathfrak{X}}_s)^c$-decides b.

Clearly there is no problem in getting Y_{k+1} starting from $[n_k, Y_k]$ by a successive application of Lemma 4.31 (5) treating each $b \in \mathcal{AR}$ such that $\operatorname{depth}_{Y_k}(b) = n_k$ and each $s \subseteq \{0, \ldots, n_k\}$. Let $X = \lim_k Y_k$. For $s \in \mathbb{N}^{[<\infty]}$, set

$$\mathcal{O}_s = \{b \in \mathcal{AR} : X \ (\widetilde{\mathfrak{X}}_s)^c\text{-accepts } b\}, \tag{4.12}$$

$$\Phi(\widetilde{\mathfrak{X}}_s) = ([a, X] \cap \widetilde{\mathfrak{X}}_s) \setminus \bigcup_{b \in \mathcal{O}_s} [b, X]. \tag{4.13}$$

Then $\Phi(\widetilde{\mathfrak{X}}_s) \supseteq [a, X] \cap \widetilde{\mathfrak{X}}_s$ and we claim that $\Phi(\widetilde{\mathfrak{X}}_s)$ serves as an envelope of the restriction of $\widetilde{\mathfrak{X}}_s$ to the basic set $[a, X]$. First of all, note that each basic set $[b, X]$ is $S(\leq X)$-Ramsey, where

$$S(\leq X) = \{Y \in S : Y \leq X\}.$$

Note also that by $A.2$ there exist only countably many $b \in \mathcal{AR}$ such that $[b, X] \neq \emptyset$. So by Lemma 4.37, we conclude that each set of the form $\Phi(\widetilde{\mathfrak{X}}_s)$ is $S(\leq X)$-Ramsey.

Claim 4.39.1 *For every $s \in \mathbb{N}^{[<\infty]}$, every $S(\leq X)$-Baire subset $\mathfrak{Y}$ of the difference $\Phi(\widetilde{\mathfrak{X}}_s) \setminus \widetilde{\mathfrak{X}}_s$ must be $S(\leq X)$-meager, and therefore, $S(\leq X)$-Ramsey null.*

Proof. Fix s and $\mathfrak{Y}$. Consider a nonempty $[b, Y]$ for some $Y \leq X$. We need to find $c \sqsupseteq b$ and $Z \leq Y$ such that $[c, Z] \neq \emptyset$ and $[c, Z] \cap \mathfrak{Y} = \emptyset$. We may assume that $b \sqsupseteq a$, or else, we are done. Pick $B \in [b, Y]$. By $A.2$ there is an $l \geq \text{length}(b)$ such that if $b' = r_l(B)$, then $\text{depth}_X(b') = n_i$ for some i such that $s \subseteq \{0, \ldots, n_i\}$. Since $\mathfrak{Y}$ is $\mathcal{S}(\leq X)$-Baire, we can find $c \sqsupseteq b'$ and $Z \leq Y$ such that $[c, Z] \neq \emptyset$ and either $[c, Z] \subseteq \mathfrak{Y}$ or $[c, Z] \subseteq \mathfrak{Y}^c$. Since the second alternative would give us the desired conclusion, we assume that $[c, Z] \subseteq \mathfrak{Y}$ and work toward a contradiction. Find a $k \geq i$ such that $\text{depth}_X(c) = n_k$. By property (3) of the fusion sequence $([n_k, Y_k])_k$ and the fact that

$$n_k = \text{depth}_X(c) = \text{depth}_{Y_k}(c),$$

we conclude that Y_{k+1} $(\widetilde{\mathfrak{X}}_s)^c$-decides c. Our assumption $[c, Z] \subseteq \mathfrak{Y}$ and the fact that $\mathfrak{Y}$ is disjoint from $(\widetilde{\mathfrak{X}}_s)$ means in particular that Z $(\widetilde{\mathfrak{X}}_s)^c$-accepts c. So by Lemma 4.31 it must be that Y_{k+1}, and therefore X, $(\widetilde{\mathfrak{X}}_s)^c$-accepts c. It follows that $c \in \mathcal{O}_s$, and therefore by Equation (4.13), we conclude that $[c, X] \cap \Phi(\widetilde{\mathfrak{X}}_s) = \emptyset$, a contradiction.   □

Note that for every $s \in \mathbb{N}^{[<\infty]}$,

$$[a, X] \cap \widetilde{\mathfrak{X}}_s = [a, X] \cap \bigcup_{n \geq \max(s)+1} \widetilde{\mathfrak{X}}_{s \cup \{n\}}. \tag{4.14}$$

It follows that

$$\mathfrak{Y}_s = \Phi(\widetilde{\mathfrak{X}}_s) \setminus \bigcup_{n \geq \max(s)+1} \Phi(\widetilde{\mathfrak{X}}_{s \cup \{n\}}) \tag{4.15}$$

is a $\mathcal{S}(\leq X)$-Ramsey set that is disjoint from $\widetilde{\mathfrak{X}}_s$, and therefore must be $\mathcal{S}(\leq X)$-Ramsey null by Claim 4.39.1. By Lemma 4.38, the $\mathcal{S}(\leq X)$-Ramsey null sets form a σ-ideal, so there is $Z \in [\text{depth}_X(a), X]$ such that

$$[a, Z] \cap \mathfrak{Y}_s = \emptyset \text{ for all } s \in \mathbb{N}^{[<\infty]}. \tag{4.16}$$

Applying the inclusion

$$\Phi(\mathfrak{X}) \setminus \mathfrak{X} \subseteq \bigcup_{s \in \mathbb{N}^{[<\infty]}} \left(\Phi(\widetilde{\mathfrak{X}}_s) \setminus \bigcup_{n \geq \max(s)+1} \Phi(\widetilde{\mathfrak{X}}_{s \cup \{n\}}) \right), \tag{4.17}$$

which follows immediately from Equation (4.14), it follows that

$$[a, Z] \cap \mathfrak{X} = [a, Z] \cap \Phi(\mathfrak{X}). \tag{4.18}$$

Since $\Phi(\mathfrak{X})$ is $\mathcal{S}(\leq X)$-Ramsey, we can find $Z_0 \in [\text{depth}_Z(a), Z]$ such that $[a, Z_0] \subseteq \Phi(\mathfrak{X})$ or $[a, Z_0] \cap \Phi(\mathfrak{X}) = \emptyset$. It follows that $Z_0 \in [\text{depth}_Y(a), Y]$ and $[a, Z_0] \subseteq \mathfrak{X}$ or $[a, Z_0] \subseteq \mathfrak{X}^c$, as required.   □

The Abstract Ramsey Theorem suggests the following definition giving us the primary concept of study in this book.

Definition 4.40 *Any structure of the form* $(\mathcal{R}, \mathcal{S}, \leq, \leq^0, r, s)$ *satisfying the conclusion of the Abstract Ramsey Theorem is called a* Ramsey space.

The main goal of this book is to present some of the most important examples of Ramsey spaces as well as to give some applications and connections between different spaces.

4.4 THE HALES-JEWETT SPACE

Let $L = \bigcup_{n=0}^{\infty} L_n$ be a given alphabet written as an increasing union of finite alphabets L_n and let $v \notin L$. Let W_L be the semigroup of words over L and let W_{Lv} be the semigroup of variable-words over L (i.e., words over $L \cup \{v\}$ in which v occurs at least once). By $W_L^{[\infty]}$ and $W_{Lv}^{[\infty]}$ we denote the collections of all infinite rapidly increasing infinite sequences of elements of W_L and W_{Lv}, respectively. (Recall that a sequence $X = (x_n)$ is rapidly increasing if $|x_n| > \sum_{i<n} |x_i|$ for all n.) For $X = (x_n) \in W_{Lv}^{[\infty]}$, let

$$[X]_L = \{x_{n_0}[\lambda_0]^\frown \ldots {}^\frown x_{n_k}[\lambda_k] \in W_L : n_0 < \ldots < n_k, \lambda_i \in L_{n_i} \ (i \le k)\}$$

and

$$[X]_{Lv} = \{x_{n_0}[\lambda_0]^\frown \ldots {}^\frown x_{n_k}[\lambda_k] \in W_{Lv} : n_0 < \ldots < n_k,$$

$$\lambda_i \in L_{n_i} \cup \{v\} \ (i \le k)\}$$

be the partial subsemigroups of W_L and W_{Lv}, respectively, generated by X. The assumption that X is rapidly increasing has the consequence that for every $w \in [X]_L$ and $x \in [X]_{Lv}$, the finite sets of integers $\{n_0 < \ldots < n_k\}$ and $\{m_0 < \ldots < m_\ell\}$ such that

$$w = x_{n_0}[\lambda_0]^\frown \ldots {}^\frown x_{n_k}[\lambda_k]$$

for some $\lambda_i \in L_{n_i} (i \le k)$ and

$$x = x_{m_0}[\mu_0]^\frown \ldots {}^\frown x_{m_\ell}[\mu_\ell]$$

for some $\mu_i \in L_{n_i} \cup \{v\} (i \le \ell)$ are unique. We denote these two finite sets by $\mathrm{supp}_X(w)$ and $\mathrm{supp}_X(x)$, respectively. For two finite sets of integers F and G, write $F < G$ if $\forall m \in F \ \forall n \in G \ m < n$. Define $\le$ on $W_{Lv}^{[\infty]}$ by letting $X = (x_n) \le Y = (y_n)$ if $X \subset [Y]_{Lv}$ and

$$\mathrm{supp}_Y(x_m) < \mathrm{supp}_Y(x_n) \quad \text{whenever } m < n. \tag{4.19}$$

Similarly, for $W = (w_n) \in W_L^{[\infty]}$ and $X = (x_n) \in W_{Lv}^{[\infty]}$, we write $W \le^\circ X$ if $W \subset [X]_L$ and

$$\mathrm{supp}_X(w_m) < \mathrm{supp}_X(w_n) \quad \text{whenever } m < n. \tag{4.20}$$

When $X = (x_n) \le Y = (y_n)$ happens, it is natural to say that X is a *block subsequence* of Y, and similarly for $W = (w_n) \le^\circ X = (x_n)$. The relations $\le$ and $\le^\circ$ come with their natural finitizations by letting

$$x = (x_m)_{m<p} \le_{\mathrm{fin}} y = (y_n)_{n<q},$$

if $x_m \in [y]_{Lv}$ for all $n < p$ and $(q-1) \in \mathrm{supp}_y(x)$ (i.e., $x \not\le_{\mathrm{fin}} y \restriction q'$ for all $q' < q$) and similarly for $w = (w_m)_{m<p} \le_{\mathrm{fin}}^\circ y = (y_n)_{n<q}$. During the course of developing axioms A.1 to A.4 in Section 4.3 we checked that

$$(W_L^{[\infty]}, W_{Lv}^{[\infty]}, \leq, \leq^\circ, r, s) \tag{4.21}$$

(where r and s are the restriction maps) satisfies all the requirements A.1 to A.4. Since $W_{Lv}^{[\infty]}$ is clearly a closed subset of $(W_{Lv}^{[<\infty]})^{\mathbb{N}}$ we have all the inputs of the Abstract Ramsey Theorem, and so we can conclude the following.

Theorem 4.41 $(W_L^{[\infty]}, W_{Lv}^{[\infty]}, \leq, \leq^\circ, r, s)$ *is a Ramsey space.*

The purpose of this section is to present some applications of this result. First of all let us state the following consequence of the Abstract Ramsey Theorem.

Theorem 4.42 (Infinite-Dimensional Hales-Jewett Theorem) *For every finite Souslin-measurable coloring of $W_L^{[\infty]}$ there is an $X = (x_n) \in W_{Lv}^{[\infty]}$ such that $[X]_L^{[\infty]}$ is monochromatic.*

Notation: $[X]_L^{[\infty]} = \{W \in W_L^{[\infty]} : W \leq^\circ X\}.$

As indicated before, various consequences of this sort of abstract Ramsey theoretic result tend to be related to each other. The following application exploits a relation between the Hales-Jewett space $(W_L^{[\infty]}, W_{Lv}^{[\infty]}, \leq, \leq^\circ, r, s)$ and the original Ellentuck space $(\mathbb{N}^{[\infty]}, \subseteq, r)$, leading to a parametrized version of the infinite-dimensional Ramsey theorem.

Theorem 4.43 *For every finite Souslin-measurable coloring of $\mathbb{N}^{[\infty]} \times \mathbb{R}^{\mathbb{N}}$, there is an $M \in \mathbb{N}^{[\infty]}$ and an infinite sequence (P_i) of nonempty perfect subsets of $\mathbb{R}$ such that the product $M^{[\infty]} \times \prod_{i=0}^{\infty} P_i$ is monochromatic. The same conclusion holds for countable Souslin-measurable colorings, provided the colors are invariant under finite changes of sets on the first coordinate.*

Proof. Let $L = \bigcup_{n=0}^{\infty} L_n$, where

$$L_n = \{\sigma \in 2^{\mathbb{N}} : \forall i > n \ \sigma(i) = 0\}. \tag{4.22}$$

Define $\varphi : W_L^{[\infty]} \to \mathbb{N}^{[\infty]}$ by

$$\varphi((w_k)) = \{|w_0| + \ldots + |w_k| : k \in \mathbb{N}\}, \tag{4.23}$$

and define $\psi : W_L^{[\infty]} \to 2^{\mathbb{N} \times \mathbb{N}}$ by

$$\psi((w_k))(n, i) = \sigma(i), \tag{4.24}$$

where σ occupies the nth place in the infinite word

$$w_0 {}^\frown w_1 {}^\frown \ldots {}^\frown w_k {}^\frown \ldots \tag{4.25}$$

Instead of $\mathbb{R}$, we shall work with the Cantor space $2^{\mathbb{N}}$, and instead of $\mathbb{R}^{\mathbb{N}}$ we work with $(2^{\mathbb{N}})^{\mathbb{N}}$, which we identify with $2^{\mathbb{N} \times \mathbb{N}}$ and make no difference between the sequence $((x_{ni})_n)_i$ of sequences and the doubly indexed sequence $(x_{ni})_{n,i}$. So let $c : \mathbb{N}^{[\infty]} \times (2^{\mathbb{N}})^{\mathbb{N}} \to \mathbb{N}$ be a given Souslin-measurable coloring that has either a finite range or has its colors invariant under changes of sets appearing on the first coordinate. Define $c^* : W_L^{[\infty]} \to \mathbb{N}$ by

$$c^*((w_n)) = c(\varphi((w_n)), \psi((w_n))). \tag{4.26}$$

By the Infinite-Dimensional Hales-Jewett Theorem (or more precisely from the fact that $(W_L^{[\infty]}, W_{Lv}^{[\infty]}, \leq, \leq^\circ, r, s)$ forms a Ramsey space), we can find $X = (x_k) \in W_{Lv}^{[\infty]}$ such that c^* is constant in

$$[X]_L^{[\infty]} = \{W = (w_k) \in W_L^{[\infty]} : W \leq^\circ X\}. \tag{4.27}$$

Let

$$M = \{|x_0| + \ldots + |x_k| : k \in \mathbb{N}\}. \tag{4.28}$$

Let $P \subseteq (2^{\mathbb{N}})^{\mathbb{N}}$ be the collection of all doubly-indexed sequences $(\varepsilon_{ni})_{n,i}$ for which one can find $(\sigma_k) \in \prod_{k=0}^{\infty} L_k$ such that when we form the infinite word

$$w_{(\sigma_k)} = x_0[\sigma_0]^\frown \ldots {}^\frown x_k[\sigma_k]^\frown \ldots, \tag{4.29}$$

then $\varepsilon_{ni} = \sigma(i)$, where σ occupies the nth place in this infinite word. Note that P is a closed subset of $(2^{\mathbb{N}})^{\mathbb{N}}$.

Claim 4.43.1 *There is an infinite sequence (P_i) of perfect subsets of the Cantor space $2^{\mathbb{N}}$ such that $\prod_{i=0}^{\infty} P_i \subseteq P$.*

Proof. Let P_i be the collection of all $\delta \in 2^{\mathbb{N}}$ that satisfies the following conditions, where x denotes the infinite variable-word $x_0^\frown \ldots {}^\frown x_n^\frown \ldots$.

(1) If at some place ℓ we find a letter $\sigma \in L$ in x, then $\delta(\ell) = \sigma(i)$.

(2) If $\ell < |x_0| + \ldots + |x_{i-1}|$ and at the ℓth place in x we find a variable, then $\delta(\ell) = 0$.

(3) If $n_{k-1} = |x_0| + \ldots + |x_{k-1}| \leq \ell < |x_0| + \ldots + |x_{k-1}| + |x_k| = n_k$ for some $k \geq i$ and we find a variable at the ℓth place in x, then $\delta(\ell) = \delta(\ell')$, where $\ell' \in [n_{k-1}, n_k)$ is the minimal place where x has a variable.

Thus, P_i has no restrictions at the minimal place of some interval of the form

$$I_k = [|x_0| + \ldots + |x_{k-1}|, |x_0| + \ldots + |x_{k-1}| + |x_k|) \tag{4.30}$$

for $k \geq i$, where a variable occurs. From this, one easily concludes that P_i is indeed perfect. Consider a sequence (δ_i) such that $\delta_i \in P_i$ for all i. For each k, choose $\sigma_k \in 2^{\mathbb{N}}$ such that

(4) $\sigma_k(i) = 0$ for $i > k$,

(5) $\sigma_k(i) = \delta_i(\ell)$ where ℓ is the minimal integer of the interval I_k where a variable occurs in x.

Clearly, $\sigma_k \in L_k$ for all k. Form the corresponding infinite word.

$$w_{(\sigma_k)} = x_0[\sigma_0]^\frown \ldots {}^\frown x_k[\sigma_k]^\frown \ldots \tag{4.31}$$

Let $(\varepsilon_{ni})_{n,i}$ be the doubly-indexed sequence such that $\varepsilon_{ni} = \sigma(i)$, where σ occupies the nth place in the infinite word $w_{(\sigma_k)}$. Tracing back the definitions from (5) to (1), we see that $\varepsilon_{ni} = \delta_i(n)$ for all n and i. It follows that $(\sigma_i) \in P$. This shows that $\prod_{i=0}^\infty P_i \subseteq P$. $\qquad\square$

Claim 4.43.2 $M^{[\infty]} \times P \subseteq (\varphi \times \psi)[[X]_L^{[\infty]}]$.

Proof. Consider $(N, (\varepsilon_{ni})) \in M^{[\infty]} \times P$. Find $(\sigma_k) \in \prod_k L_k$ such that $(\varepsilon_{ni}) = \psi((x_k[\sigma_k]))$. Thus, $\varepsilon_{ni} = \sigma(i)$ where σ occupies the nth place in the infinite word

$$w_{(\sigma_k)} = x_0[\sigma_0]^\frown \ldots {}^\frown x_k[\sigma_k]^\frown \ldots \tag{4.32}$$

Let (k_ℓ) be the increasing sequence of integers such that

$$N = \{|x_0| + \ldots + |x_{k_\ell}| : \ell \in \mathbb{N}\}. \tag{4.33}$$

For $\ell \in \mathbb{N}$, let

$$u_\ell = x_{k_{\ell-1}+1}[\sigma_{k_{\ell-1}+1}]^\frown \ldots {}^\frown x_{k_\ell}[\sigma_{k_\ell}], \tag{4.34}$$

where $k_{-1} = -1$. Then $(u_\ell) \in [X]_L^{[\infty]}$ and the infinite concatenation

$$u_0^\frown \ldots {}^\frown u_\ell^\frown \ldots$$

is still equal to the same infinite word w_{σ_k} determined by the sequence $(\sigma_k) \in \prod_k L_k$. It follows that $\psi((u_\ell)) = \psi((x_k[\sigma_k])) = (\varepsilon_{ni})$. Also $\varphi((u_\ell)) = N$, and therefore $(\varphi \times \psi)((u_\ell)) = (N, (\varepsilon_{ni}))$. $\qquad\square$

Note that this finishes the proof of the theorem, since Claim 4.43.2 yields that c is constant on $M^{[\infty]} \times \prod_{i=0}^\infty P_i$. $\qquad\square$

Corollary 4.44 (Laver) *Suppose $\{f_n\}_{n=0}^\infty$ is a uniformly bounded sequence of Baire- of Lebesgue-measurable functions from $\mathbb{R}^\mathbb{N}$ into $\mathbb{R}$. Then some subsequence of $\{f_n\}$ converges monotonically and uniformly on some product $\prod_{n=0}^\infty P_n$ of perfect subsets of $\mathbb{R}$.*

Proof. Assume first that $\{f_n\}_{n=0}^\infty$ is a uniformly bounded sequence of continuous functions. Let $\mathcal{C}$ be the subset of $\mathbb{N}^{[\infty]} \times \mathbb{R}^\mathbb{N}$ consisting of pairs $(A, \vec{t})$

for which $\{f_n(\vec{t})\}_{n \in A}$ is a monotonically convergent sequence of reals. Applying Theorem 4.43, we obtain $M \in \mathbb{N}^{[\infty]}$ and a sequence $(P_n)_{n=0}^{\infty}$ of perfect subsets of $\mathbb{R}$ such that

$$M^{[\infty]} \times \prod_{n=0}^{\infty} P_n \subseteq \mathcal{C} \quad \text{or} \quad (M^{[\infty]} \times \prod_{n=0}^{\infty} P_n) \cap \mathcal{C} = \emptyset. \qquad (4.35)$$

Note that the second alternative is impossible. Since $\mathcal{C}$ splits into two Borel pieces $\mathcal{C}^{\uparrow}$ and $\mathcal{C}^{\downarrow}$ consisting of pairs for which the convergence is nondecreasing and nonincreasing, respectively, we may further assume that, for example,

$$M^{[\infty]} \times \prod_{n=0}^{\infty} P_n \subseteq \mathcal{C}^{\uparrow}. \qquad (4.36)$$

Shrinking the product even further, we may assume that the pointwise limit of the subsequence $\{f_n\}_{n \in M}$ is continuous on $\prod_{n=0}^{\infty} P_n$. By Dini's theorem, we conclude that the convergence is in fact uniform.

The reduction of the Baire-measurable case to the assumption that the sequence $\{f_n\}$ consists of continuous functions is simple. One first finds a dense G_δ-subset $G \subset \mathbb{R}^{\mathbb{N}}$ such that $f_n \restriction G$ is continuous for all n. Then a simple fusion argument using the Fubini theorem for category gives us a sequence $(P_n)_{n=0}^{\infty}$ of perfect subsets of $\mathbb{R}$ such that $\prod_{n=0}^{\infty} P_n \subset G$.

To take care of the Lebesgue-measurable case, applying Egoroff's theorem, one first obtains a compact subset $H \subset \mathbb{R}^{\mathbb{N}}$ of positive measure such that $f_n \restriction H$ is continuous for all $n \in \mathbb{N}$. Using the Brodski-Eggleston lemma (see Theorem 9.44 and Corollary 9.45 below), we can recursively construct a sequence $(P_n)_{n=0}^{\infty}$ of perfect subsets of $\mathbb{R}$ such that $\prod_{n=0}^{\infty} P_n \subseteq H$.   □

For finite powers of $\mathbb{R}$, one can do a bit better than in Corollary 4.44

Corollary 4.45 (Harrington) *For every positive integer d and every uniformly bounded sequence $(f_n)_{n=0}^{\infty}$ of Baire- or Lebesgue-measurable functions from $\mathbb{R}^d$ into $\mathbb{R}$, there is a subsequence $\{f_{n_k}\}_{k=0}^{\infty}$ and a perfect set $P \subset \mathbb{R}$ such that $\{f_{n_k}\}_{k=0}^{\infty}$ is uniformly convergent on P^d.*

Proof. We first show by induction on d that there is a perfect set $P \subseteq \mathbb{R}$ such that $f_n \restriction P^d$ is continuous for all n. The case $d = 1$ is an immediate consequence of the fact that every Baire (Lebesgue) measurable function is continuous on a dense G_δ-subset of $\mathbb{R}$ (F_σ-subset of $\mathbb{R}$ of full measure) and that countable intersections of such sets contain nonempty perfect subsets. To see the inductive step from $d - 1$ to d, we first find a dense G_δ-set $G \subset \mathbb{R}^d$, respectively an F_σ-subset $H \subset \mathbb{R}^d$ of full measure such that $f_n \restriction G$, respectively, $f_n \restriction H$, is continuous for all n. Then we use Mycielski's theorem (see Theorem 6.40 below) to find a perfect set $P \subset \mathbb{R}$ such that the set

$$P^{(d)} = \{(x_i)_{i<d} \in P^d : x_i \neq x_j \text{ whenever } i \neq j\}$$

is included in G, respectively, in H.

Thus, without loss of generality, we may suppose that in the hypothesis of Corollary 4.45 we are in fact given a sequence $\{f_n\}$ of continuous functions on $\mathbb{R}^d$. By Corollary 4.44, using a subsequence of (f_n) and a perfect subset of P, we may assume that the sequence (f_n) is uniformly convergent on the diagonal of P^d. So we concentrate on getting uniform convergence off the diagonal of P^d. To this end, we prove by induction on d the following stronger statement:

($*$) For every uniformly bounded sequence $\{f_n\}$ of continuous real-valued functions on some finite-dimensional cube $\mathbb{R}^\ell$ and every decomposition $\ell = \ell_0 + \ell_1 + \ldots + \ell_{m-1}$ where $1 \le \ell_i \le d$ for $i < m$, there exist perfect sets $P_i \subset \mathbb{R}(i < m)$ and a subsequence $\{f_{n_k}\}$ that is uniformly convergent on $\prod_{i<m} P_i^{\ell_i}$.

Corollary 4.44 gives us the case $d = 1$ of ($*$), so let us assume ($*$)$_d$ and prove ($*$)$_{d+1}$. In fact we show only the case $m = 1$ of ($*$)$_{d+1}$, since the general case follows from an obvious modification of this argument. To this end, we construct a fusion sequence $P_\sigma(\sigma \in 2^{<\infty})$ of perfect subsets of $\mathbb{R}$ and a decreasing sequence $M_0 \supseteq M_1 \supseteq \ldots \supseteq M_k \supseteq \ldots$ of infinite subsets of $\mathbb{N}$ such that

(1) $P_{\sigma^\frown 0}, P_{\sigma^\frown 1} \subset P_\sigma$ and $P_{\sigma^\frown 0} \cap P_{\sigma^\frown 1} = \emptyset$,

(2) $\mathrm{diam}(P_\sigma) \le 2^{-|\sigma|}$,

(3) For every $k \in \mathbb{N}$ and every decomposition

$$\ell_0 + \ell_1 + \ldots + \ell_{m-1} = d + 1 \tag{4.37}$$

such that $1 \le \ell_i \le d(1 < m)$ and every one-to-one sequence $\sigma_i \in 2^k(i < m)$, the subsequence $\{f_n\}_{n \in M_k}$ is uniformly convergent or $Q = \prod_{i<m} P_{\sigma_i}^{\ell_i}$. Moreover, for all $n, m \in M_k$,

$$\|f_n \upharpoonright Q - f_m \upharpoonright Q\|_\infty \le 2^{-k}. \tag{4.38}$$

Clearly, the induction hypothesis ($*$)$_d$ is exactly what is needed to complete the construction of the fusion sequence $P_\sigma(\sigma \in 2^{<\infty})$ satisfying (1),(2), and (3).

Pick $N \in \mathbb{N}^{[\infty]}$ such that $N \backslash M_k$ is finite for all k. Let

$$P = \bigcup_{\sigma \in 2^\mathbb{N}} \bigcap_{k \in \mathbb{N}} P_{\sigma|k}. \tag{4.39}$$

Then P is a perfect subset of $\mathbb{R}$ and the subsequence $\{f_n\}_{n \in N}$ is uniformly convergent on P^{d+1}. This completes the description of the inductive step from ($*$)$_d$ to ($*$)$_{d+1}$ and finishes the proof of the corollary. $\square$

Remark 4.46 Note that the sequence of projections

$$\{\pi_n : \mathbb{R}^\mathbb{N} \to \mathbb{R}\}_{n=0}^\infty \tag{4.40}$$

contains no subsequence that is convergent on any set of the form $P^\mathbb{N}$ where P is a perfect subset of $\mathbb{R}$. In the next chapter we shall present a proof of the Rosenthal Dichotomy Theorem, which shows that the sequence $\{\pi_n\}_{n=0}^\infty$ of projection is a rather typical such example in the sense that any uniformly bounded sequence $\{f_n\}$ of continuous functions defined on some Polish space X contains either a subsequence $\{f_{n_k}\}$ that pointwise converges on the whole space X or a subsequence $\{f_{n_k}\}$ that on some copy of the Cantor space $2^{\mathbb{N}}$ behaves like the sequence $\{\pi_k : 2^{\mathbb{N}} \to 2\}_{k=0}^{\infty}$ of projections.

4.5 RAMSEY SPACES OF INFINITE BLOCK SEQUENCES OF LOCATED WORDS

We start with an alphabet $L = \bigcup_{n=0}^{\infty} L_n$ written as an increasing union of finite alphabets L_n and a variable $v \notin L$. A *located word* is a function from a finite nonempty subset of $\mathbb{N}$ into L, while a *located variable-word* is such a function with range in $L \cup \{v\}$ with the value v achieved at least once. Let FIN_L and FIN_{Lv} denote the collections of located words and located variable-words, respectively.

For $x \in \mathrm{FIN}_{Lv}$ and $\lambda \in L \cup \{v\}$, let $x[\lambda]$ be the function such that

(1) $\mathrm{dom}(x[\lambda]) = \mathrm{dom}(x)$,

(2) if $i \in \mathrm{dom}(x)$ and $x_i \in L$, then $x[\lambda]_i = x_i$,

(3) if $i \in \mathrm{dom}(x)$ and $x_i = v$, then $x[\lambda]_i = \lambda$.

A *block sequence* is a finite or infinite sequence $X = (x_n)$ of members of FIN_L or FIN_{Lv} such that

$$\mathrm{dom}(x_m) < \mathrm{dom}(x_n) \quad \text{whenever } m < n.$$

Let $\mathrm{FIN}_L^{[\infty]}$ and $\mathrm{FIN}_{Lv}^{[\infty]}$ denote the collection of all infinite block sequences of elements of FIN_L and FIN_{Lv}, respectively. We consider FIN_L and FIN_{Lv} partial semigroups under the operation of taking the union of two partial functions when their domains are disjoint. For $X = (x_n) \in \mathrm{FIN}_{Lv}^{[\infty]}$, we consider the following two partial subsemigroups of FIN_L and FIN_{Lv}, respectively, generated by X:

$$[X]_L = \{x_{n_0}[\lambda_0] \cup \ldots \cup x_{n_k}[\lambda_k] \in \mathrm{FIN}_L : n_0 < \ldots < n_k, \lambda_i \in L_{n_i} (i \leq k)\}$$

$$[X]_{Lv} = \{x_{n_0}[\lambda_0] \cup \ldots \cup x_{n_k}[\lambda_k] \in \mathrm{FIN}_{Lv} : n_0 < \ldots < n_k,$$

$$\lambda_i \in L_{n_i} \cup \{v\} (i \leq k)\}.$$

For $X = (x_n)$ and $Y = (y_n)$ from $\mathrm{FIN}_{Lv}^{[\infty]}$, we say that X is a *block-subsequence* of Y and write $X \leq Y$ if $x_n \in [Y]_{Lv}$ for all n. Similarly for $A = (a_n) \in \mathrm{FIN}_L^{[\infty]}$ and $X = (x_n) \in \mathrm{FIN}_{Lv}^{[\infty]}$, we say that $A \leq^{\circ} X$ if $a_n \in [X]_L$ for all n. We extend these relations to the families $\mathrm{FIN}_L^{[<\infty]}$ and $\mathrm{FIN}_{Lv}^{[<\infty]}$ of finite block sequences of located words and variable-words. This gives us the finitizations $\leq_{\mathrm{fin}}$ and $\leq_{\mathrm{fin}}^{\circ}$ of $\leq$ and $\leq^{\circ}$, respectively:

$$(x_n)_0^k \leq_{\text{fin}} (y_n)_0^\ell \quad \text{iff} \quad (x_n)_0^k \leq (y_n)_0^\ell \text{ and } (\forall \ell' < \ell) \ (x_n)_0^k \not\leq (y_n)_0^{\ell'},$$

$$(a_n)_0^k \leq_{\text{fin}}^o (x_n)_0^\ell \quad \text{iff} \quad (a_n)_0^k \leq (x_n)_0^\ell \text{ and } (\forall \ell' < \ell) \ (a_n)_0^k \not\leq (x_n)_0^{\ell'}.$$

$$(4.41)$$

The proofs that $\leq_{\text{fin}}$ and $\leq_{\text{fin}}^o$ are finitizations of $\leq$ and $\leq^o$ and that

$$(\text{FIN}_L^{[\infty]}, \text{FIN}_{Lv}^{[\infty]}, \leq, \leq^o, r, s)$$

satisfies $A.1 - A.3$ are straightforward. The fact that this space satisfies $A.4$ reduces to the following block sequence reformulation of Theorem 2.35.

Theorem 4.47 *For every finite coloring of $\text{FIN}_L \cup \text{FIN}_{Lv}$ and every $Y = (y_n) \in \text{FIN}_{Lv}^{[\infty]}$, there is $X \leq Y$ in $\text{FIN}_{Lv}^{[\infty]}$ such that $[X]_L$ and $[X]_{Lv}$ are both monochromatic.*

Proof. The proof is identical to the proof of Theorem 2.22 except that we now restrict to the semigroup S of cofinite ultrafilters concentrating on the union $[Y]_L \cup [Y]_{Lv}$. $\qquad\square$

Clearly, $\text{FIN}_{Lv}^{[\infty]}$ is a closed subset of the product $(\text{FIN}_{Lv}^{[<\infty]})^{\mathbb{N}}$, so applying the Abstract Ramsey Theorem gives us the following result.

Theorem 4.48 $(\text{FIN}_L^{[\infty]}, \text{FIN}_{Lv}^{[\infty]}, \leq, \leq^o, r, s)$ *is a Ramsey space.*

Corollary 4.49 (Bergelson-Blass-Hindman) *For every finite Souslin-measurable coloring of the space $\text{FIN}_L^{[\infty]}$, there is $X = (x_n) \in \text{FIN}_{Lv}^{[\infty]}$ such that $\{W \in \text{FIN}_L^{[\infty]} : W \leq^o X\}$ is monochromatic.*

The following result compares the relative strengths of two Ramsey spaces encountered so far.

Theorem 4.50 *There is a natural surjection*

$$\pi_\infty : (\text{FIN}_L^{[\infty]}, \text{FIN}_{Lv}^{[\infty]}, \leq, \leq^o, r, s) \to (W_L^{[\infty]}, W_{Lv}^{[\infty]}, \leq, \leq^o, r, s)$$

that transfers the Ramsey theoretic properties from one space onto the other.

Proof. Let

$$\pi : \text{FIN}_L \cup \text{FIN}_{Lv} \to W_L \cup W_{Lv}$$

be the mapping which associates with each located word x the word $\pi(x)$ of length $|\text{dom}(x)|$ that in the ith place has the letter x_j, where $j \in \text{dom}(x)$ occupies the ith place in the increasing enumeration of $\text{dom}(x)$. The mapping π extends to

$$\pi_\infty : \text{FIN}_L^{[\infty]} \cup \text{FIN}_{Lv}^{[\infty]} \to W_L^{[\infty]} \cup W_{Lv}^{[\infty]}$$

in the natural way,

$$\pi_\infty((x_n)) = (\pi(x_n)).$$

Note, however, that in $W_L^{[\infty]}$ and $W_{Lv}^{[\infty]}$ we have restricted ourselves to *rapidly increasing* sequences of words, i.e., sequences $X = (x_n)$ that satisfy the requirement $|x_n| > \sum_{i<n} |x_i|$ for all n. Hence, strictly speaking, we have to restrict the projection π to the Ramsey space

$$(\text{FIN}_L^{[\infty]}, \text{FIN}_{Lv}^{[\infty]}, \leq, \leq^\circ, r, s)$$

of rapidly increasing block sequences of words. In any case, it is clear that the mapping so-defined is indeed a projection, i.e., it is open, 1-Lipschitz, maps basic sets to basic sets, and therefore, transfers the Ramsey theoretic properties from the space $(\text{FIN}_L^{[\infty]}, \text{FIN}_{Lv}^{[\infty]}, \leq, \leq^\circ, r, s)$ of block sequences of located words to the Hales-Jewett space $(W_L^{[\infty]}, W_{Lv}^{[\infty]}, \leq, \leq^\circ, r, s)$.   □

It follows that the fact that $(\text{FIN}_L^{[\infty]}, \text{FIN}_{Lv}^{[\infty]}, \leq, \leq^\circ, r, s)$ is a Ramsey space is more fundamental than the fact that $(W_L^{[\infty]}, W_{Lv}^{[\infty]}, \leq, \leq^\circ, r, s)$ is a Ramsey space in the sense that any application of the latter space can also be obtained as an application of the former.

NOTES TO CHAPTER FOUR

Abstract Baire theory was initiated in the 1930's by E. Marczewski [71] and completed more recently through the work of J.C. Morgan [79], J. Pawlikowski [85], and others. One of the goals of abstract Baire theory was to find a unified approach towards the classical results of Lusin and Nykodim showing that the Lebesgue measurability and the Baire property are preserved under the Souslin operation. This theory did arrive at the elegant combinatorial proofs of these classical results based on the notion of envelope for an arbitrary subset of the index set. As we have seen above during the course of the proof of Lemma 4.39, the idea of global enveloping needed an essential adjustment to the more flexible notion of *local envelope* for an arbitrary subset of the index set. We shall encounter the idea of local enveloping again in Chapter Seven of this book when proving analogous results in the context of local Ramsey theory. The first instance of combinatorial forcing appears in a paper of Nash-Williams [82], from where we can taken the terminology "accepts", "rejects" and "decides", although our combinatorial forcing is more closely related to that of Galvin-Prikry [35]. More recently the combinatorial forcing has found one of its deepest applications in the work of Gowers [38] (see also [3]), where it is used to show than an infinite-dimensional Banach space conains either an infinite basic unconditional sequence or an infinite-dimensional hereditarily indecomposable closed subspace, a key step towards the solution of Banach's homogeneous space problem. In hindsight one can see that Baumgartner's proof ([6]) of Hindman's theorem also uses a combinatorial forcing and this suggests the natural question of whether there is an analogous proof that would establish Gowers's pigeon hole principle for FIN_k (Theorem 2.22 above), originally proved using the methods of topological dynamics. The work in set theory (see for example [97]) that deals with forcing descriptions of generic reals contains deep applications of combinatorial forcing and could serve as inspiration for further applications of this method in Ramsey theory per se. Corollary 4.49 is due to Bergelson-Blass-Hindman [8], although their argument is not sufficient to prove the full Theorem 4.48. The first parametrization of the infinite-dimensional Ramsey theorem appeared in the paper of Miller [76]. Pawlikowski [85] gives an Ellentuck-style extension of that result. The results of Harrington and Laver appear in [61] and were originally based on the Halpern-Läuchli theorem rather than the Hales-Jewett theorem. We give the Halpern-Läuchli treatment of these two results in Chapter Six of this book.

Chapter Five

Topological Ramsey Theory

5.1 TOPOLOGICAL RAMSEY SPACES

The special case of the Abstract Ramsey Theorem when $\mathcal{R} = \mathcal{S}$, when $\leq = \leq^\circ$, and when $r = s$ is of independent interest since in this case the basic sets

$$[a, B] = \{A \in \mathcal{R} : A \leq B \ \& \ (\exists n) \ r_n(A) = a\}$$

for $a \in \mathcal{AR}$ and $B \in \mathcal{R} = \mathcal{S}$ form a base for a topology on $\mathcal{R}$ that we call the *natural topology* of $\mathcal{R}$ and which extends the usual metrizable topology on $\mathcal{R}$ when we consider it a subspace of the Tychonov cube $\mathcal{AR}^{\mathbb{N}}$. The axioms A.1, A.2, A.3, and A.4 from the previous chapter reduce to the following set of axioms (still denoted the same way) about a triple

$$(\mathcal{R}, \leq, r)$$

of objects, where $\mathcal{R}$ is a nonempty set, where $\leq$ is a quasi-ordering on $\mathcal{R}$, and where

$$r : \mathcal{R} \times \omega \to \mathcal{AR}$$

is a mapping giving us the sequence $(r_n(\cdot) = r(\cdot, n))$ of approximation mappings.

A.1. (1) $r_0(A) = \emptyset$ for all $A \in \mathcal{R}$.

(2) $A \neq B$ implies $r_n(A) \neq r_n(B)$ for some n.

(3) $r_n(A) = r_m(B)$ implies $n = m$ and $r_k(A) = r_k(B)$ for all $k < n$.

A.2. There is a quasi-ordering $\leq_{\text{fin}}$ on $\mathcal{AR}$ such that

(1) $\{a \in \mathcal{AR} : a \leq_{\text{fin}} b\}$ is finite for all $b \in \mathcal{AR}$,

(2) $A \leq B$ iff $(\forall n)(\exists m) \ r_n(A) \leq_{\text{fin}} r_m(B)$,

(3) $\forall a, b \in \mathcal{AR}[a \sqsubseteq b \wedge b \leq_{\text{fin}} c \to \exists d \sqsubseteq c \ a \leq_{\text{fin}} d]$.

A.3. (1) If $\text{depth}_B(a) < \infty$ then $[a, A] \neq \emptyset$ for all $A \in [\text{depth}_B(a), B]$.

(2) $A \leq B$ and $[a, A] \neq \emptyset$ imply that there is $A' \in [\text{depth}_B(a), B]$ such that $\emptyset \neq [a, A'] \subseteq [a, A]$.

A.4. If $\text{depth}_B(a) < \infty$ and if $\mathcal{O} \subseteq \mathcal{AR}_{|a|+1}$, then there is $A \in [\text{depth}_B(a), B]$ such that $r_{|a|+1}[a, A] \subseteq \mathcal{O}$ or $r_{|a|+1}[a, A] \subseteq \mathcal{O}^c$.

Recall the notation used above

$$\text{depth}_B(a) = \min\{n : a \leq_{\text{fin}} r_n(B)\}$$

if there is an n such that $a \leq_{\text{fin}} r_n(B)$ and $\text{depth}_B(a) = \infty$, otherwise. The *length* $|a|$ of a finite approximation a is the integer n such that $a = r_n(A)$ for some $A \in \mathcal{R}$. For $a, b \in \mathcal{AR}$, we say that a *is an initial segment of* b and write $a \sqsubseteq b$ if there is $B \in \mathcal{AR}$ and $m \leq n < \omega$ such that $a = r_m(B)$ and $b = r_n(B)$; the strict version $a \sqsubset b$ corresponds to the case $m < n$. For $B \in \mathcal{R}$ and $n < \omega$, set

$$[n, B] = [r_n(B), B].$$

We shall freely carry on all other conventions and notation encountered in the previous Chapter.

Example 5.1.1 *A prototype example of a triple* $(\mathcal{R}, \leq, r)$ *satisfying axioms A.1, A.2, A.3, and A.4 is the Ellentuck space* $(\mathbb{N}^{[\infty]}, \subseteq, r)$ *already encountered above in Chapter One, where*

$$N^{[\infty]} = \{M \subseteq \mathbb{N} : M \text{ infinite }\}$$

is the set of all infinite subsets of $\mathbb{N}$, *where* $r_n(A)$ *is the initial segment of A formed by taking the first n elements of A and where the relation* $\subseteq_{\text{fin}}$ *is defined on*

$$\mathcal{AN}^{[\infty]} = \mathbb{N}^{[<\infty]} = \{a \subseteq \mathbb{N} : a \text{ finite }\}$$

(= the family of all finite subsets of $\mathbb{N}$*) as follows:*

$$a \subseteq_{\text{fin}} b \text{ iff } a = b = 0 \text{ or } a \subseteq b \text{ and } \max(a) = \max(b).$$

Note that the topology of the prototype example is equal to the topology $\mathbb{N}^{[\infty]}$ *gets as a subset of the exponential space* $\exp(\mathbb{N})$.

From now on whenever we duscuss a topological property of $(\mathcal{R}, \leq, r)$, we refer of course to the *natural*, or the *Ellentuck*, topology on $\mathcal{R}$, i.e. the one induced by the basis $[a, A]$ $(a \in \mathcal{AR}, A \in \mathcal{R})$. One example of this is the following definition.

Definition 5.1 *A subset* $\mathfrak{X}$ *of* $\mathcal{R}$ *has the* property of Baire *if* $\mathfrak{X} = \mathcal{O} \Delta \mathcal{M}$ *for some Ellentuck open* $\mathcal{O} \subseteq \mathcal{R}$ *and Ellentuck meager* $\mathcal{M} \subseteq \mathcal{R}$.

Note that when Ellentuck meager subsets of $\mathcal{R}$ are in fact nowhere dense (which will be true in all cases of interest), this notion coincides with the notion of $\mathcal{R}$-Baire subsets of $\mathcal{R}$ introduced in the previous chapter of this book. Let us recall also Definition 4.22 in this special case.

Definition 5.2 *A subset* $\mathfrak{X}$ *of* $\mathcal{R}$ *is* Ramsey *if for every* $\emptyset \neq [a, A]$ *there is a* $B \in [a, A]$ *such that* $[a, B] \subset \mathfrak{X}$ *or* $[a, B] \subset \mathfrak{X}^c$. *A subset* $\mathfrak{X}$ *of* $\mathcal{R}$ *is* Ramsey null *if for every* $\emptyset \neq [a, A]$, *there is a* $B \in [a, A]$ *such that* $[a, B] \cap \mathfrak{X} = \emptyset$.

These notions are leading us naturally to the following analog of Definition 4.40 that gives us another primary object of study in this book.

Definition 5.3 *A triple* $(\mathcal{R}, \leq, r)$ *is* a topological Ramsey space *if every property of Baire subset of* $\mathcal{R}$ *is Ramsey and if every meager subset of* $\mathcal{R}$ *is Ramsey null.*

It is now clear that the Abstract Ramsey Theorem specializes to the following result, where we recall that $(\mathcal{R}, \leq, r)$ is said to be *closed* whenever $\mathcal{R}$ is closed when identified with a subset of the Tychonov power $\mathcal{AR}^{\mathbb{N}}$ of $\mathcal{AR}$ with its discrete topology.

Theorem 5.4 (Abstract Ellentuck Theorem) *If* $(\mathcal{R}, \leq, r)$ *is closed and if it satisfies axioms A.1, A.2, A.3, and A.4, then every property of Baire subset of* $\mathcal{R}$ *is Ramsey and every meager subset is Ramsey null, or in other words, the triple* $(\mathcal{R}, \leq, r)$ *forms a topological Ramsey space.*

Proof. The Abstract Ramsey Theorem in this special case gives us in particular that Ellentuck nowhere dense subsets of $\mathcal{R}$ are all Ramsey null. Since Ramsey null subsets of $\mathcal{R}$ form a σ-ideal, it follows that Ellentuck meager subsets of $\mathcal{R}$ are in fact nowhere dense, so as explained above, the property of Baire subsets $\mathcal{R}$ relative to the Ellentuck topology are all $\mathcal{R}$-Baire in the sense of Definition 4.23. $\qquad\qquad\square$

Corollary 5.5 (Ellentuck) *The space* $(\mathbb{N}^{[\infty]}, \subseteq, r)$ *is a topological Ramsey space.*

Let us now recall a notion introduced in the previous chapter but specified here in the special case of $\mathcal{R}$ as a topological space with its Ellentuck topology.

Definition 5.6 *A subset* $\mathcal{X}$ *of* $\mathcal{R}$ *is said to be* Souslin, *or more explicitly,* Souslin-measurable, *if it belongs to the minimal field of subsets of* $\mathcal{R}$ *that contains all Ellentuck open subsets of* $\mathcal{R}$ *and that is closed under the Souslin operation.*

Using the classical result that in any topological space the property of Baire is preserved under the Souslin operation (see Corollary 4.8) we get the other part of the conclusion of the Abstract Ramsey Theorem.

Corollary 5.7 *If* $(\mathcal{R}, \leq, p)$ *is closed and if it satisfies A.1, A.2, A.3, and A.4, then every Souslin-measurable subset of* $\mathcal{R}$ *is Ramsey.*

Remark 5.8 Recall our convention that whenever we do not otherwise specify the topology on $\mathcal{R}$ to which the notion of Souslin measurability refers in this corollary, it is the Ellentuck topology of $\mathcal{R}$. In most applications, however, one needs only the Souslin measurability relative to the metrizable topology of $\mathcal{R}$, i.e., the one induced from $\mathcal{AR}^{\mathbb{N}}$.

Corollary 5.9 (Abstract Silver Theorem) *If* $(\mathcal{R}, \leq, p)$ *is closed and if it satisfies A.1, A.2, A.3, and A.4, then every metrically Souslin-measurable subset of* $\mathcal{R}$ *is Ramsey.*

Corollary 5.10 (Silver) *Every metrically Souslin-measurable subset of* $\mathbb{N}^{[\infty]}$ *is Ramsey.*

Corollary 5.11 (Abstract Galvin-Prikry Theorem) *If* $(\mathcal{R}, \leq, p)$ *is closed and if it satisfies A.1, A.2, A.3, and A.4, then every metrically Borel subset of* $\mathcal{R}$ *is Ramsey.*

Corollary 5.12 (Galvin-Prikry) *Every metrically Borel subset of* $\mathbb{N}^{[\infty]}$ *is Ramsey.*

Remark 5.13 We shall see that many Ramsey spaces are related to each other in the sense that there exist maps that transfer metric corollaries like the Galvin-Prikry or Silver theorems from one space to the other. These transfer results give us a sense that one space is "stronger" than other. The price of the "strength" is typically hidden in the fact that it is considerably harder to prove the corresponding pigeon-hole principle, A.4. We would like to note, however, that the transfer procedure rarely reduces the full strength of the Abstract Ramsey Theorem from one space to the other, since the theorem really describes a particular σ-field of sets that is typically quite different from the others. There are applications for which one does need the full description of the field of Ramsey sets. This happens, for example, when one is proving, under suitable set-theoretic assumptions, that all definable subsets of $\mathcal{R}$ are Ramsey.

As already pointed out in Chapter One, in which we consider the prototype topological Ramsey space $(\mathbb{N}^{[\infty]}, \subseteq, r)$, one cannot expect that the field of Ramsey sets contains *every* subset of the domain. The following result shows that the same phenomenon appears in essentially every other topological Ramsey space.

Theorem 5.14 *Suppose* $(\mathcal{R}, \leq, p)$ *is a closed triple that satisfies A.1, A.2, A.3, and A.4 and suppose that no basic open set of* $(\mathcal{R}, \leq, p)$ *is countable, or equivalently, that the associated topological Ramsey space has no isolated points. Then there is an* $\mathcal{X} \subseteq \mathcal{R}$ *such that* $[0, B] \cap \mathcal{X} \neq \emptyset$ *and* $[0, B] \cap \mathcal{X}^c \neq \emptyset$ *for all* $B \in \mathcal{R}$.

Proof. Fix a well-ordering $\preceq$ of $\mathcal{R}$. Let

$$\mathcal{X} = \{X \in \mathcal{R} : X \preceq Y \text{ for all } Y \in [0, X]\}.$$

Let us check that $\mathcal{X}$ is the required set. Consider an arbitrary $B \in \mathcal{R}$. Let X be the $\preceq$-minimal element of the basic open set $[0, B]$. Then X is also the $\preceq$-minimal element of $[0, X]$, and so $X \in \mathcal{X}$. It remains to show that $[0, B]$ contains an element that does not belong to $\mathcal{X}$. Otherwise, we would have that

$$(\forall X, Y \in [0, B]) \ [X \prec Y \rightarrow X \not\preceq Y]. \tag{5.1}$$

It follows in particular that $\leq$ restricted to the basic open set $[0, B]$ is antisymmetric and that $([0, B], \geq)$ is a well-founded partially ordered set. We have already noted that from $A.2$ it follows that $\leq$ is a closed relation relative to the topology of $\mathcal{R}$ induced from the product space $\mathcal{AR}^{\mathbb{N}}$. Thus, $([0, B], \geq)$ is a well-founded partially ordered set whose order relation $\leq$ is closed relative to the Polish topology of $[0, B]$. So, applying Theorem 9.53(boundedness)), the well-founded poset $([0, B], \geq)$ has countable rank. So in particular, the basic open set $[0, B]$ can be covered by countably many $\leq$-antichains. Note, however, that every $\leq$-antichain is discrete and therefore nowhere dense relative to the Ellentuck topology on $\mathcal{R}$. It follows that the basic open set $[0, B]$ can be covered by countably many nowhere dense sets. Since every nowhere dense set is Ramsey null and since the ideal of Ramsey null sets is a σ-ideal it follows that $[0, B]$ itself is a Ramsey null set, which is a contradiction. $\qquad\square$

Given a triple $(\mathcal{R}, \leq, p)$ satisfying $A.1$, $A.2$, $A.3$, and $A.4$, given a family $\mathcal{F} \subseteq \mathcal{AR}$ of finite approximations, and given $X \in \mathcal{R}$, set

$$\mathcal{F}|X = \mathcal{F} \cap \{r_n(Y) : n \in \mathbb{N} \text{ and } Y \leq X\}.$$

The following corollary of the Abstract Ellentuck Theorem is also worth pointing out.

Theorem 5.15 (Abstract Galvin Lemma) *Suppose $(\mathcal{R}, \leq, p)$ is a closed triple that satisfies $A.1$, $A.2$, $A.3$, and $A.4$. Then for every family $\mathcal{F} \subseteq \mathcal{AR}$ of finite approximations and for every $X \in \mathcal{R}$, there is $Y \leq X$ such that either $\mathcal{F}|Y = \emptyset$, or else for every $B \leq Y$ there is $n \in \mathbb{N}$ such that $r_n(B) \in \mathcal{F}$.*

Proof. This follows from the abstract Galvin-Prikry theorem applied to the open set

$$\mathcal{O} = \{B \in \mathcal{R} : (\exists n \in \mathbb{N}) \; r_n(B) \in \mathcal{F}\}.$$

$\qquad\square$

Definition 5.16 *We shall say that a family $\mathcal{F} \subseteq \mathcal{AR}$ of finite approximation is*

(1) Nash-Williams *if $a \not\sqsubseteq b$ for all $a \neq b \in \mathcal{F}$,*

(2) Sperner *if $a \not\leq_{\mathrm{fin}} b$ for all $a \neq b \in \mathcal{F}$,*

(3) Ramsey *if for every partition $\mathcal{F} = \mathcal{F}_0 \cup \mathcal{F}_1$ and every $X \in \mathcal{R}$, there is $Y \leq X$ and $i = 0, 1$ such that $\mathcal{F}_i|Y = \emptyset$.*

Theorem 5.17 (Abstract Nash-Williams Theorem) *Suppose $(\mathcal{R}, \leq, p)$ is a closed triple that satisfies $A.1$, $A.2$, $A.3$, and $A.4$. Then every Nash-Williams family of finite approximations is Ramsey.*

Proof. Applying the Abstract Galvin Lemma successively for $\mathcal{F}_0$ and then for $\mathcal{F}_1$, we get $Y \in \mathcal{R}$ such that for every $i = 0, 1$, either $\mathcal{F}_i|Y = \emptyset$, or else for every $B \leq Y$ there is $n \in \mathbb{N}$ such that $r_n(B) \in \mathcal{F}_i$. Note that we must have $\mathcal{F}_i|Y = \emptyset$ for at least one $i = 0, 1$, or else we would get a single set $B \leq Y$

and two integers m and n such that $a = r_m(B) \in \mathcal{F}_0$ and $b = r_n(B) \in \mathcal{F}_1$. Since $\mathcal{F}_0$ and $\mathcal{F}_1$ are disjoint, we must have $m \neq n$, and therefore $a \sqsubset b$ or $b \sqsubset a$, depending on whether $m < n$ or $n < m$, respectively. This contradicts the assumption that $\mathcal{F}$ is a Nash-Williams family. □

Definition 5.18 *We shall say that a family $\mathcal{F} \subseteq \mathcal{AR}$ of finite approxima-tion is a*

(1) front *on $X \in \mathcal{R}$ if $\mathcal{F}$ is a Nash-Williams family with the property that for every $Y \leq X$ there is n such that $r_n(Y) \in \mathcal{F}$.*

(2) barrier *on $X \in \mathcal{R}$ if $\mathcal{F}$ is a front on X and $\mathcal{F}$ is a Sperner family.*

Corollary 5.19 *Suppose $(\mathcal{R}, \leq, p)$ is a closed triple that satisfies A.1, A.2, A.3, and A.4 and that $\leq_{\mathrm{fin}}$ is an actual partial ordering on $\mathcal{AR}$ rather than just a quasi-ordering. Suppose further that a family $\mathcal{F}$ of finite approxima-tions is a front on some $X \in \mathcal{R}$. Then there is $Y \leq X$ such that the restric-tion $\mathcal{F}|Y$ is a barrier on Y.*

Proof. By the Abstract Nash-Williams Theorem, we know that $\mathcal{F}$ is Ramsey, so we can apply this to the partition $\mathcal{F} = \mathcal{F}_0 \cup (\mathcal{F} \setminus \mathcal{F}_0)$, where

$$\mathcal{F}_0 = \{b \in \mathcal{F} : (\forall a <_{\mathrm{fin}} b)\ a \notin \mathcal{F}\},$$

and find $Y \leq X$ such that either $(\mathcal{F} \setminus \mathcal{F}_0)|Y = \emptyset$, or else $\mathcal{F}_0|Y = \emptyset$ Note that $\mathcal{F}_0$ is a Sperner family and that the first alternative will give us the desired conclusion that $\mathcal{F}|Y = \mathcal{F}_0|Y$ is a barrier on Y. So it suffices to show that $\mathcal{F}_0|Y \neq \emptyset$. To see this, take an arbitrary $b \in \mathcal{F}|Y$ and pick an n and $B \leq Y$ such that $b = r_n(B)$. Since

$$\mathcal{F} \cap \{a \in \mathcal{AR} : a \leq_{\mathrm{fin}} b\}$$

is a finite set, it has a minimal element, call it a, relative to the partial ordering $\leq_{\mathrm{fin}}$. Clearly, $a \in \mathcal{F}_0$, but we need to show that $a \in \mathcal{F}_0|Y$. From $a \leq_{\mathrm{fin}} b = r_n(B)$, we conclude that $\mathrm{depth}_B(a) < \infty$. By A.3(1) this gives us that $[a, B] \neq \emptyset$, so in particular, there is $A \leq B \leq Y$ such that $a = r_m(A)$ for some m. So a belongs to the restriction $\mathcal{F}_0|Y$. □

The following formulation of Theorem 5.15 is also worth pointing out.

Corollary 5.20 *Suppose $(\mathcal{R}, \leq, p)$ is a closed triple that satisfies A.1, A.2, A.3, and A.4 and that the relation $\leq_{\mathrm{fin}}$ is moreover antisymmetric. Then for every family $\mathcal{F} \subseteq \mathcal{AR}$ of finite approximations and for every $X \in \mathcal{R}$, there is $Y \leq X$ such that either the restriction $\mathcal{F}|Y$ is empty or it contains a barrier on Y.*

This shows that the theory of fronts and barrier seen above in Chapter One in the case of the prototype space $(\mathbb{N}^{[\infty]}, \subseteq, r)$ allows extension to the context of general topological Ramsey spaces. Judging on the basis of the applicability of the original theory, it is reasonable to expect that the abstract theory or one of its specializations to a particular new topological Ramsey space will find interesting applications.

The rest of this chapter is devoted to particular examples of topological Ramsey spaces, their relationships to the prototype space $(\mathbb{N}^{[\infty]}, \subseteq, r)$, and their applications. We shall also devote the whole Chapter Six to a particular example of a topological Ramsey space based on the Halpern-Läuchli Theorem.

5.2 TOPOLOGICAL RAMSEY SPACES OF INFINITE BLOCK SEQUENCES OF VECTORS

For a positive integer k, set

$\mathrm{FIN}_k = \{p : \mathbb{N} \to \{0, 1, \ldots, k\} : \{n : p(n) \neq 0\}$ is finite and $k \in \mathrm{range}(p)\}$.

We consider FIN_k a partial semigroup under the partial semigroup operation of taking the sum of two disjointly supported elements of FIN_k. For $p \in \mathrm{FIN}_k$, let $\mathrm{supp}(p) = \{n : p(n) \neq 0\}$. A *block sequence* of members of FIN_k is a (finite or infinite) sequence $P = (p_n)$ such that

$$\mathrm{supp}(p_m) < \mathrm{supp}(p_n) \text{ whenever } m < n.$$

For $1 \leq d \leq \infty$, let $\mathrm{FIN}_k^{[d]}$ be the collection of all block sequences of length d. The notion of a partial subsemigroup generated by a given block sequence depends on the operation

$$T : \mathrm{FIN}_k \to \mathrm{FIN}_{k-1} \tag{5.2}$$

defined as follows:

$$T(p)(n) = \max\{p(n) - 1, 0\}. \tag{5.3}$$

Given a finite or infinite block sequence $P = (p_n)$ of elements of FIN_k and integer j ($1 \leq j \leq k$), the *partial subsemigroup* $[P]_j$ of FIN_j *generated by* P is the collection of members of FIN_j of the form

$$T^{(i_0)}(p_{n_0}) + \ldots + T^{(i_\ell)}(p_{n_\ell})$$

for some finite sequence $n_0 < \ldots < n_\ell$ of nonnegative integers and some choice $i_0, \ldots, i_\ell \in \{0, 1, \ldots, k\}$. For $P = (p_n), Q = (q_m) \in \mathrm{FIN}_k^{[\leq\infty]}$, set $P \leq Q$ if $p_n \in [Q]_k$ for all n less than the length of the sequence P. When $P \leq Q$ happens, we say that P is a *block-subsequence* of Q. Then $\leq$ is a partial ordering on $\mathrm{FIN}_k^{[\infty]}$ that allows the finitization $\leq_{\mathrm{fin}}$ to satisfy $A.2$ and $A.3$:

$$P \leq_{\mathrm{fin}} Q \text{ iff } P \leq Q \text{ and } (\forall l < \mathrm{length}(Q)) \, P \not\leq Q \restriction l. \tag{5.4}$$

The crucial axiom $A.4$ for $(\mathrm{FIN}_k^{[\infty]}, \leq, r)$ reduces to the following variation of Gowers's theorem from Section 2.3.

Theorem 5.21 *For every finite coloring of* FIN_k *and every* $Q = (q_n)$ *in* $\mathrm{FIN}_k^{[\infty]},$ *there is an infinite block subsequence* P *of* Q *such that* $[P]_k$ *is monochromatic.*

Proof. Referring to the proof of Gowers's theorem, we now work with subsemigroups $[Q]_j^*$ of FIN_j^* $(1 \leq j \leq k)$ consisting of those cofinite ultrafilters that concentrate on the sets $[Q]_j (1 \leq j \leq k)$, respectively. Note that for $1 \leq i \leq j \leq k$ the mapping

$$T^{j-i} : [Q]_j \to [Q]_i$$

is a partial homomorphism, its extension $T^{j-i} : [Q]_j^* \to [Q]_i^*$ is an actual homomorphism. So we have all the necessary inputs for the proof of Section 2.3 to give us us idempotents $\mathcal{U}_j (1 \leq j \leq k)$ such that

(1) $[Q]_j \in \mathcal{U}_j$ for all $1 \leq j \leq k$,

(2) $\mathcal{U}_i \geq \mathcal{U}_j$ whenever $1 \leq i \leq j \leq k$,

(3) $T^{(j-i)}(\mathcal{U}_j) = \mathcal{U}_i$ whenever $1 \leq i \leq j \leq k$.

Let C be the color of the given coloring such that $C \in \mathcal{U}_k$. Starting from $A_0^k = C \cap [Q]_k$, we proceed as in the proof of Theorem 2.22 and build an infinite block subsequence $P = (p_n)$ at Q and for each $1 \leq \ell \leq k$ a decreasing sequence (A_n^ℓ) of elements of $\mathcal{U}_\ell$ such that

(a) $p_n \in A_n^k$,

(b) $T^{(k-\ell)}[A_n^k] = A_n^\ell$,

(c) $(\forall 1 \leq i, j \leq k)(\mathcal{U}_k x)[T^{(k-i)}(p_n) + T^{(k-j)}(x) \in A_n^{\max\{i,j\}}]$.

When this is done, one checks as before that $[P]_k \subset C$. $\qquad\square$

Note also that $\mathrm{FIN}_k^{[\infty]}$ is a closed subset of the infinite power of $\mathrm{FIN}_k^{[<\infty]}$. So we have verified all the hypotheses of the Abstract Ellentuck Theorem, and so we obtain the following result.

Theorem 5.22 *For every positive integer* k, *the triple* $(\mathrm{FIN}_k^{[\infty]}, \leq, r)$ *is a topological Ramsey space.*

Corollary 5.23 (Milliken) $(\mathrm{FIN}_1^{[\infty]}, \leq, r)$ *is a topological Ramsey space.*

Corollary 5.24 *For every positive integer* k *and every finite Souslin-measurable coloring of* $\mathrm{FIN}_k^{[\infty]}$ *there is an infinite block sequence* P *such that the family* $[P]_k^{[\infty]}$ *of all infinite block subsequences of* P *is monochromatic.* $\qquad\square$

Corollary 5.25 *For every positive integer k and every family $\mathcal{F} \subseteq \mathrm{FIN}_k^{[<\infty]}$ of finite block sequences, there is an infinite block sequence P such that either P contains no finite block-subsequence belonging to $\mathcal{F}$ or else every infinite block-subsequence of P has an initial segment in $\mathcal{F}$.*

Proof. Apply Corollary 5.24 to the open subset $\mathcal{X}$ of all elements Q of $\mathrm{FIN}_k^{[\infty]}$ that have an initial segment in $\mathcal{F}$. □

Corollary 5.26 *For all positive integers k and d and every finite coloring of the set $\mathrm{FIN}_k^{[d]}$ of block sequences of length d of vectors from FIN_k, there is an infinite block sequence P such that $[P]_k^{[d]}$ is monochromatic.* □

Note that for $i = 0, 1, \ldots, k - 1$, the operation $T^{(i)} : \mathrm{FIN}_k \to \mathrm{FIN}_{k-i}$ extends to a projection map,

$$T^{(i)} : \mathrm{FIN}_k^{[\infty]} \to \mathrm{FIN}_{k-i}^{[\infty]} \tag{5.5}$$

defined naturally by the requirement $T^{(i)}(P) = Q$ iff $T^{(i)}(p_n) = q_n$ for all n. This allows us to deduce from Corollary 5.24 some simultaneous coloring theorems such as the following.

Theorem 5.27 *For every positive integer k and every finite Souslin-measurable coloring of $\mathrm{FIN}_1^{[\infty]} \cup \ldots \cup \mathrm{FIN}_k^{[\infty]}$, there exists $P \in \mathrm{FIN}_k^{[\infty]}$ such that $[T^{(i)}P]^{[\infty]}$ is monochromatic for all $i = 0, 1, 2, \ldots, k - 1$.*

This process can be continued indefinitely by building a single topological Ramsey space $\mathrm{FIN}_*^{[\infty]}$ that projects onto all the spaces $\mathrm{FIN}_k^{[\infty]}$. We leave the details of the construction of the space $\mathrm{FIN}_*^{[\infty]}$ to the interested reader.

We finish this section with an application of the high-dimensional Ramsey theory of block sequences to the Ramsey classification problem for equivalence relations on FIN_k. We consider here only the case $k = 1$ although similar but considerably more complex arguments will achieve the corresponding Ramsey classification for an arbitrary positive integer k.

Theorem 5.28 (Taylor) *For every equivalence relation E on FIN, there is an infinite block sequence $X = (x_i)$ such that $E \restriction [X]$ is equal to one of the five canonical equivalence relations E_c, E_min, E_max, $E_{(\mathrm{min,max})}$, or E_id.*

Here, the five canonical equivalence relations are written in notation that indicates their representing mappings, respectively: the constant mapping $x \mapsto c$, the mapping $x \mapsto \min(x)$, the mapping $x \mapsto \max(x)$, the mapping $x \mapsto (\min(x), \max(x))$, and the identity mapping $x \mapsto x$ It is also useful to write them in terms of their defining equations as follows:

$$xE_\mathrm{c}y \Leftrightarrow x = x,$$
$$xE_\mathrm{id}y \Leftrightarrow x = y,$$
$$xE_\mathrm{min}y \Leftrightarrow \min(x) = \min(y),$$
$$xE_\mathrm{max}y \Leftrightarrow \max(x) = \max(y),$$
$$xE_{(\mathrm{min,max})}y \Leftrightarrow \min(x) = \min(y) \text{ and } \max(x) = \max(y).$$

From now on, we fix an equivalence relation E on FIN.

Definition 5.29 *An E-equation in four variables v_0, v_1, v_2, v_3 (and two constants) is a formula φ of the form*

$$a \cup v_{i_0} \cup v_{i_1} \cup v_{i_2} \cup v_{i_3} \; E \; b \cup v_{j_0} \cup v_{j_1} \cup v_{j_2} \cup v_{j_3},$$

where $i_0, j_0, i_1, j_1, i_2, j_2, i_3, j_3 \in \{0, 1, 2, 3\}$ and where $a, b \in \text{FIN} \cup \{\emptyset\}$ are two constants.

Note that for two fixed constants $a, b \in \text{FIN} \cup \{\emptyset\}$, there exist only finitely many equations of the form $\varphi(a, b, v_0, v_1, v_2, v_3)$.

Definition 5.30 *For given a block sequence Z in FIN and an equation $\varphi \equiv \varphi(a, b, v_0, v_1, v_2, v_3)$ with constants a and b in $\text{FIN} \cup \{\emptyset\}$, we say that*

(1) φ *is* true *in $[Z]$ if $\varphi(a, b, a_0, a_1, a_2, a_3)$ holds for every (a_0, a_1, a_2, a_3) in $[Z/(a \cup b)]^{[4]}$,*

(2) φ *is* false *in $[Z]$ if $\varphi(a, b, a_0, a_1, a_2, a_3)$ fails for every (a_0, a_1, a_2, a_3) in $[Z/(a \cup b)]^{[4]}$, and*

(3) φ *is* decided *in $[Z]$ if either φ is true in $[Z]$ or φ is false in $[Z]$.*

Lemma 5.31 *There is an infinite block sequence Z in FIN such that all equations $\varphi(a, b, v_0, v_1, v_2, v_3)$ with $a, b \in [Z] \cup \{\emptyset\}$ are decided in $[Z]$.*

Proof. This follows from Corollary 5.26 for $k = 1$ and $d = 4$ by using the usual fusion procedure, since at the given stage of the fusion procedure we have only finitely many pairs of constants and therefore only finitely many equations to handle. $\qquad \square$

From now on, we fix an infinite block sequence $Z = (z_n)_n$ satisfying the conclusion of this lemma and we work inside the subsemigroup $[Z]$ generated by Z. Thus, we frequently suppress writing "true in $[Z]$" or "false in $[Z]$" and all equalities or inclusions between E and our five canonical equivalence relations are to be interpreted with all these relations restricted to $[Z]$. It turns out that, when restricting to $[Z]$, the truth or falsity of the following four E-identities characterizes each of the five canonical equivalence relations E_c, $E_{\min}$, $E_{\max}$, $E_{(\min,\max)}$ and E_{id}.

$$v_0 \; E \; v_1,$$
$$v_0 \cup v_1 \; E \; v_0,$$
$$v_0 \cup v_1 \; E \; v_1,$$
$$v_0 \cup v_1 \cup v_2 \; E \; v_0 \cup v_2.$$

This will leads us naturally toward the proof of Theorem 5.28. We first isolate a lemma that is quite useful when proving such characterizations.

Lemma 5.32 $v_0 \cup v_1 \; \not\!E \; v_0$ *implies* $E \subseteq E_{\max}$.

Proof. Suppose that for some $s, t \in [Z]$, we have $s \, E \, t$ and $\max s \neq \max t$. We may assume that $\max s < \max t$. Let n be such that $t = t' \cup z_n$ with $t' < z_n$ and $t' \in [Z]$. Since $s \, E \, t$, the equations $s \, E \, t' \cup v_0$ and $s \, E \, t' \cup v_0 \cup v_1$ are both true. Hence, the equation $t' \cup v_0 \cup v_1 \, E \, t' \cup v_0$ is true. This implies that $v_0 \cup v_1 \, E \, v_0$ is true, a contradiction. $\qquad\square$

Lemma 5.33 $v_0 \, E \, v_1$ *implies* $E = E_{\mathrm{c}}$.

Proof. To see this, let $s, t \in [Z]$ and pick $u \in [Z]$ such that $u > s, t$. Then $s \, E \, u$ and $t \, E \, u$, and hence, $s \, E \, t$. $\qquad\square$

Lemma 5.34 *If* $v_0 \, \not{E} \, v_1$, $v_0 \cup v_1 \, E \, v_0$ *and* $v_0 \cup v_1 \, \not{E} \, v_1$, *then* $E = E_{\min}$.

Proof. Fix $s, t \in [Z]$. Suppose first that $s \, E_{\min} \, t$, and write $s = z_n \cup s'$ and $t = z_n \cup t'$, with $s', t' \in [Z]$ and $z_n < s', t'$. Using that $v_0 \cup v_1 \, E \, v_0$ is true in Z, we obtain that $s, t \, E \, z_n$, and hence, $s \, E \, t$. Assume that $s \, E \, t$ and suppose now that $s \, \not{E} \, t$. Without loss of generality, we may assume that $\min(s) < \min(t)$. Let n be such that $s = z_n \cup s'$, $s' \in [Z]$ and $z_n < s'$. Then $s \, E \, z_n$, $z_n < t$, and hence, $z_n \, E \, t$. Since $z_n < t$, we obtain that $v_0 \, E \, v_1$ is true in Z, a contradiction. $\qquad\square$

Lemma 5.35 *If* $v_0 \, \not{E} \, v_1$, $v_0 \cup v_1 \, \not{E} \, v_0$, *and* $v_0 \cup v_1 \, E \, v_1$, *then* $E = E_{\max}$.

Proof. By Lemma 5.32, we already know that $E \subseteq E_{\max}$. To show the converse inclusion, consider s and t such that $\max(s) = \max(t)$. Then $s = s' \cup z_n$ and $t = t' \cup z_n$ for some integer n and $s', t' \in \mathrm{FIN} \cup \{\emptyset\}$. Applying the equation $v_0 \cup v_1 \, E \, v_1$, we get that $s \, E \, z_n$ and $t \, E \, z_n$. It follows that $s \, E \, t$, and this finishes the proof. $\qquad\square$

Lemma 5.36 *If* $v_0 \, \not{E} \, v_1$, $v_0 \cup v_1 \, \not{E} \, v_0$, *and* $v_0 \cup v_1 \, \not{E} \, v_1$ *but* $v_0 \cup v_1 \cup v_2 \, E \, v_0 \cup v_2$, *then* $E = E_{(\min,\max)}$.

Proof. To prove the inclusion $E_{(\min,\max)} \subseteq E$, consider s and t such that $\min(s) = \min(t)$ and $\max(s) = \max(t)$. Then $s = z_m \cup s' \cup z_n$ and $t = z_m \cup t' \cup z_n$ for some $m \leq n$ and $s', t' \in \mathrm{FIN} \cup \{\emptyset\}$ such that $z_m < s' < z_n$ and $z_m < t' < z_n$. Using the equation $v_0 \cup v_1 \cup v_2 \, E \, v_0 \cup v_2$, we get that $s \, E \, z_m \cup z_n$ and $t \, E \, z_m \cup z_n$. It follows that $s \, E \, t$, as required.

Suppose that the other inclusion, $E \subseteq E_{(\min,\max)}$, fails. Then, by our hypothesis $v_0 \cup v_1 \, \not{E} \, v_0$ and Lemma 5.32, there exist s and t such that $s \, E \, t$ and $\max s = \max t$ but $\min s \neq \min t$. We may assume $\min s < \min t$. Write $s = z_{n_0} \cup s' \cup z_n$, $t = z_{m_0} \cup t' \cup z_n$ with $z_{n_0} < s' < z_n$, $z_{m_0} < t' < z_n$ and $n_0 < m_0 \leq n$. Since the equation $v_0 \cup v_1 \cup v_2 \, E \, v_0 \cup v_2$ is true, we may assume that $s' = t' = \emptyset$. Since $n_0 < m_0 \leq n$, one of the equations $v_0 \cup v_2 \, E \, v_1 \cup v_2$ or $v_0 \cup v_1 \, E \, v_1$ must be true. The second case is impossible by the hypothesis of the lemma. In the first case, we obtain that $v_0 \cup v_3 \, E \, v_1 \cup v_2 \cup v_3$ and $v_0 \cup v_3 \, E \, v_2 \cup v_3$ are true in Z and hence, so is $v_0 \, E \, v_0 \cup v_1$, a contradiction. This finishes the proof $\qquad\square$

Lemma 5.37 *If* $v_0 \, \not{E} \, v_1$, $v_0 \cup v_1 \, \not{E} \, v_0$, $v_0 \cup v_1 \, \not{E} \, v_1$, *and* $v_0 \cup v_1 \cup v_2 \, \not{E} \, v_0 \cup v_2$, *then* $E = E_{\mathrm{id}}$.

Proof. Suppose that $s \, E \, t$ and that $s \neq t$. Since $v_0 \cup v_1 \, E \, v_0$ is false in Z, we obtain that $\max s = \max t$ (by Lemma 5.32). Write $s = \bigcup_{i \in F} z_i$, $t = \bigcup_{i \in G} z_i$ with $\max F = \max G$. By our assumption $s \neq t$, so F and G are two different finite subsets of $\mathbb{N}$. Let n be the maximum of the symmetric difference of the two finite sets F and G. Without loss of generality we assume that $n \in F \setminus G$. This implies that $s = s' \cup z_n \cup s''$ and $t = t' \cup s''$ with $s', t' < z_n < s''$, $s'' \neq \emptyset$ and all in $[Z]$. Therefore the equation $s' \cup v_0 \cup v_1 \, E \, t' \cup v_1$ is true in Z, which implies that $s' \cup v_0 \cup v_1 \cup v_2 \, E \, t' \cup v_2$ and $s' \cup v_0 \cup v_2 \, E \, t' \cup v_2$ are both true in Z. So, the equation $v_0 \cup v_1 \cup v_2 \, E \, v_0 \cup v_2$ is true in Z, a contradiction. $\square$

Corollary 5.38 *For every positive integer m there is a positive integer n such that for every equivalence relation E on $\mathcal{P}(\{0, 1, \ldots, n-1\})$,[1] there is a block sequence $z_0 < z_1 <, \ldots, < z_{m-1}$ of nonempty subsets of $\{0, 1, \ldots, n-1\}$ such that E restricted to the sublattice*

$$\left\{ \bigcup_{i \in I} z_i : \emptyset \neq I \subseteq \{0, 1, \ldots, m-1\} \right\}$$

of nonempty subsets gennerated by $(z_i)_{(i<m)}$ is equal to one of the five canonical equivalence relations E_c, $E_{\min}$, $E_{\max}$, $E_{(\min,\max)}$, or E_{id}.

If one does not require the sequence $(z_i)_{(i<m)}$ to be a block sequence, we obtain a different result, which is really a finite form of an infinitary Ramsey classification result belonging to a different category (see Theorem 5.82).

Theorem 5.39 (Prömel-Voigt) *For every positive integer m there is a positive integer n such that for every equivalence relation E on the power set $\mathcal{P}(\{0, 1, \ldots, n-1\})$ there is sequence $z_0, z_1, \ldots, z_{m-1}$ of nonempty subsets of $\{0, 1, \ldots, n\}$ such that the restriction of E to the sublattice*

$$\left\{ \bigcup_{i \in I} z_i : \emptyset \neq I \subseteq \{0, 1, \ldots, m-1\} \right\}$$

of nonempty sets generated by $(z_i)_{i<m}$ is equal to one of the three canonical equivalence relations E_c, $E_{\min}$, or E_{id}.

Remark 5.40 Similar ideas can be used to extend Theorem 5.28 and obtain the corresponding classification result for an arbitrary FIN_k. This is indeed possible although the complexity of the corresponding classification problem increases rapidly with k. For example, the number of canonical equivalence relations on FIN_2 is equal to 43, that on FIN_3 is equal to 619, that on FIN_4 is equal to 13829, etc. The following formula involving the incomplete gamma function is one of the formulas for the number of canonical equivalence relations on the general FIN_k,

$$e^2 [k[\Gamma(k, 1) - \Gamma(k+1, 1)]^2 + \Gamma(k+1, 1)^2].$$

---

[1] The collection of all subsets of $\{0, 1, \ldots, n-1\}$.

5.3 TOPOLOGICAL RAMSEY SPACES OF INFINITE SEQUENCES OF VARIABLE WORDS

As in the previous section, $L = \bigcup_{n=0}^{\infty} L_n$ is a fixed alphabet written as an increasing union of finite alphabets L_n and $v \notin L$ is a symbol that we call *variable*. By W_{Lv} we denote the semigroup of *variable words* over L, i.e., finite nonempty strings of elements of $L \cup \{v\}$ in which v occurs. $W_{Lv}^{[\infty]}$ is the set of all infinite *rapidly increasing* sequences $X = (x_n)$ of variable-words, i.e., sequences such that $|x| > \sum_{i<n} |x_n|$ for all n. For $X = (x_n) \in W_{Lv}^{[\infty]}$, let

$$[X]_{Lv} = \{x_{n_0}[\lambda_0]\,^\frown \ldots \,^\frown x_{n_k}[\lambda_k] \in W_{Lv} : k \in \mathbb{N}, n_0 < \ldots < n_k,$$

$$\lambda_i \in L_{n_i} \cup \{v\} (i \le k)\}$$

denote the combinatorial subspace of W_{Lv} generated by X. From the fact that X is rapidly increasing, we conclude that for every $x \in [X]_L$ the set $\{n_0 < \ldots < n_k\}$ such that

$$x = x_{n_0}[\lambda_0]\,^\frown \ldots \,^\frown x_{n_k}[\lambda_k] \tag{5.6}$$

for some choice of $\lambda \in L_{n_i} \cup \{v\} (i \le k)$ is unique, and so we let $\mathrm{supp}_X(x)$ denote $\{n_0, \ldots, n_k\}$. This helps us to define an ordering $\le$ on $W_{Lv}^{[\infty]}$ by letting $X = (x_n) \le Y = (y_n)$ iff $x_n \in [Y]_{Lv}$ for all n, and

$$\mathrm{supp}_Y(x_n) < \mathrm{supp}_Y(x_n) \qquad \text{whenever } n < m. \tag{5.7}$$

Thus $X \le Y$ iff X is a *block subsequence of* Y. The ordering $\le$ has a natural finitization on the set $W_{Lv}^{[<\infty]}$ of finite rapidly increasing sequences: $(x_m)_0^{k-1} \le_{\mathrm{fin}} (y_n)_0^{\ell-1}$ iff $(x_n)_0^{\ell-1}$ is a block subsequence of $(y_n)_0^{\ell-1}$ but not a block subsequence of any $(y_n)_0^{\ell'-1}$ for $\ell' < \ell$ (of course, we let $\emptyset \le_{\mathrm{fin}} \emptyset$). In proving the Abstract Infinite-Dimensional Ramsey Theorem in Sections 4.2 and 4.3, we verified that $(W_{Lv}^{[\infty]}, \le, r)$ satisfies A.1 to A.4, and since $W_{Lv}^{[\infty]}$ is a closed subset of $(W_{Lv}^{[<\infty]})^{\mathbb{N}}$, we have the following consequence.

Theorem 5.41 (Carlson) $(W_{Lv}^{[\infty]}, \le, r)$ *is a topological Ramsey space.*

Corollary 5.42 *For every Souslin-measurable coloring of the space W_{Lv}^{∞} of all infinite sequences of variable-words over the alphabet $L = \bigcup_{n=0}^{\infty} L_n$, there is an infinite rapidly increasing sequence $X = (x_n)$ of variable-words over L such that the set $[X]_{Lv}^{[\infty]}$ of all infinite block subsequences of X is monochromatic.*

Corollary 5.43 *For every positive integer d and every finite coloring of the set W_{Lv}^d of all d-tuples of variable-words over the alphabet $L = \bigcup_{n=0}^{\infty} L_n$, there is an infinite rapidly increasing sequence $X = (x_n)$ of variable-words over L such that the set $[X]_{Lv}^{[d]}$ of all block d-subsequences of X is monochromatic.*

To give a particular application of this corollary, we concentrate on the case of a finite alphabet L and, as in the previous sections, we let W_L denote the semigroup of all words over L, i.e., the family of all finite strings of symbols from L. For a positive integer d, a *d-dimensional combinatorial subspace* of W_L is a subset of W_L of the form

$$S[x_0, ..., x_{d-1}] = \{x_0[a_0]^\frown ...^\frown x_{d-1}[a_{d-1}] : a_i \in L \quad \text{for} \quad i < d\}, \quad (5.8)$$

where $(x_0, ..., x_{d-1})$ is a sequence of a variable-words over L. An *infinite-dimensional combinatorial subspace* of W_L has in fact already been defined above (although not named like this), it is the partial subsemigroup $[X]_L$ of W_L generated by some infinite sequence of $X = (x_n)$ of variable-words in the following way:

$$[X]_L = \{x_{n_0}[a_0]^\frown ...^\frown x_{n_k}[a_k] : k \in \mathbb{N}, n_0 < ... < n_k, a_i \in L(i \leq k)\}.$$

Theorem 5.44 *For a finite alphabet $|L| \geq 2$, for every positive integer d, and every finite coloring of the family of all d-dimensional combinatorial subspaces of W_L, there exists an infinite-dimensional combinatorial subspace S of W_L such that the family of all d-dimensional combinatorial subspaces of W_L that are included in S is monochromatic.*

Proof. Color the set W_{Lv}^d of all d-tuples $(x_0, ..., x_{d-1})$ of variable-words by the color of the corresponding combinatorial subspace $S[x_0, ..., x_{d-1}]$. By Corollary 5.43, there is an infinite rapidly increasing sequence $X = (x_n)$ of variable-words such that the family $[X]_L^{[d]}$ of all block d-subsequences of X is monochromatic. We claim that $S = [X]_L$ satisfies the conclusion of the theorem. So suppose $S[y_0, ..., y_{d-1}]$ is a d-dimensional combinatorial subspace of W_L such that $S[y_0, ..., y_{d-1}] \subset [X]_L$. Note that elements of $S[y_0, ..., y_{d-1}]$ all have the same length ℓ. Since X is rapidly increasing any element w of $[X]_L$ has a support relative to X, $\text{supp}_X(w)$, the unique finite set $n_0 < ... < n_k$ such that

$$w = x_{n_0}[\lambda_0]^\frown ...^\frown x_{n_k}[\lambda_k]$$

for some choice $\lambda_i \in L(i \leq k)$. For the same reason, if for two $w_0, w_1 \in [X]_L$ we have that

$$\text{supp}_X(w_0) \neq \text{supp}_X(w_1),$$

then w_0 and w_1 have different lengths. From all this it follows that there is a fixed finite set $n_0 < ... < n_k$ such that for all $w \in S[y_0, ..., y_{d-1}]$,

$$\text{supp}_X(w) = \{n_0, ..., n_k\}.$$

It follows that, in particular

$$S[y_0, ..., y_{d-1}] \subset S[x_{n_0}, ..., x_{n_k}] \subset [X]_L.$$

Choose a sequence (v_i) of new variables and consider the following two multivariable-words over L:

$$y = y_0[v_0]^\frown ...^\frown y_{d-1}[v_{d-1}], \quad x = x_{n_0}[v_0]^\frown ...^\frown x_{n_k}[v_k].$$

Then every instance $y[\lambda_0, \ldots, \lambda_{d-1}](\lambda_i \in L, i < d)$ of y is also an instance $x[\mu_0, \ldots, \mu_k](\mu_i \in L_{n_i}, i \leq k)$. Moreover, x and y have the same length ℓ. Note that if λ is a letter from L that occupies some position $j < \ell$ in x, then at the jth place in y we must find a nonvariable (i.e., a letter from L), and therefore the jth position in y must be occupied by the letter λ itself. It follows that if at some position $j < \ell$ we find a variable in y, then the jth place in x must also be occupied by some variable. Moreover, no two positions can be occupied by the same variable in x and different variables in y. So y is obtained from x by substituting its variables by either letters from L or variables from $\{v_0, \ldots, v_{d-1}\}$. For $i < d$ set

$$J_i = \{j \leq k : v_j \text{ is replaced by } v_i \text{ when forming } y \text{ from } x\}.$$

It follows that if $i < i' < d$ then $J_i < J_{i'}$. So we can choose a pairwise disjoint sequence $\bar{J}_i \supseteq J_i$, $(i < d)$, of intervals of integers such that

$$\{0, \ldots, k\} = \bigcup_{i < d} \bar{J}_i.$$

For $i < k$, let $\bar{J}_i = \{p, p+1, \ldots, p+q\}$ and let

$$x_{n_p}[v_p]^\frown \ldots {}^\frown x_{n_{p+q}}[v_{p+q}]$$

be the segment of x that gets transferred to a segment $t_i(v_i)$ of y when $v_p, \ldots, v_{p+q}$ are replaced either by letters from L or by v_i (with at least one replacement by v_i). Then $\langle t_0[v], \ldots, t_{d-1}[v] \rangle$ is a block subsequence of X such that

$$y_0[v_0]^\frown \ldots {}^\frown y_{d-1}[v_{d-1}] = t_0[v_0]^\frown \ldots {}^\frown t_{d-1}[v_{d-1}],$$

and therefore $S[y_0, \ldots, y_{d-1}] = S[t_0, \ldots, t_{d-1}]$. This shows that the subspace $S[y_0, \ldots, y_{d-1}]$ is of the color equal to the color of the block d-subsequence $\langle t_0, \ldots, t_{d-1} \rangle$ of X. Since $X = (x_n)$ is chosen to have $[X]_L^{[d]}$ monochromatic, this finishes the proof.   □

We finish this section with an application that in some sense corresponds to the application given in Theorem 4.43 of the Hales-Jewett space developed above in the Chapter Four. Recall that FIN is the collection of all finite nonempty subsets of $\mathbb{N}$ and that $\text{FIN}^{[\infty]}$ is the collection of all infinite block sequences of members of FIN i.e., sequences $X = (x_n)$ such that $x_m < x_n$ whenever $m < n$. For $X = (x_n) \in \text{FIN}^{[\infty]}$, the set

$$[X] = \{x_{n_0} \cup \ldots \cup x_{n_k} : k \in \mathbb{N}, n_0 < \ldots < n_k\}$$

is the sublattice of FIN generated by X. For two sequences $X = (x_n)$ and $Y = (y_n)$ from $\text{FIN}^{[\infty]}$, we say that X is a *block subsequence* of Y and write $X \leq Y$ if

$$x_n \in [Y] \text{ for all } n.$$

Let $[Y]^{[\infty]}$ denote the collection of all infinite block subsequence of Y.

Theorem 5.45 (Parametrized Milliken Theorem) *For every finite Souslin-measurable coloring of the product* $\mathrm{FIN}^{[\infty]} \times \mathbb{R}^\infty$ *there is a* $B \in \mathrm{FIN}^{[\infty]}$ *and an infinite sequence* (P_i) *of nonempty perfect subsets of* $\mathbb{R}$ *such that* $[B]^{[\infty]} \times \prod_{i=0}^\infty P_i$ *is monochromatic.*

Proof. Let $L = \bigcup_{n=0}^\infty L_n$, where

$$L_n = \{\sigma \in 2^\infty : \forall i > n \ \sigma(i) = 0\}. \tag{5.9}$$

Define $\varphi : W_{Lv}^{[\infty]} \to \mathrm{FIN}^{[\infty]}$ by letting $\varphi((x_k)) = (a_k)$, where

$$a_k = \{|x_0| + \ldots + |x_{k-1}| + l : v \text{ occupies the } l\text{th place in } x_k\}. \tag{5.10}$$

Define $\psi : W_{Lv}^{[\infty]} \to 2^{\mathbb{N} \times \mathbb{N}}$, by

$$\psi((x_k))(n, i) = \sigma(i), \tag{5.11}$$

where σ occupies the nth place in the infinite word

$$x_0 ^\frown x_1 ^\frown \ldots ^\frown x_k ^\frown \ldots,$$

and where we put $\psi((x_k))(n, i) = 0$ if at the nth place of this infinite word we find the variable v.

Let $c : \mathrm{FIN}^{[\infty]} \times (2^{\mathbb{N}})^{\mathbb{N}} \to F$ be a given Souslin-measurable coloring with range a finite subset F of $\mathbb{N}$. Define $c^* : W_{Lv}^{[\infty]} \to F$ by

$$c^*((x_n)) = c(\varphi((x_n)), \psi((x_n))), \tag{5.12}$$

where we identify $(2^{\mathbb{N}})^{\mathbb{N}}$ with $2^{\mathbb{N} \times \mathbb{N}}$ via the mapping

$$((\varepsilon_{n,i})_n)_i \to (\varepsilon_{n,i})_{(n,i)}.$$

By Corollary 5.42, there is a $Y = (y_k) \in W_{Lv}^{[\infty]}$ such that c^* is monochromatic on the set $[Y]_{Lv}^{[\infty]} = \{X \in W_{Lv}^{[\infty]} : X \leq Y\}$.

Let $B = (b_k) \in \mathrm{FIN}^{[\infty]}$ be defined as follows:

$$b_k = \{|y_0| + \ldots + |y_{2k}| + \ell : \text{ the } \ell\text{th place in } y_{2k+1} \text{ is occupied by } v\}.$$

Let P be subset of $(2^{\mathbb{N}})^{\mathbb{N}}$ consisting of all doubly indexed sequences (ε_{ni}) for which we can find $(\sigma_{2k}) \in \prod_{k=0}^\infty L_{2k}$ that determines (ε_{ni}) in the following way: $\varepsilon_{ni} = \sigma(i)$ if and only if the nth place of the infinite word

$$y_{(\sigma_{2k})} = y_0[\sigma_0] ^\frown y_1 ^\frown \ldots ^\frown y_{2k}[\sigma_{2k}] ^\frown y_{2k+1} ^\frown \ldots$$

is occupied by the letter σ; $\varepsilon_{ni} = 0$ if the nth place is occupied by the variable v. Equivalently, a doubly indexed sequence (ε_{ni}) belongs to P if we can find $(\sigma_{2k}) \in \prod_{k=0}^\infty L_{2k}$ such that

$$(\varepsilon_{ni}) = \psi((y_{2k}[\sigma_{2k}] ^\frown y_{2k+1})).$$

Going back though the definition of ψ we see that the ψ-image of $[Y]_{Lv}^{[\infty]}$ contains P.

Claim 5.45.1 *There is an infinite sequence* (P_i) *of perfect subsets of the Cantor space* $2^{\mathbb{N}}$ *such that* $\prod_{i=0}^\infty P_i \subseteq P$.

Proof. Let P_i be the collection of all $\delta \in 2^{\mathbb{N}}$ that satisfies the following conditions, where y denotes the infinite variable-word $y_0{}^\frown \dots {}^\frown y_n{}^\frown \dots$.

(1) If at some place n we find a letter $\sigma \in L$ in y, then $\delta(n) = \sigma(i)$.

(2) If $n < |y_0| + \dots + |y_{2i-1}|$ and at nth place in y we find a variable, then $\delta(n) = 0$.

(3) If $n_{2k-1} = |y_0| + \dots + |y_{2k-1}| \leq n < |x_0| + \dots + |y_{2k-1}| + |y_{2k}| = n_{2k}$ for some $k \geq i$ and we find a variable at nth place in x, then $\delta(n) = \delta(n')$, where $n' \in [n_{2k-1}, n_{2k})$ is the minimal place where y has variable.

Thus, P_i has no restrictions at the minimal place of some interval of the form $I_{2k-1} = [n_{2k-1}, n_{2k})$ for $k \geq i$ where a variable occurs. From this one easily concludes that P_i is indeed perfect. Consider a sequence $(\delta_i) \in \prod_{i=0}^\infty P_i$.. Let $(\varepsilon_{ni})_{n,i}$ be the doubly-indexed sequence such that $\varepsilon_{ni} = \delta_i(n)$. For each k choose $\sigma_{2k} \in L_{2k}$ such that $\sigma_{2k}(i) = \delta_i(n)$, where n is the minimal integer of the interval I_{2k-1} where a variable occurs in y. Consider the corresponding infinite word $y_{(\sigma_{2k})} = y_0[\sigma_0]{}^\frown y_1{}^\frown \dots {}^\frown y_{2k}[\sigma_{2k}]{}^\frown y_{2k+1}{}^\frown \dots$. Then $\varepsilon_{ni} = \sigma(i)$ if and only if the nth place of the infinite word $y_{(\sigma_{2k})}$ is occupied by the letter σ; $\varepsilon_{ni} = 0$ if the nth place is occupied by the variable v. Referring to the definition of P, we see that $(\delta_i) \in P$. This shows that $\prod_{i=0}^\infty P_i \subseteq P$. $\square$

Claim 5.45.2 $[B]^{[\infty]} \times P \subseteq (\varphi \times \psi)[[Y]^{[\infty]}_{Lv}]$.

Proof. Consider $(A, (\varepsilon_{ni})) \in [B]^\infty \times P$ and fix $(\sigma_{2k}) \in \prod_k L_{2k}$ such that

$$(\varepsilon_{ni}) = \psi((y_{2k}[\sigma_{2k}]{}^\frown y_{2k+1}). \tag{5.13}$$

Since $A = (a_\ell)$ is a block subsequence of B there is block sequence (F_ℓ) of finite subsets of $\mathbb{N}$ such that

$$a_\ell = \bigcup_{k \in F_\ell} b_k$$

for all l. Note that if for every k, we let $x_k = y_{2k}[\sigma_k]{}^\frown y_{2k+1}$, then we get an infinite sequence $X = (x_k)$ of variable-words such that $\varphi(X) = B$. For a given ℓ, let $\{p, p+1, \dots, p+q\}$ enumerate the interval $(\max(F_{\ell-1}), \max(F_\ell)]$, where $\max(F_{\ell-1}) = -1$ if $\ell = 0$, and let

$$z_\ell = x_p[\lambda_p]{}^\frown \dots {}^\frown x_{p+1}[\lambda_{p+q}], \tag{5.14}$$

where $\lambda_{p+i}(0 \leq i \leq q)$ is either equal to v or to the constantly 0 member of 2^∞ depending whether $p + i \in F_\ell$ or not. Clearly, $Z = (z_\ell)$ is a block subsequence of $X = (x_k)$, and therefore, a block subsequence of $Y = (y_k)$. The choice of $Z = (z_\ell)$ is of course made to ensure that $\varphi(Z) = A$, so it remains only to check that the we did not lose (ε_{ni}), i.e., that we still have the equality $\psi((z_\ell)) = (\varepsilon_{ni})$. This equality depends on which letter we find in some place n of the infinite word $z = z_0{}^\frown \dots {}^\frown z_k{}^\frown \dots$. Tracing back the definitions, we get that

$$z = y_0[\sigma_0]{}^\frown y[\lambda_1]{}^\frown \dots {}^\frown y_{2k}[\sigma_{2k}]{}^\frown y_{2k+1}[\lambda_{2k+1}]{}^\frown \dots,$$

where $(\sigma_{2k}) \in \prod_k L_{2k}$ was the sequence that gave us (ε_{ni}) from $Y = (y_k)$ and where the λ_{2k+1} are either equal to v or to the constantly equal to 0 member of L. Thus the only difference between the infinite word $y_{(\sigma_{2k})}$ and z is that at some places where $y_{(\sigma_{2k})}$ has a variable z, has the constantly equal to 0 member of L. Referring to the reading of $\psi((z_\ell))$, we realize that it gives us the original double-indexed sequence (ε_{ni}). This finishes the proof of Claim 5.45.2. $\square$

It is clear that the two claims finish the proof of Theorem 5.45. $\square$

Recall now the definitions given at the beginning of Section 4.5. We start with an alphabet $L = \bigcup_{n=0}^{\infty} L_n$ written as an increasing union of finite alphabets L_n and a variable $v \notin L$. A *located variable-word* is a finitely supported map from $\mathbb{N}$ into $L \cup \{v\}$ with the value v achieved at least once. Let FIN_{Lv} denote the partial semigroup of located variable-words with the partial semigroup operation of taking the union of two disjointly supported located variable-words. For $x \in \mathrm{FIN}_{Lv}$ and $\lambda \in L \cup \{v\}$, let $x[\lambda]$ be the located word or located variable-word obtained from x by replacing every occurrence of v by λ. A *block sequence* is a finite or infinite sequence $X = (x_n)$ of members of FIN_{Lv} such that

$$\mathrm{dom}(x_m) < \mathrm{dom}(x_n) \quad \text{whenever } m < n.$$

Let $\mathrm{FIN}_{Lv}^{[\infty]}$ denote the collection of all infinite block sequences of elements of FIN_{Lv}. For $X = (x_n) \in \mathrm{FIN}_{Lv}^{[\infty]}$, we consider the following the partial subsemigroup of FIN_{Lv} generated by X:

$$[X]_{Lv} = \{x_{n_0}[\lambda_0] \cup \ldots \cup x_{n_k}[\lambda_k] \in \mathrm{FIN}_{Lv} : k \in \mathbb{N}, \, n_0 < \ldots < n_k,$$

$$\lambda \in L_{n_i} \cup \{v\} (i \leq k)\}.$$

For $X = (x_n)$ and $Y = (y_n)$ from $\mathrm{FIN}_{Lv}^{[\infty]}$, we say that X is a *block subsequence* of Y and write $X \leq Y$ if $x_n \in [Y]_{Lv}$ for all n. We extend these relations to the families $\mathrm{FIN}_{Lv}^{[<\infty]}$ of finite block sequences of located words and variable-words. This will give us the finitizations $\leq_{\mathrm{fin}}$ of $\leq$ satisfying axioms $A.2$ and $A.3$. That $(\mathrm{FIN}_{Lv}^{[\infty]}, \leq, r)$ also satisfies axiom $A.4$ follows from Theorem 4.47 above, so we have the following result.

Theorem 5.46 $(\mathrm{FIN}_{Lv}^{[\infty]}, \leq, r)$ *is a topological Ramsey space.*

Corollary 5.47 (Milliken) $(\mathrm{FIN}_{\emptyset v}^{[\infty]}, \leq, r)$ *is a topological Ramsey space.*

Proof. This is the case $L = \emptyset$ of the previous theorem, for which $A.4$ reduces to Hindman's theorem. $\square$

The projection mapping π of Theorem 4.50 also establishes the following result, showing that Theorem 5.41 is a consequence of Theorem 5.46.

Theorem 5.48 *There is a natural projection from the topological Ramsey space* $(\mathrm{FIN}_{Lv}^{[\infty]}, \leq, r)$ *onto the topological Ramsey space* $(W_{Lv}^{[\infty]}, \leq, r)$ *transferring the Ramsey theoretic properties of* $\mathrm{FIN}_{Lv}^{[\infty]}$ *to* $W_{Lv}^{[\infty]}$.

5.4 PARAMETRIZED VERSIONS OF ROSENTHAL DICHOTOMIES

The purpose of this section is to point out the potential for applications of the Ramsey spaces constructed so far. We do this by listing some of the typical applications of Milliken's space $(\text{FIN}^{[\infty]}, \leq r)$, the simplest space in the category of Ramsey spaces on infinite sequences of words.

First of all, note that Milliken's space naturally connects to the Ellentuck space $\mathbb{N}^{[\infty]}$ via the following two mappings.

$$\textbf{min} : \text{FIN}^{[\infty]} \to \mathbb{N}^{[\infty]} \text{ and } \textbf{uni} : \text{FIN}^{[\infty]} \to \mathbb{N}^{[\infty]},$$

defined as follows:

$$\textbf{min}((x_n)) = (\min(x_n)) \text{ and } \textbf{uni}((x_n)) = \bigcup_{n=0}^{\infty} x_n.$$

Note that $\textbf{min}$ is an open 1-Lipschitz projection from Milliken's space onto the Ellentuck space and that $\textbf{uni}$ is continuous with respect to the metric as well as to the Ellentuck topologies on these two spaces. The proof of the following result points out the strength of Milliken's space over Ellentuck's and we note that its conclusion has already been obtained when the number of colors is finite (see Theorem 5.45 above).

Theorem 5.49 (Parametrized Perfect Set Theorem)   *For every Souslin measurable coloring of the product $\mathbb{N}^{[\infty]} \times \mathbb{R}$ with finitely many colors, or countably many colors, provided they are invariant under finite changes on the first coordinate, there exist $M \in \mathbb{N}^{[\infty]}$ and perfect $P \subseteq \mathbb{R}$ such that $M^{[\infty]} \times P$ is monochromatic.*

Proof. Since we are interested in getting a perfect set on the second co-ordinate, we may replace $\mathbb{R}$ with the set $\mathbb{N}^{[\infty]}$ equipped with its natural metric topology, which is as we know homeomorphic to $\mathbb{R} \setminus \mathbb{Q}$. So let $c : \mathbb{N}^{[\infty]} \times \mathbb{N}^{[\infty]} \to \mathbb{N}$ be a given Souslin-measurable coloring. Assume first that c has finite range in $\mathbb{N}$. Define $c^* : \text{FIN}^{[\infty]} \to \mathbb{N}$ by

$$c^*(X) = c(\textbf{min}((X_{\text{odd}})), \textbf{uni}((X_{\text{even}}))). \tag{5.15}$$

Clearly, c^* is also Souslin-measurable. By Milliken's theorem, there is a $Y = (y_n) \in \text{FIN}^{[\infty]}$ such that c^* is constant on the set $[Y]^{[\infty]}$ of all infinite block subsequences of Y. Let

$$M = \{\min(y_{3n+2}) : n \in \mathbb{N}\}, \tag{5.16}$$

$$P = \left\{ \bigcup_{n=0}^{\infty} y_{3n+\varepsilon(n)} : \varepsilon \in 2^{\mathbb{N}} \right\}. \tag{5.17}$$

Clearly, $M \in \mathbb{N}^{[\infty]}$ and P is a perfect subset of $\mathbb{N}^{[\infty]}$. To show that c is constant on $M^{[\infty]} \times P$, it suffices to prove that

$$M^{[\infty]} \times P \subseteq \{(\textbf{min}((X_{\text{odd}})), \textbf{uni}((X_{\text{even}})) : X \leq Y\}. \tag{5.18}$$

Consider an $N \in M^{[\infty]}$ and $t = \bigcup_{n=0}^{\infty} Y_{3n+\varepsilon(n)} \in P$. Let (n_i) be the strictly increasing sequence such that $N = \{\min(y_{3n_i+2}) : i \in \mathbb{N}\}$ and define an element $X = (x_n)$ of $\text{FIN}^{[\infty]}$ as follows:

$$x_{2i} = \bigcup_{n=n_{i-1}+1}^{n_i} Y_{3n+\varepsilon(n)}, \tag{5.19}$$

$$x_{2i+1} = y_{3n_i+2}. \tag{5.20}$$

(Here, $n_1 = -1$.) Then X is a block sequence of Y and

$$\langle N, t \rangle = \langle \mathbf{min}((X_{\text{odd}})), \mathbf{uni}((X_{\text{even}})) \rangle, \tag{5.21}$$

as required. If c has infinite range, the construction would give us a perfect set $P \subseteq \mathbb{N}^{[\infty]}$ and an Ellentuck basic open set $[a, M]$ such that c is constant on the product $[a, M] \times P$. Since we are now assuming that c is invariant under finite changes on the first coordinate, we conclude that c is actually constant on $M^{[\infty]} \times P$. This finishes the proof. $\square$

The following is a typical application of the Parameterized Perfect Set Theorem to the geometry of Banach spaces.

Theorem 5.50 (Parametrized Rosenthal ℓ_1-Theorem)[2] *Suppose that $\{x_s : s \in 2^{<\infty}\}$ is a bounded sequence of elements of a separable Banach space X indexed by the complete binary tree $2^{<\infty}$. Then there is a perfect set $P \subseteq 2^{\mathbb{N}}$ and an infinite increasing sequence $\{n_k\}$ of positive integers such that either*

(1) *$\{x_{a \restriction n_k}\}$ is a weakly Cauchy[3] sequence in X for all $a \in P$, or*

(2) *there is $\delta > 0$ such that*

$$\left\| \sum_0^m \lambda_k x_{a \restriction n_k} \right\| \geq \delta \sum_0^m |\lambda_k| \tag{5.22}$$

for every $a \in P$ and every finite sequence $\lambda_0, .., \lambda_m$ of scalars.

Proof. Define $c : \mathbb{N}^{[\infty]} \times 2^{\infty} \to \mathbb{N} \cup \{\infty\}$ as follows:

(a) $c(M, a) = \infty$ if $\{x_{a \restriction k}\}_{k \in M}$ is weakly Cauchy.

(b) $c(M, a) = n > 0$ if $\|\sum_0^m \lambda_k x_{a \restriction n_k}\| \geq (1/n) \sum_0^m |\lambda_k|$ for every finite sequence $\lambda_0, \ldots, \lambda_m$ of scalars and $n_0 < \ldots < n_m$ in M.

(c) $c(M, a) = 0$ if neither (a) nor (b) is true.

---

[2]The "ℓ_1" here refers to the fact that the second alternative of this theorem means that for each $a \in P$, the mapping $e_k \mapsto x_{a \restriction n_k}$ extends to an isomorphic embedding $T_a : \ell_1 \to X$ of norm $\leq \delta^{-1}$. Here, of course, ℓ_1 is the Banach space of absolutely converging series and (e_k) is its canonical basis.

[3]Here "weakly Cauchy" refers to the fact that $(f(x_{a \restriction n_k}))_k$ is a Cauchy sequence of scalars for every bounded linear functional f on X.

Since the sets of all pairs (M, a) satisfying (a) or (b) are coanalytic subsets of $\mathbb{N}^{[\infty]} \times 2^\infty$, this defines a Souslin-measurable coloring of $\mathbb{N}^{[\infty]} \times 2^\infty$. By the Parameterized Perfect Set Theorem, there exist $M \in \mathbb{N}^{[\infty]}$ and a perfect set $P \subseteq 2^\infty$ such that c is constant on the product $M^{[\infty]} \times P$. Clearly, we are done if the constant value is > 0, so let us show that the constant value cannot be 0. For this, it suffices to show that for a fixed $a \in P$, there is an $N \in M^{[\infty]}$ such that the pair (a, N) falls into case (a) or case (b) above. Fixing a, let S^* be the unit sphere of the dual space X^* of X and let $\{n_k\}$ be the increasing enumeration of M. We define

$$\{f_k : S^* \to \mathbb{R}\} \tag{5.23}$$

as follows: $f_k(x^*) = x^*(x_{a \restriction n_k})$. So, it suffices to show the following fact which is clearly just a reformulation of Rosenthal's original ℓ_1-theorem. $\qquad \square$

Theorem 5.51 (Rosenthal's ℓ_1-theorem) *Suppose $(f_n : S \to \mathbb{R})_{n=0}^\infty$ is a bounded sequence of functions defined on some set S. Then either there is a subsequence (f_{n_k}) such that*

(1) *(f_{n_k}) is pointwise convergent on S, or*

(2) *there is a $\delta > 0$ such that*

$$\left\| \sum_0^m \lambda_k \, f_{n_k} \right\|_\infty \geq \delta \sum_0^m |\lambda_k| \tag{5.24}$$

for every finite sequence $\lambda_0, .., \lambda_m$ of scalars.

The proof will be given in a sequence of lemmas, but first we need a definition that introduces a combinatorial version of the second alternative of Theorem 5.51.

Definition 5.52 *We say that an infinite sequence $(\langle X_n, Y_n \rangle)$ of pairs of subsets of some set S is* independent *if*

(i) *$X_n \cap Y_n = \emptyset$ for all n, and*

(ii) *$(\bigcap_{n \in K} X_n) \cap (\bigcap_{n \in L} Y_n) \neq \emptyset$ for every choice of disjoint finite nonempty sets K and L of integers.*

The relevance of this notion to our situation here stems from the following simple observation.

Lemma 5.53 *Suppose $(g_n : S \to \mathbb{R})_{n=0}^\infty$ is a bounded sequence of functions defined on some set S such that for some rationals $p < q$, the sequence[4]*

$$(S(g_n < p), S(g_n > q))_{n=0}^\infty$$

---

[4]In what follows, for $f : S \to \mathbb{R}$ and $p \in \mathbb{R}$, we let $S(f < p) = \{x \in S : f(x) < p\}$ (and similarly, for any of the other relations $>$, $\leq$, and $\geq$ in place of $<$.

of pairs of disjoint subsets of S is independent. Then for $\delta = (q - p)/2$, we have that

$$\left\| \sum_1^m \lambda_n\, g_n \right\|_\infty \geq \delta \sum_1^m |\lambda_n|. \tag{5.25}$$

for every choice $\lambda_0, .., \lambda_n$ of scalars.

Proof. Let $K = \{n \leq m : \lambda_n \geq 0\}$ and $L = \{n \leq m : \lambda_n < 0\}$. By independence, we can choose

$$x \in \left(\bigcap_{n \in K} S(g_n < p) \right) \cap \left(\bigcap_{n \in L} S(g_n > q) \right),$$

$$y \in \left(\bigcap_{n \in L} S(g_n < p) \right) \cap \left(\bigcap_{n \in K} S(g_n > q) \right).$$

Then

$$\sum_1^m \lambda_n\, g_n(x) \leq p \sum_{n \in K} |\lambda_n| - q \sum_{n \in L} |\lambda_n| \tag{5.26}$$

$$\sum_1^m \lambda_n g_n(y) \geq -p \sum_{n \in L} |\lambda_n| + q \sum_{n \in K} |\lambda_n|. \tag{5.27}$$

Subtracting Equation (5.26) from Equation (5.27), we get

$$\sum_1^m \lambda_n g_n(y) - \sum_1^m \lambda_n g_n(x) \geq (q - p) \sum_1^m |\lambda_n|. \tag{5.28}$$

Since the left-hand side of Equation (5.28) is bounded by $2\left\| \sum_1^m \lambda_n g_n \right\|$, we have verified the estimate in Equation (5.25) above. $\qquad \square$

The proof of Theorem 5.51 is complete once we show the following.

Lemma 5.54 *A pointwise bounded sequence (f_n) of real-valued functions defined on some set S either contains an infinite pointwise convergent subsequence, or an infinite subsequence (f_{n_k}) such that for some rationals $p < q$ the sequence of pairs $(S(f_{n_k} < p), S(f_{n_k} > q))$ is independent.*

A simple diagonal sequence procedure taking into account all possible pairs $p < q$ of rational numbers reduces Lemma 5.54 to the following purely combinatorial fact.

Lemma 5.55 *Suppose $(\langle X_n, Y_n \rangle)$ is a given infinite sequence of pairs of disjoint subsets of some set S. Then there is an infinite subsequence $(\langle X_{n_k}, Y_{n_k} \rangle)$ such that either*

(i) *$(\langle X_{n_k}, Y_{n_k} \rangle)$ is an independent sequence of pairs of subsets of S, or*

(ii) *there is no point of S that belongs to infinitely many X_{n_k} as well as to infinitely many Y_{n_k}.*

Proof. Color an $N = (n_k) \in \mathbb{N}^{[\infty]}$ by Red iff

$$\left(\bigcap_{k=0}^{\ell} X_{n_{2k}} \right) \cap \left(\bigcap_{k=0}^{\ell} Y_{n_{2k+1}} \right) \neq \emptyset \qquad (5.29)$$

for all $\ell \in \mathbb{N}$; otherwise, color M by Green. Clearly, this is a Borel coloring. By the Galvin-Prikry theorem, there is a $M \in \mathbb{N}^{[\infty]}$ such that $M^{[\infty]} \subseteq$ Red or $M^{[\infty]} \subseteq$ Green. Note that in the second case, $M^{[\infty]} \subseteq$ Green, the corresponding subsequence $(\langle X_{m_k}, Y_{m_k} \rangle)$, where $M = (m_k)$ is the increasing enumeration of M, satisfies the second alternative of the lemma. So suppose $M^{[\infty]} \subseteq$ Red. Let $N = (m_{2k+1})$ be the subset of M formed by taking the odd indexed members of M. We claim that $(\langle X_m, Y_m \rangle)_{m \in N}$ is an independent sequence of pairs of disjoint subsets of S. To see this, consider two disjoint finite sets $K, L \subseteq N$. Since K and L are subsets of the subset of M consisting of its odd indexed elements, we can extend K and L by two disjoint finite supersets $\bar{K} \supseteq K$ and $\bar{L} \supseteq L$ such that $\bar{K} \cup \bar{L}$ forms an initial segment of some infinite subset C of M such that $K \subseteq C_{\text{even}}$ and $L \subseteq C_{\text{odd}}$. Since $C \in M^{[\infty]} \subseteq$ Red, we have that

$$\left(\bigcap_{m \in K} X_m \right) \cap \left(\bigcap_{m \in L} Y_m \right) \supseteq \left(\bigcap_{m \in \bar{K}} X_m \right) \cap \left(\bigcap_{m \in \bar{L}} Y_m \right) \neq \emptyset. \qquad (5.30)$$

This finishes the proof. $\qquad\qquad\qquad\qquad\qquad\qquad\qquad\qquad\qquad\quad\square$

A closely related result to Rosenthal's ℓ_1-theorem is his dichotomy about pointwise convergence of sequences of continuous real-valued functions defined on some Polish space P, which could be stated as follows.

Theorem 5.56 (Rosenthal Dichotomy Theorem) *Suppose (f_n) is a pointwise bounded sequence of continuous functions on some Polish space P. Then (f_n) contains either a pointwise convergent subsequence or a subsequence (f_{n_k}) such that $\mathcal{U} \mapsto \lim_{k \to \mathcal{U}} f_{n_k}$ is a homeomorphical embedding of $\beta \mathbb{N}$ into the topological closure of (f_{n_k}) inside the product space $\mathbb{R}^P$.*

Proof. By Lemmas 5.54 and 5.55, and using a subsequence of (f_n), we may assume that there is a pair $\varepsilon < \delta$ of reals such that for every infinite $M \subseteq \mathbb{N}$ there exists $x \in P$ such that the sets

$$M(x < \varepsilon) = \{n \in M : f_n(x) < \varepsilon\} \text{ and } M(x > \delta) = \{n \in M : f_n(x) > \delta\}$$

are both infinite. For a given infinite subset M of $\mathbb{N}$, let $P(M)$ be the topological closure of the set of all $x \in P$ for which both sets $M(x < \varepsilon)$ and $M(x > \delta)$ are infinite. Note that $P(M)$ is a nonempty closed subset of P with no isolated points and that $P(M) \subseteq P(M')$ whenever $M \setminus M'$ is finite. So, going to a subsequence, we may assume that $P(M) = P(\mathbb{N})$ for all infinite $M \subseteq \mathbb{N}$. As noted above, the set $P(\mathbb{N})$ has no relatively isolated points, so we can pick two countable disjoint subsets D and E that are dense in $P(\mathbb{N})$ in such a way that the family of sets

$$\mathbb{N}(x < \varepsilon) \cap \mathbb{N}(y > \delta) \quad (x \in D, y \in E) \qquad (5.31)$$

has the finite intersection property. Pick a perfect set $Q \subseteq P(\mathbb{N})$ such that $Q \cap D$ and $Q \cap E$ are dense in Q and pick infinite $M \subseteq \mathbb{N}$, which is almost included in every member of the family

$$\mathbb{N}(x, < \varepsilon) \cap \mathbb{N}(y, > \delta) \quad (x \in E, y \in D). \tag{5.32}$$

Then no subsequence of $(f_n)_{n \in M}$ is convergent on Q, since any pointwise cluster point g of $(f_n)_{n \in M}$ is pointwise bounded by ε on $Q \cap D$ and pointwise dominates δ on $Q \cap E$. So in particular, it cannot be a Baire Class 1 function by the Baire characterization theorem of such functions (see [54]). Applying Lemma 5.54 to $S = Q$ and $(f_n)_{n \in M}$, we get a subsequence (f_{n_k}) of $(f_n)_{n \in M}$ and $\bar{\varepsilon} < \bar{\delta}$ such that the sequence of pairs

$$Q_k^0 = Q(f_{n_k} \leq \bar{\varepsilon}) \text{ and } Q_k^1 = Q(f_{n_k} \geq \bar{\delta}) \tag{5.33}$$

is independent. Since Q is compact, for each $\sigma \in 2^{\mathbb{N}}$, we can pick a point

$$q_\sigma \in \bigcap_{k=0}^{\infty} Q_k^{\sigma(k)}. \tag{5.34}$$

For an ultrafilter $\mathcal{U}$ on $\mathbb{N}$, let $g_{\mathcal{U}} = \lim_{k \to \mathcal{U}} f_{n_k}$. Then $\mathcal{U} \mapsto g_{\mathcal{U}}$ is a continuous map from $\beta\mathbb{N}$ into the pointwise closure of (f_{n_k}) in $\mathbb{R}^P$. We claim that this mapping is 1-1. To see this pick, $\mathcal{U} \neq \mathcal{V}$ in $\beta\mathbb{N}$ and find $\sigma \in 2^{\mathbb{N}}$ such that

$$\{k : \sigma(k) = 0\} \in \mathcal{U} \text{ and } \{k : \sigma(k) = 1\} \in \mathcal{V}. \tag{5.35}$$

Then $g_{\mathcal{U}}(q_\sigma) \leq \bar{\varepsilon}$, while $g_{\mathcal{V}}(q_\sigma) \geq \delta$, so in particular $g_{\mathcal{U}} \neq g_{\mathcal{V}}$. This completes the proof. □

Corollary 5.57 (Sequential Compactness Theorem) *Every pointwise compact set K of Baire Class 1 functions defined on some Polish space[5] X is sequentially compact.*

Proof. Consider a sequence (f_n) of elements of K. Increasing the topology of X by adding to it countably many Borel sets and still keeping it Polish, we may assume that all f_n are actually continuous functions. Applying Rosenthal's Dichotomy Theorem to this sequence, we must get the first alternative since the second one is impossible for $\beta\mathbb{N}$ cannot be embedded into K because K has cardinality not bigger than the continuum. □

Corollary 5.58 (Parametrized Sequential Compactness Theorem)   *Let f_s ($s \in 2^{<\mathbb{N}}$) be a relatively compact subset of Borel functions defined on some Polish space X. Then there are a perfect set $P \subseteq 2^{\mathbb{N}}$ and a strictly increasing sequence (n_k) of integers such that for every $a \in P$, the corresponding sequence $(f_{a \restriction n_k})$ is pointwise convergent on X.*

Proof. As before, we may assume that f_s ($s \in 2^{<\mathbb{N}}$) is a sequence of continuous functions on X. Let

$$\mathcal{C} = \{(a, M) \in 2^{\mathbb{N}} \times \mathbb{N}^{[\infty]} : (f_{a \restriction m})_{m \in M} \text{ is pointwise convergent on } X\}.$$

---

[5]Recall that a *Polish space* is any nonempty separable completely metrizable topological space.

Clearly, $\mathcal{C}$ is a Souslin-measurable subset of the product $2^{\mathbb{N}} \times \mathbb{N}^{[\infty]}$. Applying the Parameterized Perfect Set Theorem, we get a perfect set $P \subseteq 2^{\mathbb{N}}$ and infinite $M \subseteq \mathbb{N}$ such that $P \times M^{[\infty]} \subseteq \mathcal{C}$, or else $P \times M^{[\infty]} \cap \mathcal{C} = \emptyset$. Note that by Corollary 5.57, the second alternative is not possible, so we are left with the first. This gives us the conclusion of the corollary.   □

Remark 5.59 Note that if, for $a \in P$, we let $f_a = \lim_k f_{n_k}$, and if we define $\Phi : P \times X \to \mathbb{R}$ by $\Phi(a, x) = f_a(x)$, then we get a Borel-measurable map. This is a phenomenon that appears in similar constructions below and at some point we will even be in a position to exploit it.

5.5 RAMSEY THEORY OF SUPERPERFECT SUBSETS OF POLISH SPACES

Recall that the space of irrationals $\mathbb{R} \backslash \mathbb{Q}$ (or its homeomorphic version, the Baire space $\mathbb{N}^{\mathbb{N}}$) is an example of a Polish space X that is not σ-compact, and in fact, that is nowhere σ-compact in the sense that the interior of every σ-compact subset of X is empty. The following classical result (see Theorem 9.51) goes a bit deeper in explaining the property of σ-compactness in the realm of Polish spaces.

Theorem 5.60 (Hurewicz) *A Polish space X is σ-compact if and only if it does not contain a closed subset homeomorphic to the irrationals, or equivalently, the Baire space $\mathbb{N}^{\mathbb{N}}$.*

Since the purpose of this section is to give some applications of the Milliken space $\mathrm{FIN}^{[\infty]}$ to the Ramsey theory in the context of Polish space, we need to disclose the notions of largeness relevant to this theory.

Definition 5.61 *We shall say that a subset P of some Polish space X is* perfect *if it is nonempty and closed and it has no isolated points. A subset P of X will be said to be* superperfect *if it is nonempty, closed, without isolated points, and not σ-compact.*

Remark 5.62 Clearly, every perfect subset P of some Polish space X contains a homeomorphic copy of the Cantor space $2^{\mathbb{N}}$, which itself is a perfect Polish space. So if perfectness is the notions of largeness in our Ramsey theory where we, in order to simplify a given coloring, need to go to large subsets of large sets, we shall sometimes think of *perfect* as simply a *topological copy of the Cantor set*. Similarly, by Hurewicz's theorem a superperfect subset P of a Polish space X always contains a closed copy of the irrationals, so we think of *superperfect* as simply a *closed copy of the irrationals*.

We are now ready to state and prove our first result in this new Ramsey theory on Polish spaces.

Theorem 5.63 (Superperfect Set Theorem) *Suppose that the infinite power* $X^{\mathbb{N}}$ *of a non-σ-compact Polish space* X *is colored by countably many Souslin-measurable colors. Then there is a sequence* (P_i) *of perfect subsets of* X *such that* P_0 *and* P_1 *are superperfect and such that the product* $\prod_{i=0}^{\infty} P_i$ *is monochromatic.*

Proof. By Hurewicz's theorem (Theorem 5.60 above), every non-σ-compact Polish space contains a closed copy of the Baire space $\mathbb{N}^{\mathbb{N}}$. Therefore we may assume that $X = \mathbb{N}^{[\infty]}$ with its usual metric topology and that we are given a Souslin-measurable coloring

$$c : \mathbb{N}^{[\infty]} \times \mathbb{N}^{[\infty]} \times (2^{\mathbb{N}})^{\mathbb{N}} \to \mathbb{N} \tag{5.36}$$

and concentrate on producing two superperfect subsets S_0 and S_1 of $\mathbb{N}^{[\infty]}$ and a sequence (P_i) of perfect subsets of $2^{\mathbb{N}}$ such that c is monochromatic on $S_0 \times S_1 \times \prod_{i=0}^{\infty} P_i$. Let

$$c^* : \mathrm{FIN}^{[\infty]} \times (2^{\mathbb{N}})^{\mathbb{N}} \to \mathbb{N} \tag{5.37}$$

be defined now as follows

$$c^*(A, x) = c\left(\bigcup_{k=0}^{\infty} a_{2k}, \bigcup_{k=0}^{\infty} a_{2k+1}, x \right). \tag{5.38}$$

By the proof of Theorem 5.45, we can find a basic open set $[2\ell, B]$ of the Milliken space and a sequence (P_i) of perfect sets such that the product $[2\ell, B] \times \prod_{i=0}^{\infty} P_i$ is monochromatic for c^*. Let B_0 be obtained from B by removing its $(2\ell + 1)$st element $b_{2\ell+1}$, and let B_1 be obtained from B by removing its (2ℓ)th element $b_{2\ell}$. Let

$$S_0 = \{ \bigcup_{k=0}^{\infty} a_{2k} : A = (a_n) \in [2\ell + 1, B_0] \},$$
$$S_1 = \{ \bigcup_{k=0}^{\infty} a_{2k+1} : A = (a_n) \in [2\ell + 1.B_1] \} \tag{5.39}$$

So suppose we are given a pair $C = (c_n) \in [2\ell + 1, B_0]$ and $D = (d_n) \in [2\ell + 1, B_1]$ representing a pair (s, t) from the product $S_0 \times S_1$. Thus C agrees with B up to 2ℓ and its (2ℓ)th element is equal to $b_{2\ell}$, and D agrees with B up to 2ℓ and its (2ℓ)th element is equal to $b_{2\ell+1}$. Consider now the following two disjoint subsets of $\mathbb{N}$:

$$\bar{s} = \bigcup_{n \geq 2\ell} c_n \quad \text{and} \quad \bar{t} = \bigcup_{n \geq 2\ell} d_n. \tag{5.40}$$

Define the equivalence relation $\sim$ on $\bar{s} \cup \bar{t}$ by letting $n \sim m$ iff the closed interval of the set $\bar{s} \cup \bar{t}$ determined by n and m (i.e., the set $[n, m] \cap (\bar{s} \cup \bar{t})$ or the set $[m, n] \cap (\bar{s} \cup \bar{t})$ depending on whether $n \leq m$ or $m \leq n$, respectively) is included either in $\bar{s}$ or in $\bar{t}$. Note that indeed this is an equivalence relation on $\bar{s} \cup \bar{t}$ and that equivalence classes form a block subsequence of B. Note also that the first class is included in $\bar{s}$ as it contains $b_{2\ell}$ which $\bar{t}$ misses. Define $A = (a_n) \in [2\ell, B]$ by the following rules:

(1) $A \upharpoonright 2\ell = B \upharpoonright 2\ell$,

(2) $\{a_{2k}\}_{k=\ell}^{\infty}$ is an increasing enumeration of $\bar{s}/\sim$,

(3) $\{a_{2k+1}\}_{k=\ell}^{\infty}$ is an increasing enumeration of $\bar{t}/\sim$.

Note that

$$\bigcup_{k=0}^{\infty} a_{2k} = \bigcup_{k=0}^{\infty} c_{2k} = s \text{ and } \bigcup_{k=0}^{\infty} a_{2k+1} = \bigcup_{k=0}^{\infty} d_{2k+1} = t. \tag{5.41}$$

It follows that $c(s,t,x) = c^*(A,x)$ for every $x \in \prod_{i=0}^{\infty} P_i$. This shows that c is monochromatic on $S_0 \times S_1 \times \prod_{i=0}^{\infty} P_i$. $\qquad\square$

Corollary 5.64 (Spinas) *Suppose that X is a non-σ-compact Polish space and that X^2 is colored by countably many Souslin-measurable colors. There exist two superperfect subsets S_0 and S_1 of X such that $S_0 \times S_1$ is monochromatic.*

Remark 5.65 It should be noted that the number of superperfect sets in the conclusion of Theorem 5.63 cannot exceed 2. This is a consequence of the following result which appears below as Corollary 6.37.

Theorem 5.66 *There is a continuous map $c : (\mathbb{R}\backslash\mathbb{Q})^{[3]} \to \mathbb{N}$ that takes all the values from $\mathbb{N}$ on the cube of any superperfect subset of $\mathbb{R}\backslash\mathbb{Q}$.*

This result suggests the question about the validity of the symmetric case of the two-dimensional superperfect set theorem. It turns out that Theorem 5.63 does allow the symmetric variation, and this is our next result, which can be seen as yet another contribution to parametrized Ramsey theory, a theory that receives full attention in later chapters of this book.

Theorem 5.67 (Symmetric Version of the Superperfect Set Theorem) *Suppose that X is a non-σ-compact Polish space and that $X^{[2]} \times X^{\mathbb{N}}$ is colored by finitely many Souslin-measurable colors. Then there are a superperfect set $S \subseteq X$ and an infinite sequence (P_n) of perfect subsets at X such that the product $S^{[2]} \times \prod_{i=0}^{\infty} P_i$ is monochromatic.*

Proof. Applying Hurewicz's theorem (Theorem 5.60), we may assume that that $X = \mathbb{N}^{[\infty]}$. Similarly, we may replace the second coordinate of the colored product $X^{[2]} \times X^{\mathbb{N}}$ by the infinite power $(2^{\mathbb{N}})^{\mathbb{N}}$ of the Cantor cube $2^{\mathbb{N}}$. Thus we may assume to have given a finite Souslin measurable coloring

$$c : (\mathbb{N}^{[\infty]})^{[2]} \times (2^{\mathbb{N}})^{\mathbb{N}} \to F.$$

To this c, we associate a Souslin-measurable coloring

$$c^* : \text{FIN}^{[\infty]} \times (2^{\mathbb{N}})^{\mathbb{N}} \to F \tag{5.42}$$

as follows:

$$c^*(A,x) = c\left(\left\{a_0 \cup \bigcup_{k=0}^{\infty} a_{2k+1}, \bigcup_{k=0}^{\infty} a_{2k}\right\}, x\right). \tag{5.43}$$

By Theorem 5.45, there exist $B = (b_n)_0^\infty \in \text{FIN}^{[\infty]}$ and a sequence (P_i) of perfect subsets of the Cantor space $2^\mathbb{N}$ such that c^* is monochromatic on the product $[B]^\infty \times \prod_{i=0}^\infty P_i$. Let

$$S = \left\{ b_0 \cup \bigcup_{k=1}^\infty b_{p_1 \ldots p_k} : (p_k)_1^\infty \in \text{PR}^{[\infty]} \right\},$$

where PR denotes the set of all primes. Thus S is equal to the set of all unions of subsequences of the sequence $B = (b_n)_0^\infty$ of the form

$$b_0, b_{p_1}, b_{p_1 p_2}, b_{p_1 p_2 p_3} \ldots, b_{p_1 \ldots p_k}, \ldots$$

for some strictly increasing sequence $(p_k)_1^\infty$ of positive prime numbers. This particular choice of subsequences of B is made for two purposes: first, to make S a superperfect subset of $\mathbb{N}^{[\infty]}$, and, second, to have that for every two different subsequences (b_{n_k}) and (b_{m_k}) of B forming an unordered pair of elements of S there is an $\ell \geq 0$ such that

(1) $b_{n_k} = b_{m_k}$ for all $k \leq \ell$,

(2) $b_{n_k} \cap b_{n_{k'}} = \emptyset$ for all $k \neq k' > \ell$.

Consider two such subsequences (b_{n_k}) and (b_{m_k}) of B and $\ell \geq 0$ satisfying (1) and (2). For concreteness assume that $n_{\ell+1} < m_{\ell+1}$. Let

$$s = \bigcup_{k=0}^\infty b_{n_k} \text{ and } t = \bigcup_{k=0}^\infty b_{m_k}. \tag{5.44}$$

There is a natural equivalence relation $\sim$ on the symmetric difference $(s \backslash t) \cup (t \backslash s)$ of s and t : $i \sim j$ iff the interval of integers determined by i and j contains only points of one of the sets $s \backslash t$ or $t \backslash s$. Note that (1) and (2) imply that an equivalence class of $\sim$ is a finite union of the form

$$\bigcup_{k \in I} b_{n_k} \text{ or } \bigcup_{k \in J} b_{m_k} \tag{5.45}$$

for some intervals I and J of integers $> \ell$. Define an $A = (a_n) \leq B$ by the following conditions:

(3) $a_0 = \bigcup_{k=0}^\ell b_{n_k}$,

(4) $(a_{2k+1})_{k=0}^\infty$ is an increasing enumeration of $s \backslash t / \sim$,

(5) $(a_{2k})_{k=0}^\infty$ is an increasing enumeration of $t \backslash s / \sim$.

It follows that for any $x \in \prod_{i=0}^\infty P_i$,

$$c(\{s, t\}, x) = c^*(A, x).$$

Since $\{s, t\}$ was an arbitrary unordered pair of elements of S, this shows that c is monochromatic on $S^{[2]} \times \prod_{i=0}^\infty P_i$. $\qquad \square$

Corollary 5.68 (Spinas) *Suppose that the symmetric square $X^{[2]}$ of some non-σ-compact Polish space X has been colored by finitely many Souslin-measurable colors. Then there is a superperfect $S \subseteq X$ such that $S^{[2]}$ is monochromatic.*

Note that the conclusion of this corollary is false for infinite colorings. This is most easily seen in the case $X = \mathbb{N}^{\mathbb{N}}$. For example, given an unordered pair $\{x, y\}$ of elements of $\mathbb{N}^{\mathbb{N}}$ such that $x <_{\text{lex}} y$, set

$$c(\{x, y\}) = \min\{n : x(n) > y(n)\},$$

where we use the convention $\min \emptyset = 0$. Then c realizes infinitely many values on any square $S^{[2]}$ of a superperfect subset S of $\mathbb{N}$.

5.6 DUAL RAMSEY THEORY

Let $\mathcal{E}_\infty = \mathcal{E}_\infty(\mathbb{N})$ be the collection of all *equivalence relations* E on $\mathbb{N}$ whose quotients $\mathbb{N}/E$ are infinite. It is convenient to think of E as a partition of $\mathbb{N}$ into infinitely many *equivalence classes*. Each class $[x]_E$ of E has a minimal representative. Let $p(E)$ be the set of all minimal representatives of classes of E. Let $\{p_n(E)\}_{n=0}^\infty$ be the increasing enumeration of $p(E)$. Note that $0 \in p(E)$ for all $E \in \mathcal{E}_\infty$, so we have that $p_0(E) = 0$ for all $E \in \mathcal{E}_\infty$.

For $E, F \in \mathcal{E}_\infty$ we say that E *is coarser than* F and write $E \leq F$ if every class of E can be represented as the union of certain set of classes of F. The *nth approximation $r_n(E)$* to some $E \in \mathcal{E}_\infty$ is defined as follows:

$$r_n(E) = E \restriction p_n(E). \tag{5.46}$$

Thus, $r_n(E)$ is simply the restriction of the equivalence relation E to the finite set $\{0, 1, \ldots, p_n(E) - 1\}$ of integers. Each approximation $a \in \mathcal{AE}_\infty$ has its *length* $|a|$, the integer n such that $a = r_n(E)$ for some $E \in \mathcal{E}_\infty$ (or equivalently, the number of equivalence classes of a) and its *domain*, the integer $p_{|a|}(E) = \{0, 1, \ldots, p_{|a|}(E) - 1\}$, where E is some member of $\mathcal{E}_\infty$ such that $a = r_{|a|}(E)$. The relation $\leq$ of $\mathcal{E}_\infty$ allows a natural finitization $\leq_{\text{fin}}$ on $\mathcal{AE}_\infty$ satisfying A.2 and A.3: $a \leq_{\text{fin}} b$ if $\text{dom}(a) = \text{dom}(b)$ and a is coarser than b.

To show that $(\mathcal{E}_\infty, \leq, r)$ is a topological Ramsey space, it remains to show that it satisfies A.4.

Lemma 5.69 *Let $[a, E]$ be a nonempty basic set, let n be the length of a and let $\mathcal{O}$ be a family of members of $\mathcal{AE}_\infty$ of length $n + 1$. Then there is an $F \in [a, E]$ such that $r_{n+1}[a, F]$ is either contained in or disjoint from $\mathcal{O}$.*

Proof. Using the amalgamation property, we may replace E and assume that $a = r_n(E)$, or in other words, that a is an equivalence relation on the set $p_n(E) = \{0, 1, \ldots, p_n(E) - 1\}$ and that it is equal to the restriction of E on that set. An end-extension $b \in r_{n+1}[a, E]$ of a is an equivalence relation on a set of the form $p_m(E) = \{0, 1, \ldots, p_m(E) - 1\}$ for some $m > n$, which has one more class $[p_n(E)]_b$ and that joins some classes of E (or more precisely, the classes of E restricted to the set of integers below $p_m(E)$), with minimal representatives among $p_n(E), p_{n+1}(E), \ldots, p_{m-1}(E)$ to either one of the classes with representatives $< p_n(E)$ or to the new class $[p_n(E)]_E$. Hence, any such end-extension b of a can be coded as a word w^b in the alphabet $L = \{0, 1, \ldots, n\}$ such that the length of w^b is equal to $m - n$ and such that for $i < m - n$, we have $w_i^b = j$ iff the class of $p_{n+i}(E)$ has been joined with the class of $p_j(E)$. Conversely, any word w of W_L that has n as its first letter is of the form w^b for some $b = b(w)$ in $r_{n+1}[a, E]$. So the coloring by $\mathcal{O}$ and $\mathcal{O}^c$ induces the coloring of W_L. By Theorem 2.37 there is a word w_0 that starts with the letter n and an infinite sequence $X = (x_k)$ of left variable words such that the translate $w_0{}^\frown[X]_L$ is monochromatic, or in other words:

(1) $b(w) \in \mathcal{O}$ for every $w \in w_0{}^\frown[X]_L$, or

(2) $b(w) \notin \mathcal{O}$ for every $w \in w_0{}^\frown[X]_L$.

Form the infinite word $x = w_0{}^\frown x_0{}^\frown x_1{}^\frown \ldots {}^\frown x_k{}^\frown \ldots$ out of w_0 and X and use it to define an equivalence relation $F \in [a, E]$ as follows. To define F, it suffices to say how it acts on the set $p(E)$ of minimal representatives of classes of E. If at place i of x we find a letter $j \in L$, we let $p_j(E)$ and $p_{n+i}(E)$ be F-equivalent. If for some $k \geq 0$ and

$$ i, j \in [|w_0| + |x_0| + \ldots + |x_{k-1}|, |w_0| + |x_0| + \ldots + |x_{k-1}| + |x_k|), $$

we find v on both places i and j of x, then we put $p_{n+i}(E)$ and $p_{n+j}(E)$ to be F-equivalent. (Here, $x_{-1} = \emptyset$.) We put no other connections in F. Then this F satisfies the conclusion of the lemma. $\qquad\square$

Since we have established axioms A.1 to A.4 for the triple $(\mathcal{E}_\infty, \leq, r)$ and since $\mathcal{E}_\infty$ is a closed subset of the cube $\mathcal{AE}_\infty^{\mathbb{N}}$, the Abstract Ellentuck Theorem gives us the following result.

Theorem 5.70 (Carlson-Simpson) *The space $(\mathcal{E}_\infty, \leq, r)$ is a topological Ramsey space.*

Before we state the usual corollaries, note that $\mathcal{E}_\infty$ has three topologies. The first one is of course the one induced by the basic open sets of the form $[n, E]$, which makes $\mathcal{E}_\infty$ a topological Ramsey space. The second one is the metrizable topology induced from the Tychonov cube $\mathcal{AE}_\infty^{\mathbb{N}}$. The third one is also a metrizable topology induced from the Cantor cube $2^{\mathbb{N} \times \mathbb{N}}$ when members of $\mathcal{E}_\infty$ are identified with subsets of $\mathbb{N} \times \mathbb{N}$. In the following corollary, "Souslin-measurable" can be based on any of the three topologies.

Corollary 5.71 (Dual Silver Theorem) *Suppose that c is a finite Souslin-measurable coloring of the family $\mathcal{E}_\infty$ of all equivalence relations on $\mathbb{N}$ with infinitely many classes. Then there is $E \in \mathcal{E}_\infty$ such that the family $\mathcal{E}_\infty|E$ of all $F \in \mathcal{E}_\infty$ that are coarser than E is c-monochromatic.*

For a positive integer k, let $\mathcal{E}_k = \mathcal{E}_k(\mathbb{N})$ be the collection of all equivalence relations on $\mathbb{N}$ with exactly k classes. It is natural to identify $\mathcal{E}_k$ with the subset of the Tychonoff cube $\{0, 1, \ldots, k-1\}^{\mathbb{N}}$ consisting of surjective maps f with the property that

$$\min f^{-1}(i) < \min f^{-1}(j) \text{ for } i < j < k. \tag{5.47}$$

Note that in this identification, $\mathcal{E}_k$ is an open subset of the Tychonov cube.

Corollary 5.72 (Dual Ramsey Theorem) *Suppose that for some positive integer k, we are given a finite coloring c of $\mathcal{E}_k$ that is Baire-measurable relative to the topology induced from the Tychonoff cube $\{0, 1, \ldots, k-1\}^{\mathbb{N}}$. Then there is an $E \in \mathcal{E}_{\infty}$ such that the family $\mathcal{E}_k|E$ of all $F \in \mathcal{E}_k$ that are coarser than E is c-monochromatic.*

Proof. Assume first that c is actually Souslin-measurable. Define $\pi : \mathcal{E}_{\infty} \to \mathcal{E}_k$ by letting $\pi(E)$ be the equivalence relation that makes every $p_n(E)$ for any $n \geq k$ equivalent to 0. Then the lifted coloring $c \circ \pi$ of $\mathcal{E}_{\infty}$ is Souslin-measurable and so by Corollary 5.71, there is an $E \in \mathcal{E}_{\infty}$ such that $c \circ \pi$ is monochromatic on $\mathcal{E}_{\infty}|E$. Let $E^* \in \mathcal{E}_{\infty}$ be a coarsening of E that has the property that every class of E^* contains infinitely many classes of E. Then for every $F \in \mathcal{E}_k|E^*$, there is a $G \in \mathcal{E}_{\infty}|E$ such that $\pi(G) = F$. If follows that $\mathcal{E}_k|E^*$ is c-monochromatic.

Assume now that c is merely Baire-measurable. Then there is a dense G_{δ}-subset $\mathcal{G}$ of $\mathcal{E}_k$ on which the coloring c is actually continuous. Consider a finite approximation b to a member of $\mathcal{E}_k$ considered as a subset of $\{0, 1, \ldots, k-1\}^{\mathbb{N}}$. Thus b is simply a surjection from $\{0, 1, \ldots, |b|-1\}$ onto $\{0, 1, \ldots, k-1\}$ satisfying the condition in Equation (5.47). Note that the full basic open set

$$[b] = \{f \in \{0, 1, \ldots, k-1\}^{\mathbb{N}} : f \upharpoonright |b| = b\}$$

of the Tychonov cube is included in $\mathcal{E}_k$. By standard representation of dense G_{δ} subsets of Tychonov cubes like $\{0, 1, \ldots, k-1\}^{\mathbb{N}}$ (see Lemma 9.34), there is an infinite block sequence (D_i) of finite subsets of $\mathbb{N}$ and for each i a surjective mapping $f_i : D_i \to \{0, 1, \ldots, k-1\}$ such that every $f \in \{0, 1, \ldots, k-1\}^{\mathbb{N}}$ that extends b and infinitely many of the f_i belong to $\mathcal{G}$. Let b_{ℓ} ($\ell \in \mathbb{N}$) be a list of all such finite approximations to elements of $\mathcal{E}_k$, each appearing with infinite repetition, and let f_i^{ℓ} ($i \in \mathbb{N}$) be the corresponding block sequence of partial maps guaranteeing the membership in $\mathcal{G}$. We build an equivalence relation $E \in \mathcal{E}_{\infty}$ by recursively deciding its restrictions $E \upharpoonright n_{\ell}$ ($\ell \in \mathbb{N}$) for some strictly increasing sequence (n_{ℓ}) of positive integers as follows. At odd stages we choose $E \upharpoonright n_{2\ell+1}$ in a way that ultimately guarantee that E has infinitely many infinite classes. At even stages, if b_{ℓ} happens to be coarser than $E \upharpoonright |b_{\ell}|$, we choose a finite partial mapping $f_{i_{\ell}}^{\ell}$ from the family of functions f_i^{ℓ} ($i \in \mathbb{N}$) associated with b_{ℓ} such that $D_{\ell} = \text{dom}(f_{i_{\ell}}^{\ell})$ lies above $n_{2\ell-1}$. Put $n_{2\ell} = \max(D_{\ell}) + 1$ and define an extension $E \upharpoonright n_{2\ell}$ of $E \upharpoonright n_{2\ell-1}$ that for each $j < k$ makes each point of the preimage $f_{i_{\ell}}^{-1}(j)$ equivalent to the minimal point of the preimage $b_{\ell}^{-1}(j)$ and makes no other commitments (subject of course to $E \upharpoonright n_{2\ell}$ being an equivalence relation).

Consider an arbitrary equivalence relation $F \in \mathcal{E}_k | E$, i.e., an equivalence relation that is coarser than E and that has exactly k classes. Let $p_F(0), p_F(1), \ldots, p_F(k-1)$ be the list of the minimal elements of the classes of F. Let $n = p_F(k-1) + 1$ and let b be the surjective map from n onto k that is constant on each class of $F \upharpoonright n$ and satisfies the requirement in Equation (5.47). Recall that the full equivalence relation F is identified with a surjective map $h_F : \mathbb{N} \to \{0, 1, \ldots, k-1\}$ satisfying 5.47. Then by our construction of E, the map h_F extends b as well as $f_{i_\ell}^\ell$ for every ℓ such that $b_\ell = b$. It follows that h_F extends b and infinitely many of the partial mappings f_i^ℓ ($i \in \mathbb{N}$) and that it therefore belongs to the set $\mathcal{G}$. From this, we conclude that the restriction of our coloring c to $\mathcal{E}_k | E$ is continuous, and so we are done with first part of the proof. □

We finish the discussion of the dual Ramsey theorem with the finite version of this result.

Theorem 5.73 (Finite Dual Ramsey Theorem) *For every triple k, ℓ and n of positive integers, there is a positive integer m such that for every ℓ-coloring of the family $\mathcal{E}_k(m)$ of all equivalence relations on $m = \{0, 1, \ldots, m-1\}$ with exactly k classes there is an equivalence relation E on m with exactly n classes such that $\{F \in \mathcal{E}_k(m) : F \text{ is coarser than } E\}$ is monochromatic.*

Proof. Suppose that for some triple k, ℓ, and n such an m cannot be found. So, for each $m \in \mathbb{N}$, we can fix

$$c_m : \mathcal{E}_k(m) \to \{0, \ldots, \ell - 1\} \tag{5.48}$$

such that no equivalence relation E on m with exactly n classes has the property that the set $\{F \in \mathcal{E}_k(m) : F \text{ is coarser than } E\}$ is monochromatic. Define $c : \mathcal{E}_{k+1} \to \{0, \ldots, \ell - 1\}$ by

$$c(F) = c_m(F \upharpoonright m), \tag{5.49}$$

where m is the largest integer that is equal to the minimum of some class of F. Clearly, c is a continuous map. By the Dual Ramsey Theorem there is an $E \in \mathcal{E}_\infty$ such that $\mathcal{E}_{k+1} | E$ is monochromatic. Let m be the $(n+1)$st member of $p(E)$, the set of all integers that are minimal members of various classes of E. Then $E \upharpoonright m$ has exactly n classes and the set

$$\{F \in \mathcal{E}_k(m) : F \text{ is coarser than } E \upharpoonright m\}$$

is monochromatic relative to the coloring c, which is a contradiction. This finishes the proof. □

Fix a finite (possibly) empty set L disjoint from $\mathbb{N}$. Let $\mathcal{E}_\infty^L = \mathcal{E}_\infty^L(\mathbb{N})$ be the collection of all *equivalence relations* E on $L \cup \mathbb{N}$ whose quotients $(L \cup \mathbb{N})/E$ are infinite and which have the property that $E \upharpoonright L$ is the equality relation on L. We order $\mathcal{E}_\infty^L$ as before, $E \leq F$ if E is coarser than F, meaning that every class of E is the union of classes of F.

Given $E \in \mathcal{E}_\infty^L$, each equivalence class $[x]_E$ of E that is disjoint from L has a minimal representative. Let $p(E)$ be the set of all minimal representatives of classes of E that are disjoint from L. Let $\{p_n(E)\}_{n=0}^{\infty}$ be the increasing enumeration of $p(E)$. Let $r_0(E) = \emptyset$ and for $n \in \mathbb{N}$, let

$$r_{n+1}(E) = E \upharpoonright (L \cup \{0, 1, \ldots, p_{n+1}(E) - 1\}). \tag{5.50}$$

Each approximation $a \in \mathcal{AE}_{\infty}^L$ has its *length* $|a|$ defined to be 0 if $a = \emptyset$ or the integer $n > 0$ such that $a = r_n(E)$ for some $E \in \mathcal{E}_{\infty}^L$ (or equivalently, the number of equivalence classes of a that are disjoint from L) and its *domain*, the set $L \cup \{0, 1, \ldots, p_{|a|}(E) - 1\}$, where E is any member of $\mathcal{E}_{\infty}^L$ such that $a = r_{|a|}(E)$. The relation $\leq$ of $\mathcal{E}_{\infty}^L$ still allows a natural finitization $\leq_{\mathrm{fin}}$ on $\mathcal{AE}_{\infty}$ satisfying A.2 and A.3: $a \leq_{\mathrm{fin}} b$ if $\mathrm{dom}(a) = \mathrm{dom}(b)$ and a is coarser than b. Working as in the proof of Lemma 5.69 (which amounts to using Theorem 2.37 at the crucial point) one shows that $(\mathcal{E}_{\infty}^L, \leq, r)$ satisfies A.4 as well.

Lemma 5.74 *Let $[a, E]$ be a nonempty basic set of the space $(\mathcal{E}_{\infty}^L, \leq, r)$ and let n be the length of a. Let $\mathcal{O}$ be a family of members of $\mathcal{AE}_{\infty}^L$ of length $n + 1$. Then there is $F \in [a, E]$ such that $r_{n+1}[a, F]$ is either contained in or disjoint from $\mathcal{O}$.* $\square$

This establishes the following variation on Theorem 5.70.

Theorem 5.75 (Carlson-Simpson) *For every finite alphabet L, the triple $(\mathcal{E}_{\infty}^L, \leq, r)$ forms a topological Ramsey space.*

Note that Theorem 5.70 is the case $L = \emptyset$ of Theorem 5.75. For a positive integer k, let $\mathcal{E}_k^L = \mathcal{E}_k^L(\mathbb{N})$ be the collection of all equivalence relations E on $L \cup \mathbb{N}$ whose restrictions to L is the equality relation of L and that have exactly k classes disjoint from L.

Corollary 5.76 *For every finite Borel coloring c of $\mathcal{E}_k^L$, there is $E \in \mathcal{E}_{\infty}^L$ such that c is constant on the set $\mathcal{E}_k^L | E$ of all members of $\mathcal{E}_k^L$ that are coarser than E.*

Remark 5.77 It should be noted that Corollary 5.76 is true for the wider class of Baire-measurable colorings, but the proof now is a bit more difficult than the proof of Corollary 5.72.

We also have a finitary version of this result, which appears to be just a re-formulation of the well-known Graham-Rothschild n-parameter set theorem.

Corollary 5.78 (Graham-Rothschild) *For every finite alphabet L, for every triple k, l, m of positive integers such that $k < m$, there is a positive integer n such that for every coloring[6] $c : \mathcal{E}_k^L(n) \to l$ there is $E \in \mathcal{E}_m^L(n)$ such that c is constant on the set of all equivalence relations of $\mathcal{E}_k^L(n)$ that are coarser than E.*

---

[6]Recall the notation used here, $n = \{0, 1, \ldots, n - 1\}$ and $l = \{0, 1, \ldots, l - 1\}$.

One of the reasons for presenting the Ramsey space $(\mathcal{E}_\infty^L, \leq, r)$ of Theorem 5.75 is that it admits a natural reformulation in terms of variable-words similar to the Ramsey spaces of Section 5.3 above. To see this, let $W_\infty^L(\omega)$ be the collection of all words x of length ω over $L \cup \{v_n : n \in \mathbb{N}\}$ in which each variable occurs in such a way that if $m < n$, then the first occurrence of v_m is before the first occurrence of v_n. Elements of $W_\infty^L(\omega)$ are are called ω-variable-words. For $x, y \in W_\infty^L(\omega)$, set $x \leq y$ if x is obtained from y by substituting variables with letters or variables, keeping of course the rule about occurrences of variables. Let $r_0(x) = \emptyset$, and let

$$r_{n+1}(x) = x \restriction m, \quad \text{where } m = \min\{l : x(l) = v_n\}. \tag{5.51}$$

It should be clear that is just another way of representing the space of Theorem 5.75, so we have the following result, worth noting.

Theorem 5.79 (Carlson-Simpson) *For every finite alphabet L the space $(W_\infty^L(\omega), \leq, r)$ of all ω-variable-words is a topological Ramsey space.*

Given a positive integer k, if by $W_k^L(\omega)$,, we denote the collection of all infinite k-variable-words over $L \cup \{v_i : i < k\}$, i.e., words of length ω in which each variable occurs in such a way that if $i < j$ then the first occurrence of v_i is before the first occurrence of v_j, then we have the following corollary.

Corollary 5.80 *For every finite alphabet L, every positive integer k, and every finite Borel coloring c of $W_k^L(\omega)$, there is an ω-variable-word x such that c is monochromatic on the set $W_k^L(\omega)|x$ of all k-variable reductions of the word x.*

Finally, for positive integers k, n, let $W_k^L(n)$ denote the collection of all words of length n over $L \cup \{v_i : i < k\}$ in which each variable occurs in such a way that if $i < j$, then the first occurrence of v_i is before the first occurrence of v_j. Then we have the following formulation of the Graham-Rothschild n-parameter set theorem.

Corollary 5.81 (Graham-Rothschild) *Let L be a finite alphabet and let k, l, m be positive integers such that $k < m$. Then there is an integer n such that for every coloring $c : W_k^L(n) \to l$ there is $x \in W_m^L(n)$, an m-variable-word of length n, such that c is constant on the set $W_k^L(n)|x$ of all k-variable reductions of the word x.*

We finish this section by mentioning two Ramsey classification results for Borel equivalence relations in the context of dual Ramsey theory. Recall Theorem 1.58 which gives a Ramsey classification of smooth Borel equivalence relations for the space $\mathbb{N}^{[\infty]}$ of all infinite subsets of $\mathbb{N}$. This should be compared with our next result which gives a Ramsey classification of the same class of Borel equivalence relations relative to the dual Ramsey space considered in this section.

Theorem 5.82 (Prömel-Simpson-Voigt) *For every smooth Borel equivalen-ce relation E on the powerset $\mathcal{P}(\mathbb{N})$ there is an infinite sequence (M_k) of pairwise disjoint nonempty subsets of $\mathbb{N}$ such that one of the following three conditions holds for all nonempty $X, Y \subseteq \mathbb{N}$,*

(1) $(\bigcup_{k \in X} M_k) \ E \ (\bigcup_{k \in Y} M_k)$ *if and only if $X = X$,*

(2) $(\bigcup_{k \in X} M_k) \ E \ (\bigcup_{k \in Y} M_k)$ *if and only if $X = Y$,*

(3) $(\bigcup_{k \in X} M_k) \ E \ (\bigcup_{k \in Y} M_k)$ *if and only if $\min(X) = \min(Y)$.*

Theorem 5.83 (Prömel-Simpson-Voigt) *For every Borel map φ from the power set $\mathcal{P}(\mathbb{N})$ into some metric space, there exist two sets $M \subseteq N \subseteq \mathbb{N}$ with $N \setminus M$ infinite such that the restriction of φ to the corresponding complete Boolean sublattice $\{X : M \subseteq X \subseteq N\}$ isomorphic to $\mathcal{P}(\mathbb{N})$ is a continuous map that is either constant or one-to-one.*

5.7 A RAMSEY SPACE OF INFINITE-DIMENSIONAL VECTOR SUBSPACES OF $F^{\mathbb{N}}$

Let F be a fixed finite field. All matrices considered in this section are matrices over F, i.e., mappings of the form

$$A : n \times m \to F,$$

where $m, n \in \mathbb{N} \cup \{\mathbb{N}\}$. The *$i$th column* of A is the mapping $A_i : m \to F$ defined by

$$A_i(j) = A(i, j).$$

The *jth row* of A is the mapping $A^j : n \to F$ defined by

$$A^j(i) = A(i, j).$$

We say that an $n \times m$-matrix is in *reduced echelon form* if

(1) every column of A has a nonzero entry,

(2) if p_i is the first nonzero entry of A_i, then $A_i(p_i) = 1$ and further $A(j, p_i) = 0$ for $j \neq i$,

(3) $p_i < p_j$ whenever $i < j$.

Let $\mathcal{M}_\infty$ be the collection of all reduced echelon $\mathbb{N} \times \mathbb{N}$-matrices. For $A \in \mathcal{M}_\infty$ and $n \in \mathbb{N}$, set

$$p_n(A) = \min\{j : A_n(j) \neq 0\}. \tag{5.52}$$

Let $p(A) = \{p_n(A) : n \in \mathbb{N}\}$. Thus the sequence $(p_n(A))$ gives us an increasing enumeration of the infinite set $p(A)$. This leads us to the notion of a sequence $(r_n(A))$ of *finite approximations* to A:

$$r_0(A) = \emptyset \text{ and } r_{n+1}(A) = A \restriction (n \times p_n(A)). \tag{5.53}$$

Note that $\mathcal{M}_\infty$ is a closed subset of $\mathcal{AM}_\infty^{\mathbb{N}}$.

For $A, B \in \mathcal{M}_\infty$, we put

$$A \le B \text{ iff } (\forall i \in \mathbb{N}) \ A_i \in \overline{\text{span}}\{B_n : n \in \mathbb{N}\},$$

or in other words, every column of A belongs to the closure taken in $F^{\mathbb{N}}$ of the linear span of the set $\{B_n : n \in \mathbb{N}\}$ of columns of B. Thus, if we identify matrices from $\mathcal{M}$ with closed linear subspaces the ordering $\le$ between them corresponds to the inclusion ordering on the set of all closed linear subspaces of $F^{\mathbb{N}}$. We extend this quasi-ordering on the set $\mathcal{AM}_\infty$ of finite approximations in the most natural way, $a \le_{\text{fin}} b$ if and only if

(4) a and b have the same number of rows, say, n and

(5) every column of a belongs to the subspace of F^n generated by the columns of b.

Note that $\le_{\text{fin}}$ is indeed a finitization of $\le$, i.e., that A.2 is satisfied.

Lemma 5.84 *Suppose that $A \le B$ for some $A, B \in \mathcal{M}_\infty$. Then $p(A) \subseteq p(B)$, and if $I = \{i : p_i(B) \notin p(A)\}$ then for every $\bar{n} \in \mathbb{N}$ there exist a unique $n \in \mathbb{N}$ and a sequence $\lambda_i (i \in I \backslash n)$ of elements of F such that $A_{\bar{n}} = B_n + \sum_{i \in I \backslash n} \lambda_i B_i$.*

Proof. That the representation $A_{\bar{n}} = \sum_{i=0}^{\infty} \lambda_i B_i$ exists and that it is unique follows easily from the definitions. Note that if $n = \min\{i : \lambda_i \ne 0\}$, then $p_{\bar{n}}(A) = p_n(B)$ and $\lambda_n = 1$. Note that $i > n$ and $p_i(B) \in p(A)$ imply that $\lambda_i = 0$, or else we would contradict condition (3) for the matrix A. $\square$

Clearly, the conclusion of Lemma 5.84 also holds for the relation $a \le_{\text{fin}} b$ between members of $\mathcal{AM}_\infty$. From this, one easily concludes that A.3(1) holds for $(\mathcal{M}_\infty, \le, r)$, i.e., that $\text{depth}_B(a) \ge 0$ implies $[a, B] \ne \emptyset$ for every $a \in \mathcal{AM}_\infty$ and $B \in \mathcal{M}_\infty$. The following lemma gives us the other part of A.3.

Lemma 5.85 *Suppose that $\ell = \text{depth}_B(a) \ge 0$ and that $A \in [a, B]$. Then there is an $A' \in [\ell, B]$ such that $\emptyset \ne [a, A'] \subseteq [a, A]$.*

Proof. Let $\ell = \text{depth}_B(a)$ and let $\bar{\ell} = \text{depth}_A(a)$. Then $a = r_{\bar{\ell}}(A)$ and $0 \le \bar{\ell} \le \ell$. Moreover, $p_{\bar{\ell}-1}(A) = p_{\ell-1}(B) = m$ and a is an $(\bar{\ell} - 1) \times m$-matrix. If $\ell = 0$ there is nothing to prove, so let us assume that $\ell > 0$. Let $I = \{i : p_i(B) \notin p(A)\}$. We define A' by letting its columns A'_n be determined as follows. Let $A'_{(\ell-1)+n} = A_{(\bar{\ell}-1)+n}$ for $n \ge 0$. If $n < \ell - 1$ and $n \in I$, let $A'_n = B_n$. If $n < \ell - 1$ and $n \notin I$ then by Lemma 5.84 there is an $\bar{n} < \bar{\ell} - 1$ and a sequence $(\lambda_i)_{i \in I \backslash n}$ such that $A_{\bar{n}} = B_n + \sum_{i \in I \backslash n} \lambda_i B_i$. Let

$$A'_n = B_n + \sum_{i \in I \backslash (\ell-1)} \lambda_i B_i. \tag{5.54}$$

Clearly, $A' \in [\ell, B]$. It remains to check that every $C \in [a, A']$ belongs to $[a, A]$. This will follow if we can show that $C \leq A$. Let

$$J = \{i : p_i(A') \notin p(C)\}.$$

Note that

$$J \cap [0, \ell - 1) = I \cap [0, \ell - 1).$$

By Lemma 5.84, the relation $C \leq A'$ gives us that for every $\bar{n} \in \mathbb{N}$ there are a unique $n \in J$ and a sequence $\gamma_i (i \in J \backslash n)$ of elements of F such that $C_{\bar{n}} = A'_n + \sum_{i \in J \backslash n} \gamma_i A'_i$. If $n \geq \ell - 1$, by our definition of A', this equation is the same as the equation $C_{\bar{n}} = A_n + \sum_{i \in J \backslash n} \gamma_i A_i$, which we want. So let us assume $n < \ell - 1$. Going back to the definition of A'_n, we get that

$$C_{\bar{n}} = A_{\bar{n}} + \sum_{i \in I \cap (n, \ell-1)} (-\lambda_i) B_i + \sum_{i \in I \cap (n, \ell-1)} \gamma_i A'_i + \sum_{i \in J \backslash (\ell-1)} \gamma_i A_i. \quad (5.55)$$

Referring to the definition of the A'_i, we conclude that

$$C_{\bar{n}} = A_{\bar{n}} + \sum_{i \in I \cap (n, \ell-1)} \xi_i B_i + \sum_{i \in J \backslash (\ell-1)} \gamma_i A_i \quad (5.56)$$

for some scalars $\xi_i \in F$ $(i \in I \cap (n, \ell - 1))$. Taking the restriction of each column appearing in this equation to m, we get that

$$C_{\bar{n}} \upharpoonright m = A_{\bar{n}} \upharpoonright m + \sum_{i \in I \cap (n, \ell-1)} \xi_i B_i \upharpoonright m. \quad (5.57)$$

Since $C_{\bar{n}} \upharpoonright m = A_{\bar{n}} \upharpoonright m$ and since the columns $B_i \upharpoonright m$ $(i < \ell - 1)$ are linearly independent, we get that $\xi_i = 0$ for all $i \in I \cap (n, \ell - 1)$. It follows that $C_{\bar{n}} = A_{\bar{n}} + \sum_{i \in J \backslash (\ell-1)} \gamma_i A_i$. This finishes the verification of $C \leq A$ and the proof of the lemma. $\square$

It remains to check that $(\mathcal{M}_\infty, \leq, r)$ satisfies A.4.

Lemma 5.86 *Suppose $[a, B]$ is a nonempty basic open set, ℓ is the length of a and $\mathcal{O}$ is a set of approximations of length $\ell + 1$. Then there is an $A \in [\mathrm{depth}_B(a), B]$ such that $r_{\ell+1}[a, A]$ is either contained in or is disjoint from $\mathcal{O}$.*

Proof. By A.3, we may assume that $a = r_\ell(B)$, i.e., that $[a, B] = [\ell, B]$. We shall apply Theorem 2.35 for the alphabet $L = F^\ell$. A word $w \in W_L$ of length k is seen as an $\ell \times k$ matrix (w_{ij}). To any such word, we associate an $\ell \times m$-matrix $b = b(w)$, where $m = p_{\ell+k}(B)$, by letting

$$b_i = (B_i \upharpoonright m) + \sum_{j < k} w_{ij} (B_{\ell+j} \upharpoonright m) \quad (5.58)$$

for $i < \ell$. Note that any such $b(w)$ belongs to $r_{\ell+1}[\ell, B]$, i.e., is equal to $r_{\ell+1}(A)$ for some $A \in [\ell, B]$. Conversely, every $r_{\ell+1}(A)$ for $A \in [\ell, B]$ has the form $b(w)$ for some $w \in W_L$. The w is determined as follows using Lemma

5.84. The length of w is the integer k such that $p_\ell(A) = p_{\ell+k}(B)$ and the w_{ij} is equal to the jth scalar appearing in the unique representation

$$A_i = B_i + \sum_{j \in I \setminus i} \lambda_{ij} B_j. \tag{5.59}$$

Note that $[\ell, \ell+k] \subset I$, so we have that $w_{ij} = \lambda_{i(\ell+j)}$ for all $i < \ell$ and $j < k$. Consider the coloring of W_L by the set $\mathcal{O}^* = \{w \in W_L : b(w) \in \mathcal{O}\}$ and its complement. Applying Theorem 2.37, we obtain a word t and an infinite sequence (x_n) of left variable words over L such that finite concatenations of the form $t^\frown x_0[\lambda_0]^\frown \ldots ^\frown x_n[\lambda_n]$, where $\lambda_i \in L$ $(i \le n)$, are monochromatic relative to this coloring. We use (x_n) to define $A \in [\ell, B]$ as follows. We first let x denote the infinite multi variable-word $t^\frown x_0[v_0]^\frown \ldots ^\frown x_n[v_n]^\frown \ldots$ and use the notation x_{ij} to denote the ith coordinate of the letter λ that occupies the jth place in x, provided of course $\lambda \in L$. Let J be the set of all positions of x occupied by some letter, and for $n \in \mathbb{N}$, let I_n be the set of positions of x occupied by the variable v_n. For $i < \ell$, let

$$A_i = B_i + \sum_{j \in J} x_{ij} B_{\ell+j}, \tag{5.60}$$

and for $n \ge 0$,

$$A_{\ell+n} = \sum_{j \in I_n} B_{\ell+j}. \tag{5.61}$$

To show that this A satisfies the conclusion of the lemma it suffices to verify that every $b \in r_{\ell+1}[\ell, A]$ is of the form $b(w)$ for some $w \in W_L$ that is equal to some finite concatenation of the form $t^\frown x_0[\lambda_0]^\frown \ldots ^\frown x_n[\lambda_k]$, where $\lambda_i \in L$ $(i \le k)$. To see this, pick $C \in [\ell, A]$ and let $b = r_{\ell+1}(C)$. Let $m = p_\ell(C)$, then $m = p_{\ell+\bar{k}}(A)$ for some $\bar{k} > 0$. Note that $p_{\ell+\bar{k}}(A) = p_{\ell+k}(B)$, where k is the length of the word $t^\frown x_0^\frown \ldots ^\frown x_{\bar{k}-1}$. So reading $C \le A \le B$ via Lemma 5.84, we infer that for all $i < \ell$,

$$C_i = B_i + \sum_{i \in J} x_{ij} B_{\ell+j} + \sum_{n < k} \lambda_{in} \left(\sum_{j \in I_n} B_{\ell+j} \right). \tag{5.62}$$

For $n < k$, let λ_n denote the element $(\lambda_{in})_{i<\ell}$ of our alphabet $E^\ell = L$. Let $w = t^\frown x_0[\lambda_0]^\frown \ldots ^\frown x_{k-1}[\lambda_{k-1}]$. Tracing back the definitions, one checks that $C \restriction (\ell \times m) = b(w)$ which was to be shown. $\square$

Applying the Abstract Ellentuck Theorem, we get the following result.

Theorem 5.87 (Carlson) $(\mathcal{M}_\infty, \le, r)$ *is a topological Ramsey space.*

For $A \in \mathcal{M}_\infty$, let

$$V(A) = \overline{\text{span}}\{A_i : i \in \mathbb{N}\}, \tag{5.63}$$

i.e., the closed linear span of the set of columns of A in the space $F^\mathbb{N}$ equipped with the usual product topology. A routine argument shows that

$$A \mapsto V(A)$$

is a bijection between $\mathcal{M}_\infty$ and the collection of all closed infinite-dimensional subspaces of $F^\mathbb{N}$. Moreover, by the definition of $\leq$, we have that $A \leq B$ iff $V(A) \subseteq V(B)$. The most natural topology on the collection $\mathcal{V}_\infty(F)$ of all closed infinite-dimensional subspaces of $F^\mathbb{N}$ is the Vietoris hyperspace topology. In fact, it can easily be seen that $A \mapsto V(A)$ is a homeomorphism between $\mathcal{M}_\infty(F)$ and $\mathcal{V}_\infty(F)$, provided we take the metric topology on $\mathcal{M}_\infty(F)$ induced from $\mathcal{A M}_\infty(F)^\mathbb{N}$. Then Theorem 5.87 has the following reformulation.

Corollary 5.88 *For every finite Souslin-measurable partition of the space $\mathcal{V}_\infty(F)$ of all closed infinite-dimensional subspaces of $F^\mathbb{N}$, there is a $V \in \mathcal{V}_\infty(F)$ such that the collection of all closed infinite-dimensional subspaces of V is monochromatic.*

It is also worth pointing out the finite-dimensional version of this result.

Theorem 5.89 (Voigt) *For every positive integer k and every finite Baire-measurable coloring of the space $\mathcal{V}_k(F)$ of all k-dimensional subspaces of $F^\mathbb{N}$, there is $V \in \mathcal{V}_\infty(F)$ such that the collection of all k-dimensional subspaces of V is monochromatic.*

Proof. Clearly, it suffices to prove the theorem by working with the collection $\mathcal{M}_k = \mathcal{M}_k(F)$ of all reduced echelon $k \times \mathbb{N}$ matrices with no zero columns, rather than the collection $\mathcal{V}_k(F)$. The restriction mapping $\pi_k : \mathcal{M}_\infty \longrightarrow \mathcal{M}_k$ defined by

$$\pi_k(A) = A \restriction (k \times \mathbb{N}) \tag{5.64}$$

is continuous and onto. So if the given coloring of $\mathcal{M}_k$ is Souslin-measurable rather than just Baire-measurable, we get the desired conclusion using Carlson's theorem. Since every finite Baire-measurable coloring of $\mathcal{M}_k$ is continuous on a dense G_δ-subset G of $\mathcal{M}_k$, it suffices to produce a matrix $B \in \mathcal{M}_\infty$ such that $\pi_k(A) \in G$ for all $A \leq B$. Consider a finite reduced echelon $k \times m_i$-matrix $a : k \times m_i \to F$ with no zero rows. Then

$$\{M \in \mathcal{M}_k : M \restriction k \times m = a\} = \{X \in 2^{k \times \mathbb{N}} : X \restriction k \times m_0 = a_0\}. \tag{5.65}$$

By a standard fact about dense G_σ-subsets of the Cantor cube (see Lemma 9.34), for every such matrix a, we can fix an infinite sequence (x_i^a) of finite mappings $x_i^a : k \times J_i^a \to F$, where (J_i^a) is a strictly increasing sequence of finite intervals of $\mathbb{N}$, such that any $M \in \mathcal{M}_k$ that extends a as well as infinitely many of the x_i^a belongs to G. As in the proof of Corollary 5.72, we perform a simple recursive construction of an increasing infinite sequence (b_n) of finite row-reduced echelon matrices such that $B = \bigcup_{n=0}^\infty b_n$ is a member of $\mathcal{M}_\infty$ and has the following property: for every $M \in \mathcal{M}_k$ with $M \leq B$, if m is the minimal integer such that $a = M \restriction k \times m$ is a reduced echelon matrix with no zero columns then there are infinitely many i such that M extends x_i^a. It follows that

$$\{M \in \mathcal{M}_k : M \leq B\} \subseteq G, \tag{5.66}$$

as required. This finishes the proof. □

We finish the section with a finite version of the Ramsey theorem for vector spaces.

Theorem 5.90 (Graham-Leeb-Rothschild) *For every triple k, ℓ, and n of positive integers, there is a positive integer m such that for every ℓ-coloring of the family of all k-dimensional subspaces of F^m, there is an n-dimensional subspace V of F^m such that the family of all k-dimensional subspaces of V is monochromatic.*

Proof. Suppose that for some k, ℓ, and n, such an m cannot be found. Then for each m we can fix an ℓ-coloring c_m of the family $\mathcal{V}_k(F^m)$ of all k-dimensional subspaces of F^m with no n-dimensional subspace V of F^m such that $\mathcal{V}_k(V) = \{W \in \mathcal{V}_k(F^m) : W \subseteq V\}$ is c_m-monochromatic. Define a coloring

$$c : \mathcal{M}_{k+1} \rightarrow \{0, 1, \ldots, \ell - 1\} \tag{5.67}$$

by letting

$$c(M) = c_m(V(M \upharpoonright (k \times m))),$$

where m is the first nonzero entry of the last column of M and where $V(M \upharpoonright (k \times m))$ denotes the subspace of F^m generated by the restrictions of the first k columns of M. By Voigt's theorem, there is an $A \in \mathcal{M}_\infty$ such that c is monochromatic on

$$\{M \in \mathcal{M}_{k+1} : M \leq A\}. \tag{5.68}$$

Let m be the first nonzero entry of the $(n+1)$st column of A and let V be the subspace of F^m generated by the restrictions of the first n columns of A. Tracing back the definitions, we see that $\mathcal{V}_k(V)$ is c_m-monochromatic, which is a contradiction. This completes the proof. □

NOTES TO CHAPTER FIVE

Ellentuck invented his topological Ramsey theorem in order to re-prove Silver's result that the field of Ramsey subsets of $\mathbb{N}^{[\infty]}$ is closed under the Souslin operation, since the original proof involved sophisticated metamathematical ideas. In other words, by putting a natural topology on $\mathbb{N}^{[\infty]}$ and by proving that Ramsey notions correspond to Baire-category notions, Ellentuck was able to deduce Silver's result from the classical fact that the Baire property in any topological space is preserved under the Souslin operation. It is therefore not surprising that the difficulties in the proof of the Abstract Ramsey Theorem compared to those encountered in the proof of the Abstract Ellentuck Theorem lie exactly in the preservation of the Ramsey property under the Souslin operation. Milliken's space $\text{FIN}^{[\infty]}$, appearing in his paper [77] was the first space built after that of Ellentuck [27] and also the first space that was built on the basis of a substantially different pigeon hole principle. Its power over the Ellentuck space was not fully realized until relatively recently, after Gowers's successful applications of the "block Ramsey theory" when treating some problems from Banach space geometry (see [37] and [38]). It should be noted that Gowers's block Ramsey theory has two new features not present in the classical theory, a feature with great potential for applications. The first of these new features is the "approximate metric Ramsey theory". We have already presented a result from this theory in Section 2.4 above during the course of showing that the sphere of the Banach space c_0 is oscillation stable. In this theory one has a metric on the space of all block sequences under consideration and one is interested in getting ε-close to a given color rather than inside a given color. The second feature is the "strategic Ramsey theory" where for a given finite coloring one looks for a strategy for building an infinite block sequence of a given color rather than an infinite block sequence all of whose block subsequences are monochromatic. Regarding the original Milliken's theorem, it should be noted that $FIN^{[\infty]}$ was anticipated earlier by A. D. Taylor [104], who was in need of the higher-dimensional version of Hindman's theorem for his beautiful Theorem 5.28. The extension of Taylor's classification theorem to all FIN_k's mentioned in Remark 5.40 is due to Lopez-Abad [65]. It should also be mentioned that the extension of Taylor's Ramsey classification result in another direction was done by Lefmann [63], who was able to identify all canonical equivalence relations on any finite power $FIN^{[d]}$. For example, there are 26 canonical equivalence relations on $FIN^{[2]}$. The corresponding infinite-dimensional result for Borel equivalence relations on $FIN^{[\infty]}$ is due to Klein-Spinas [56]. There are no known analogous Ramsey classification results for FIN_k when $k \geq 2$. Even the list of canonical equivalence relations on $FIN_2^{[2]}$ is unknown. Rosenthal's ℓ_1-theorem and Rosenthal's dichotomy appear in his papers [95] and [96]. A slightly weaker form of the parametrized Rosenthal's ℓ_1-theorem appears in a paper of Stern [103] as a positive answer to a question of Brunel and Sucheston. The two theorems of Spinas appear in his paper [102]. The first version of the Abstract Ellentuck Theorem appeared in a paper of Carlson-Simpson [16], using a slightly different axiomatization and different proofs. The infinite dual Ramsey theory was developed by Carlson-Simpson [15], extending the famous finite dual Ramsey theorem due to Graham-Rothschild [40]. The corresponding extension of the n-parameter-set theorem of Graham-Rothschild [40] is also due to Carlson and Simpson although its proof appeared first in the paper of Carlson [13]. This extension could perhaps be better seen in terms of a topological Ramsey space of infinite words with infinitely many variables (v_k) enumerated according to their first occurrences. The extension of Corollary 5.76 to Baire-measurable colorings is due to Prömel-Voigt [90]. The description of canonical Borel equivalence relations in the context of the Carlson-Simpson space (the space of Theorem 5.70) was given in the paper [88] of Prömel,

Simpson, and Voigt. More Ramsey classification results of this type and in the finite case can be found in the papers [113], [112], [89], and [114]. A version of the space $(W_{Lv}^{[\infty]}, \leq, r)$ was first developed by Carlson [14] in a paper that represents a major contribution to this area of Ramsey theory after those of Ellentuck and Milliken. A version of Theorem 5.44 appears in the paper of Furstenberg-Katznelson [32] although its present form is from the paper of Blass-Bergelson-Hindman [8]. The Ramsey space of all closed infinite-dimensional vector subspaces of $F^{\mathbb{N}}$ was also developed by Carlson [13] extending another famous result of finite Ramsey theory due to Graham-Leeb-Rothschild [39]. The finite-dimensional infinitary extension of the Graham-Leeb-Rothschild theorem is due to Voigt [114].

Chapter Six

Spaces of Trees

6.1 A RAMSEY SPACE OF STRONG SUBTREES

In this section, unless otherwise specified, by a *tree* we always mean a rooted finitely branching tree of some height $\leq \omega$. Given a tree T and $n \in \omega$, let $T(n)$ denote the nth *level* of T. Thus, the *height* of T is simply the minimal $n \leq \omega$ such that $T(n) = \emptyset$. For a set $A \subseteq \omega$, let $T(A) = \bigcup_{n \in A} T(n)$. Throughout most of this chapter, we let U be a *fixed* rooted finitely branching tree of height ω with no terminal nodes and we study its subtrees. Recall that a *subtree* of U is simply a subset T of U that, with the induced ordering, is a rooted tree that in general can be of finite height. A *regular subtree* of U is a subtree T with the property that every level $T(n)$ of T is a subset of some level $U(m)$ of U. We say that T is a *strong subtree* of U if T is a special kind of regular subtree of U, or more precisely, for which we can find a set $A \subseteq \omega$ such that

(str1) $T \subseteq U(A)$ and $T \cap U(n) \neq \emptyset$ for all $n \in A$,

(str2) if $m < n$ are two successive elements of the set A, then for every $s \in T \cap U(m)$, every immediate successor of s in U has *exactly one* extension in $T \cap U(n)$.

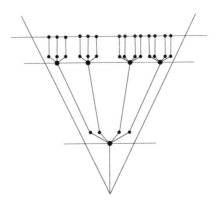

Figure 6.1 The tree U and one of its strong subtrees T of height 3.

Note that a regular subtree of a regular subtree is a regular subtree and that a strong subtree of a strong subtree is a strong subtree. Let $\mathcal{S}_\infty(U)$

denote the collection of all strong subtrees of U of infinite height, and for a positive integer k, let $\mathcal{S}_k(U)$ denote the collection of all strong subtrees of U of height k. For $T \in \mathcal{S}_\infty(U)$ the sequence $(r_n(T))$ of finite approximations (restrictions) is defined as follows:

$$r_n(T) = T \upharpoonright n \left(= \bigcup_{m<n} T(m) \right). \tag{6.1}$$

Thus, the set of finite approximations to elements of $\mathcal{S}_\infty(U)$ is the set

$$\mathcal{S}_{<\infty}(U) = \bigcup_{n=0}^{\infty} \mathcal{S}_n(U) \tag{6.2}$$

of strong subtrees of U of finite heights. Note that $\mathcal{S}_\infty(U)$ becomes a closed subset of the Tychonov product $\prod_{n=0}^{\infty} \mathcal{S}_n(U)$ when $T \in \mathcal{S}_\infty(U)$ is identified with its sequence $(r_n(T))$ of finite approximations. The inclusion order on $\mathcal{S}_\infty(U)$ is finitized as follows:

$$a \subseteq_{\text{fin}} b \text{ iff } a = b = \emptyset \text{ or } a \subseteq b \text{ and } a(\max) \subseteq b(\max), \tag{6.3}$$

where $a(\max)$ and $b(\max)$ denote the maximal levels of the tree a and b, respectively. Finitized this way, the space $(\mathcal{S}_\infty(U), \subseteq, r)$ is easily seen to satisfy the requirements A.1, A.2, and A.3 from Section 4.2. So, it remains to check the crucial requirement A.4 for this space.

Lemma 6.1 *Let $[n, T]$ be a basic open set of $(\mathcal{S}_\infty(U), \subseteq, r)$ and let $\mathcal{O}$ be a subset of $\mathcal{S}_{n+1}(U)$. Then there is an $S \in [n, T]$ such that $r_{n+1}[n, S]$ is either included in or disjoint from $\mathcal{O}$.*

Proof. Let $u_0, \ldots, u_{d-1}$ be a 1-1 enumeration of the set of nodes of U that happen to be immediate successors of some node of the level $T(n-1)$ of T. (If $n = 0$, we put $d = 1$ and $u_0 = \text{root}(U)$). For $i < d$, let

$$T_i = \{t \in T : u_i \le t\}. \tag{6.4}$$

Note that every $\vec{t} = (t_0, \ldots, t_{d-1})$ from

$$\bigcup_{k=0}^{\infty} \prod_{i<d} T_i(k) \tag{6.5}$$

determines the strong subtree

$$b(\vec{t}) = (T \upharpoonright n) \cup \{t_0, \ldots, t_{d-1}\} \tag{6.6}$$

of U of height $n + 1$. Define

$$\mathcal{O}^* = \left\{ \vec{t} \in \bigcup_{k=0}^{\infty} \prod_{i<d} T_i(k) : b(\vec{t}) \in \mathcal{O} \right\}. \tag{6.7}$$

By the strong subtree version of the Halpern-Läuchli theorem (see Theorem 3.2 in Section 3.1), there is a sequence S_i $(i < d)$ of trees and a strictly increasing infinite sequence $(l_n)_n$ of nonnegative integers such that

(1) S_i is a strong subtree of T_i for all $i < d$,

(2) $S_i(n) \subseteq T_i(l_n)$ for all $i < d$ and $n < \omega$,

(3) $\bigcup_{n=0}^{\infty} \prod_{i<d} S_i(n)$ is a subset of $\mathcal{O}^*$ or its complement.

Let

$$S = (T \upharpoonright n) \cup \bigcup_{i<d} S_i. \tag{6.8}$$

Then S is a strong subtree of U that belongs to the basic set $[n, T]$ and such that $r_{n+1}[n, S]$ is included either in $\mathcal{O}$ or its complement, depending on which of the two alternatives of (3) hold. $\square$

Applying the Abstract Ellentuck Theorem to $(\mathcal{S}_\infty, \subseteq, r)$, we obtain the following result.

Theorem 6.2 (Milliken)   *For every rooted finitely branching tree U of height ω with no terminal nodes, the triple $(\mathcal{S}_\infty(U), \subseteq, r)$ forms a topological Ramsey space.*

Corollary 6.3 (Ellentuck) $(\mathbb{N}^{[\infty]}, \subseteq, r)$ *is a Ramsey space.*

Proof. Take U to be ω with the usual ordering. $\square$

Corollary 6.4 *For every rooted finitely branching tree U of height ω with no terminal nodes and for every finite Souslin-measurable coloring of the set $\mathcal{S}_\infty(U)$ there exists a strong subtree T of U such that $\mathcal{S}_\infty(T)$ is monochromatic.*

Corollary 6.5 *Let U be a rooted finitely branching tree of height ω with no terminal nodes, and let k be a positive integer. Then for every finite coloring of the set $\mathcal{S}_k(U)$ of all strong subtrees of U of height k there is a strong subtree T of U of infinite height such that $\mathcal{S}_k(T)$ is monochromatic.*

Corollary 6.6 *Let U be a rooted finitely branching tree of height ω with no terminal nodes, and let k, l, and n be a given triple of positive integers. Then there is a positive integer m such that for every l-coloring of the set $\mathcal{S}_k(U \upharpoonright m)$ of strong subtrees included in the first m levels of U, there is a strong subtree T of $U \upharpoonright m$ of height n such that $\mathcal{S}_k(T)$ is monochromatic.*

Proof. Otherwise, for each positive integer m we can fix a coloring

$$c_m : \mathcal{S}_k(U \upharpoonright m) \to l, \tag{6.9}$$

which would violate the conclusion of the corollary. Define

$$c : \mathcal{S}_{k+1}(U) \to l \tag{6.10}$$

by

$$c(T) = c_m(T \upharpoonright k), \tag{6.11}$$

where m is such that $T(k) \subseteq U(m)$. By Corollary 6.5, there is $T \in \mathcal{S}_\infty(U)$ such that c is constant on $\mathcal{S}_{k+1}(T)$. Let $S = T \upharpoonright n$ and let m be such that $T(n) \subseteq U(m)$. Then $\mathcal{S}_k(S)$ is c_m-monochromatic, a contradiction. $\square$

Corollary 6.7 *Let U be a rooted finitely branching tree of height ω with no terminal nodes. Then, for every $\mathcal{F} \subseteq \mathcal{S}_{<\infty}(U)$ there is an infinite strong subtree T of U such that either*

(a) $\mathcal{S}_{<\infty}(T) \cap \mathcal{F} = \emptyset$, *or*

(b) *for every $S \in \mathcal{S}_{\infty}(T)$ there is n such that $S \restriction n \in \mathcal{F}$.*

Proof. Color elements of $\mathcal{S}_{\infty}(U)$ according to whether they have a restriction in $\mathcal{F}$ or not. This is clearly a Borel coloring. Now apply Corollary 6.4. □

Corollary 6.8 *Let U be a rooted finitely branching tree of height ω with no terminal node and let $\mathcal{F} \subseteq \mathcal{S}_{<\infty}(U)$ be a family that contains no two subtrees one of which is a strict initial part of the other. Then, for every finite coloring of $\mathcal{F}$, there is an infinite strong subtree T of U such that $\mathcal{S}_{<\infty}(T) \cap \mathcal{F}$ is monochromatic.*

Proof. Apply the previous corollary successively to each of the colors. □

Corollary 6.9 (Galvin) *For every family $\mathcal{F}$ of finite subsets of $\mathbb{N}$ there exist infinite $B \subseteq \mathcal{F}$ such that*

(a) *either B contains no member of $\mathcal{F}$, or*

(b) *for every infinite $A \subseteq B$ there is an n such that $A \cap n \in \mathcal{F}$.*

Proof. Apply Corollary 6.7 to the tree $U = (\omega, <)$. □

Corollary 6.10 (Nash-Williams) *Suppose $\mathcal{F}$ is a family of nonempty finite subsets of $\mathbb{N}$ such that no member of $\mathcal{F}$ is a strict initial segment of some other member of $\mathcal{F}$. Then for every finite coloring of $\mathcal{F}$ there is infinite $B \subseteq \mathbb{N}$ such that $\mathcal{F}|B$ is monochromatic.*

Proof: Apply Corollary 6.8 to the tree $U = (\omega, <)$. □

6.2 APPLICATIONS OF THE RAMSEY SPACE OF STRONG SUBTREES

Fix a finitely branching rooted tree U of height ω. In this section we show how to use Milliken's space of strong subtrees to get the corresponding Ramsey theoretic results about *arbitrary* subsets of the tree U. The key to all these application is the notion of the *strong subtree envelope* of a given subset of U. This notion is most easily seen if we consider our base tree U as a downward closed finitely branching subtree of $\mathbb{N}^{<\infty}$ with no terminal nodes. Thus, root$(U) = \emptyset$, the ordering is the end-extension $\sqsubseteq$, and the height of a node $t \in U$ is simply equal to the integer $|t| \in \mathbb{N}$ such that $t : |t| \to \mathbb{N}$. For $s, t \in U$, set

$$s \wedge t = \max\{u \in U : u \sqsubseteq s \text{ and } u \sqsubseteq t\}. \tag{6.12}$$

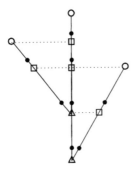

Figure 6.2 A set $A = \{\bigcirc, \bigcirc, \bigcirc\}$ its $\wedge$-closure $A^\wedge = \{\bigcirc, \bigcirc, \bigcirc, \triangle, \triangle\}$ and its embedding type, the structure $(A^\wedge, \sqsubseteq, \wedge, \bullet\bullet, \bullet\bullet, \bullet, \bullet, \bullet, \bullet)$.

The $\wedge$-*closure* of a subset A of U is the set

$$A^\wedge = \{s \wedge t : s, t \in A\}. \tag{6.13}$$

Note that $A \subseteq A^\wedge$, and that $A^\wedge$ is a rooted subtree of U.

Definition 6.11 *For $A, B \subseteq U$ we say that A and B have the same embedding type in U and write A Em B if there is a bijection $f : A^\wedge \to B^\wedge$ such that for all $s, t \in A^\wedge$,*

(a) $s \sqsubseteq t \Leftrightarrow f(s) \sqsubseteq f(t)$,

(b) $|s| < |t| \Leftrightarrow |f(s)| < |f(t)|$,

(c) $s \in A \Leftrightarrow f(s) \in B$,

(d) $t(|s|) = f(t)(|f(s)|)$ *whenever* $|s| < |t|$.

Clearly, Em is an equivalence relation on the power set of U, and its classes are called the *embedding types*, which can also be identified with the isomorphism types of the structure appearing in Figure 6.2. The black dots appearing in this structure represent the integers that code immediate successors of a triangle point and a square point that have extensions in $A^\wedge$. The square points, while not part of the structure, are important when one tries to construct the strong tree envelope that we introduce below.

For $A \subseteq U$, set

$$\|A\| = |\{|s \wedge t| : s, t \in A\}|, \tag{6.14}$$

the number of levels U which $A^\wedge$ intersects. Clearly, A Em B implies that $\|A\| = \|B\|$. For $A \subseteq U$, let the *strong subtree envelope* of A be the following subset of $\mathcal{S}_{\|A\|}(U)$:

$$\mathcal{C}_A(U) = \{T \in \mathcal{S}_{\|A\|}(U) : T \supseteq A^\wedge\}. \tag{6.15}$$

The following property of the cover functor will be the key in transferring colorings of arbitrary subsets of U into colorings of strong subtrees of U.

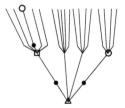

Figure 6.3 A set $A = \{\bigcirc, \bigcirc\}$, its $\wedge$-closure $A^\wedge = \{\bigcirc, \bigcirc, \triangle\}$, its embedding type $(A^\wedge, \sqsubseteq, \wedge, \bullet\bullet, \bullet)$, and one of the members of its strong subtree envelope.

Lemma 6.12 *If $A \operatorname{Em} B$ and $\mathcal{C}_A(U) \cap \mathcal{C}_B(U) \neq \emptyset$, then $A = B$.*

Proof. Choose $T \in \mathcal{C}_A(U) \cap \mathcal{C}_B(U)$ and choose a bijection $f : A^\wedge \to B^\wedge$ witnessing that $A \operatorname{Em} B$. We shall show by induction on $|t|$ that $f(t) = t$ for all $t \in A^\wedge$. First of all, note that

$$\min(A^\wedge) = \operatorname{root}(T) = \min(B^\wedge),$$

so $f(t) = t$ for $t = \operatorname{root}(A^\wedge)$. Suppose that for some $t \in A^\wedge$, the mapping f is equal to the identity on $A^\wedge \cap (U \restriction |t|)$. Since $B^\wedge$ must have a node at every level of T, condition (b) of Definition 6.11 requires that $|f(t)| = |t|$, or in other words, that t and $f(t)$ lie on the same level of T (and U). Let s be the maximal strict predecessor of t in $A^\wedge$. Then, by hypothesis, $f(s) = s$ is the maximal strict predecessor of $f(t)$ in $B^\wedge$. By Condition (d) of Definition 6.11, $t(|s|) = f(t)(|s|) = i$, or equivalently, $s^\frown i \sqsubseteq t$ and $s^\frown i \sqsubseteq f(t)$. Thus, t and $f(t)$ are two nodes of the strong subtree T of U lying on the same level and extending the same immediate successor $s^\frown i$ of s. Applying the condition (str2) from the definition of strong subtree, we conclude that we must have $t = f(t)$. This completes the proof. □

Now we are ready to state and prove the first basic result of this section.

Theorem 6.13 *Suppose U is a nonempty downward closed finitely branching subtree of $\mathbb{N}^{<\infty}$ with no terminal nodes and let A be an arbitrary subset of U. Then, for every finite Souslin-measurable coloring of the class $[A]_{\operatorname{Em}}$ of all subsets of U realizing the same embedding type as A, there is a strong subtree T of U such that the restriction of $[A]_{\operatorname{Em}}$ to T is monochromatic.*

Proof. Color each $S \in \mathcal{S}_{\|A\|}(U)$ with the color of the unique subset that belongs to $[A]_{\operatorname{Em}}$, if there is one; otherwise, S gets a color different from those appearing on $[A]_{\operatorname{Em}}$. From Lemma 6.12 we infer that this is a well-defined Souslin-measurable coloring. By Corollary 6.4, there is a strong subtree T of U such that $\mathcal{S}_{\|A\|}(T)$ is monochromatic relative to the coloring we have just defined. It follows that $[A]_{\operatorname{Em}} \restriction T$ is monochromatic relative to the given coloring of $[A]_{\operatorname{Em}}$. □

Corollary 6.14 (Milliken) *Suppose U and A satisfy the hypothesis of the previous theorem. Assume, moreover, that A is finite. Then for every finite* *coloring of the equivalence class* $[A]_{\mathrm{Em}}$ *of all subsets of* U *of the same embedding type in* U *as* A, *there is a strong subtree* T *of* U *such that the restriction* $[A]_{\mathrm{Em}}|T$ *is monochromatic.*

The rest of the section is devoted to similar partition theorems where the equivalence relation Em is relaxed at the necessary expense of shrinking the output subtree T of U. The general procedure of reducing a given coloring to a coloring of a family of strong subtrees of U and then applying Corollary 6.4 will however remain unchanged.

We now assume that our base tree U is particulary placed as a downward closed subtree of $\mathbb{N}^{<\infty}$, i.e., that it satisfies the following condition:

(1) $(\forall t \in U)(\exists d = \deg_U(t) \geq 1)(\forall i)\ t^\frown(i) \in U \Leftrightarrow i < d.$

Fix also a downward closed subtree A of U satisfying a similar condition:

(2) $(\forall t \in A)(\exists d = \deg_A(t) \geq 0)(\forall i)\ t^\frown(i) \in A \Leftrightarrow i < d.$

In other words, we are assuming that the immediate successors of a node of U or A is formed out of $t^\frown(i)$ for i running through an initial segment of non-negative integers. While we impose that $\deg_U(t) \geq 1$, or in other words, that U has no terminal nodes, the tree A can have terminal nodes, i.e., nodes of degree 0. We wish U to be rich with "copies" of A, so it is natural to impose the following restriction on U:

(3) $(\forall t \in U)(\forall s \in A)(|t| \geq |s| \Rightarrow \deg_U(t) \geq \deg_A(s)).$

Let $[A]_{\mathrm{Eb}}$ be the collection of all $B \subseteq U$ for which there is a bijection $f : A \to B$ such that

(4) $s \sqsubseteq t \Leftrightarrow f(s) \sqsubseteq f(t),$

(5) $s^\frown(i) \sqsubseteq t \Leftrightarrow f(s)^\frown(i) \subseteq f(t),$

(6) $(\forall m < n)(\forall s \in A(m))(\forall t \in A(n))|f(s)| < |f(t)|,$

(7) $(\forall m)(\forall s, t \in A(m))(s <_{\mathrm{lex}} t \to |f(s)| \leq |f(t)|).$

Thus, $[A]_{\mathrm{Eb}}$ is the equivalence class of a relation Eb on $\mathcal{P}(U)$ that is a bit less restrictive that the relation Em considered above in Definition 6.11. While mappings witnessing $A\,\mathrm{Em}\,B$ must map levels into levels, this is no longer true about Eb.

A *degree preserving* subtree of U is a nonempty (and therefore infinite) rooted subtree T of U such that:[1]

(8) $(\forall t \in T)(\forall i < \deg_U(t))(\exists! s \in \mathrm{Imsucc}_T(t))\ t^\frown(i) \subseteq s.$

Condition (3) above ensures that every degree preserving subtree T of U is also rich with Eb-copies of the tree A. We are now ready to state and prove the second basic result of this section.

---

[1] Here $\mathrm{Imsucc}_T(t)$ denotes the set of all immediate successors of t in T.

Theorem 6.15 *For U and A as above and every finite Souslin-measurable coloring of $[A]_{\mathrm{Eb}}$ there is a degree preserving subtree T of U such that $[A]_{\mathrm{Eb}}|T$ is monochromatic.*

Proof. We start with a particularly placed Eb-copy of the tree A inside $\mathbb{N}^{<\infty}$, i.e., a subtree A_0 of $\mathbb{N}^{<\infty}$ for which there is a bijection $f_0 : A \to A_0$ satisfying conditions $(4) - (7)$ above, together with the following two additional conditions:

(9) $(\forall n)|A_0 \cap \mathbb{N}^n| \leq 1$,

(10) $(\forall t \in A_0)(\forall s \notin A_0)(s \sqsubseteq t \Rightarrow s^\frown(0) \sqsubseteq t)$.

Clearly $[A_0]_{\mathrm{Em}} \subseteq [A]_{\mathrm{Eb}}$, so the Souslin-measurable coloring also acts on $[A_0]_{\mathrm{Em}}$. By Theorem 6.13, there is a strong subtree S of U such that the restriction $[A_0]_{\mathrm{Em}}|S$ is monochromatic. Let $(m_k)_k$ be the increasing enumeration of the infinite subset of $\mathbb{N}$ where levels of S appear, i.e., such that $S(k) \subseteq U(m_k)$ for all k. Choose a degree preserving subtree T of S (and therefore a degree preserving subtree of U) such that

(11) $(\forall k) |T \cap S(k)| = 1$,

(12) $(\forall k < l)(\forall t \in T \cap S(l))[t \upharpoonright m_k \notin T \Rightarrow t(m_k) = 0]$,

(13) $(\forall k < l)(\forall s \in T(k))(\forall t \in T(l))|s| < |t|$.

Choose $B \in [A]_{\mathrm{Eb}}$ such that $B \subseteq T$. Then there is a bijection $f : A \to B$ satisfying $(4) - (7)$ above. Let $g = f \circ f_0^{-1} : A_0 \to B$. We claim that g witnesses that $A_0 \operatorname{Em} B$, i.e., that g satisfies conditions $(a) - (d)$ of Definition 6.11. Since Eb-copies of A are all $\wedge$-closed, the conditions (a) and (c) follow from (4) for f and f_0. To check (b) pick $s, t \in A_0$ such that $|s| < |t|$. We may restrict ourselves to the case that $s, t \in A_0(m)$ for some m. Then $g(s), g(t) \in B(m)$ and $f^{-1}(s), f^{-1}(t) \in A(m)$. By (7) for f_0, we must have that $f_0^{-1}(s) <_{\mathrm{lex}} f_0^{-1}(t)$, and therefore by (7) for f, we have that $|g(s)| \leq |g(t)|$. However, since B is a subset of T, it has no two different nodes of the same length, so we must have $|g(s)| < |g(t)|$. The converse implication $|g(s)| < |g(t)| \Rightarrow |s| < |t|$ is also immediate and is due to the fact that A_0 has no two different nodes of the same length. When $s \sqsubseteq t$ condition (d) follows from (5) for f and f_0, so let us suppose that s and t are two nodes of A_0 such that $|s| < |t|$ but $s \not\sqsubseteq t$. By the property (b) of g, which we have just established, we know that $|g(s)| < |g(t)|$ and $g(s) \not\sqsubseteq g(t)$. Thus for some $k < l, g(t) \in T \cap S(l)$ and $g(s) \in T \cap S(k)$ and $g(t) \upharpoonright m_k \neq g(s)$, which by (11) implies $g(t) \upharpoonright m_k \notin T$, and so (12) gives us that $g(t)(|g(s)|) = 0$. By property (10) of f_0, A_0 and A, we have that $t(|s|) = 0$ as well. This checks (d) for g and finishes the proof that $A_0 \operatorname{Em} B$. Since B was an arbitrary member of $[A]_{\mathrm{Eb}}$ included in T, we have shown that

$$[A]_{\mathrm{Eb}}|T \subseteq [A_0]_{\mathrm{Em}}|T \subseteq [A_0]_{\mathrm{Em}}|S. \tag{6.16}$$

This finishes the proof.   □

Corollary 6.16 *Let U be a finitely branching rooted tree with no terminal nodes. Then for every Souslin-measurable partition of the set $U^{[\infty]}$ of all infinite chains of U, there is a degree preserving infinite subtree T of U such that $T^{[\infty]}$ is monochromatic.*

Proof. Take $A = \{0^{(n)} : n \in \mathbb{N}\}$. $\square$

Corollary 6.17 (Stern) *For every finite Souslin-measurable partition of the set $Ch_\infty(2^{<\infty})$ of all infinite chains of the complete binary tree $2^{<\infty}$, there is a perfect subtree T of $2^{<\infty}$ such that $Ch_\infty(T)$ is monochromatic.*

Corollary 6.18 (Silver) *For every finite Souslin-measurable partition of the set $\mathbb{N}^{[\infty]}$ of all infinite subsets of $\mathbb{N}$, there is an infinite $M \subseteq \mathbb{N}$ such that $M^{[\infty]}$ is monochromatic.*

Proof. Take $A = U = \{0^{(n)} : n \in \mathbb{N}\}$. $\square$

6.3 PARTITION CALCULUS ON FINITE POWERS OF THE COUNTABLE DENSE LINEAR ORDERING

In this section we give another application of the Ramsey space of strong subtrees, this time to the partition calculus of the countable dense linear ordering. It turns out that there are numerical invariants $t_k (k \geq 1)$ that characterize the Ramsey theoretic properties of the countable dense linear ordering $(\mathbb{Q}, <_\mathbb{Q})$ in a very precise sense. The numbers t_k are some sort of Ramsey degrees that measure the complexity of an arbitrary finite coloring of the set $Q^{[k]}$ of all k-element subsets of $\mathbb{Q}$ modulo, of course, restricting to $X^{[k]}$ for some appropriately chosen dense linear subordering X of $\mathbb{Q}$. It turns out that these Ramsey degrees of $\mathbb{Q}$ are most easily described if[2] one works with $\mathbb{Q} = 2^{<\infty}$ and $<_\mathbb{Q} = <_{\text{lex}}$. A k-tuple A of elements of the complete binary tree $2^{<\infty}$ determines a subtree

$$A^\wedge = \{s \wedge t : s, t \in A\} \tag{6.17}$$

which has as its *embedding type* the equivalence class $[A]_{\text{Em}}$. It turns out that order-isomorphic copies of $\mathbb{Q}$ in $2^{<\infty}$ can omit certain embedding types so one has to isolate the types that are really essential.

Definition 6.19 *A finite set $A \subseteq 2^{<\infty}$ of some size $k \geq 1$ realizes a Devlin embedding type iff:*

(i) *A is the set of terminal nodes of $A^\wedge$ that therefore has size exactly $2k - 1$.*

---

[2]Here, the $<_{\text{lex}}$ is the usual (see Appendix) lexicographical ordering of finite binary sequences defined by letting $s <_{\text{lex}} t$ if either $s \sqsubset t$ or else s and t are incomparable in the ordering $\sqsubset$ of end-extension and $s(k) < t(k)$ for k the minimal integer i such that $s(i)$ and $t(i)$ are both defined and different.

(*ii*) $|s| \neq |t|$ *for every pair* $s \neq t$ *from* $A^{\wedge}$.

(*iii*) $t(|s|) = 0$ *for all* $s, t \in A^{\wedge}$ *such that* $|s| < |t|$ *and* $s \not\sqsubseteq t$.

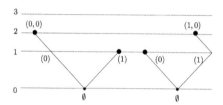

Figure 6.4 Devlin embedding types for $k = 2$.

The interest in Devlin embedding types is based on the following lemma.

Lemma 6.20 *Every strongly embedded subtree* T *of* $2^{<\infty}$ *contains an antichain* X *such that*

(*a*) $(X, <_{\text{lex}})$ *contains a dense linear ordering,*

(*b*) *Every finite subset of* X *realizes a Devlin embedding type,*

(*c*) *For every Devlin embedding type* $[A^{\wedge}]_{\text{Em}}$ *and every* $Y \subseteq X$ *order-isomorphic to* X, *there exists* $B \subseteq Y$ *such that* $[B^{\wedge}]_{\text{Em}} = [A^{\wedge}]_{\text{Em}}$.

Proof. Clearly, we may assume $T = 2^{<\infty}$. Let S be the $\wedge$-closed subtree of $2^{<\infty}$ uniquely determined by the following properties:

(1) $\text{root}(S) = \emptyset$,

(2) $|S \cap 2^{3n}| = 1$ and $S \cap 2^{3n+1} = S \cap 2^{3n+2} = \emptyset$ for all n,

(3) S is isomorphic to $2^{<\infty}$,

(4) $(\forall m)(\forall s, t \in S(m))(s <_{\text{lex}} t \Rightarrow |s| < |t|)$,

(5) $(\forall m < n)(\forall s \in S(m))(\forall t \in S(n))\, |s| < |t|$,

(6) $(\forall s \in S)(\forall t \notin S)(t \sqsubseteq s \Rightarrow t^{\wedge}(0) \sqsubseteq s)$.

Let $X = \{s^{\wedge}(0, 1) : s \in S\}$. We claim that this X has the properties $(a)-(c)$. Clearly $(X, <_{lex})$ contains a dense linear ordering and this is true for every $Y \subseteq X$ with the property that $Y^{\wedge}$ includes a perfect ($\wedge$-closed) subtree of S. The properties $(1) - (4)$ ensure that every finite $A \subseteq X$ realizes a Devlin embedding type. So it remains to show that for every non-scattered $Y \subseteq X$ and every Devlin embedding type $[A^{\wedge}]_{\text{Em}}$, there is a $B \subseteq Y$ such that $[B^{\wedge}]_{\text{Em}} = [A^{\wedge}]_{\text{Em}}$. Pick a $\wedge$-closed perfect subtree U of $Y^{\wedge}$ and a map

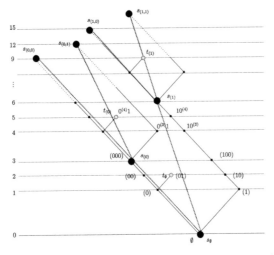

Figure 6.5 $S = \{s_f : f \in 2^{<\infty}\}$ and $X = \{t_f : f \in 2^{<\infty}\}$

$f : U \to Y$ such that $|f(s)| < |t|$ for every pair $s, t \in U$ such that $|s| < |t|$. Note that it suffices to find a finite antichain $C \subseteq U$ such that $C^\wedge$ Em $A^\wedge$, since then

$$B = \{f(s) : s \in C\} \tag{6.18}$$

is as required, i.e., $B^\wedge$ Em $A^\wedge$. By induction on the size $k = |A|$, we show that there exists an infinitely branching $\subseteq$-tree

$$W_k \subseteq \bigcup_{l=0}^{2k-1} U^{[l]} \tag{6.19}$$

such that

(7) $\emptyset \in W_k$,

(8) $F \in W_k$ implies that either $|F| = 2k - 1$ or else there exist infinitely many $s \in U$ such that $F \cup \{s\} \in W_k$.

(9) Every $F \in W_k$ of size $2k - 1$ is $\wedge$-closed and has the property that F Em $A^\wedge$.

For $k = 1$ this is clear. For the inductive step, let $s_0 = \text{root}(A^\wedge)$ and for $i = 0, 1$, let

$$A_i = \{t \in A : s_0^\wedge(i) \subseteq t\}. \tag{6.20}$$

Let $k_i = |A_i|$ for $i = 0, 1$. Let $u_0 = \text{root}(U)$ and let

$$U_i = \{u \in U : u_0^\wedge(i) \subseteq u\}. \tag{6.21}$$

By the induction hypothesis, there exist

$$W_{k_i} \subseteq \bigcup_{l=0}^{2k_i-1} U_i^{[l]} \quad (i = 0, 1) \tag{6.22}$$

satisfying $(7)-(9)$. Note that the length function $t \to |t|$ gives us an ordering $<_{A^\wedge}$ of $A^\wedge$ that shows us how $A_0^\wedge$ and $A_1^\wedge$ are intertwined to form $A^\wedge$. We use this information together with W_{k_0} and W_{k_1} to form an infinitely branching subtree

$$W_k \subseteq \bigcup_{l=0}^{2k-1} U^{[l]} \tag{6.23}$$

satisfying $(7)-(9)$. □

For a positive integer k, let t_k denote the number of different equivalence classes $[A^\wedge]_{\text{Em}}$, where $A \subseteq 2^{<\infty}$ is an antichain of size k realizing a Devlin embedding type.

Lemma 6.21 $t_k = \sum_{l=1}^{k-1} \binom{2k-2}{2l-1} t_l \cdot t_{k-l}$ *with the initial value* $t_1 = 1$.

Proof. Let $k > l \geq 1$. For $i = 0, 1$, set

$$S_i = \{s \in S : i \text{ is the first digit of } s\}. \tag{6.24}$$

Let $X_i = \{s^\frown 01 : s \in S_i\}$ for $i = 0, 1$. By the proof of the previous lemma, we have trees

$$W_l \subseteq \bigcup_{j=0}^{2l-1} S_0^{[j]} \text{ and } W_{k-l} \subseteq \bigcup_{j=0}^{2k-2l-1} S_1^{[j]} \tag{6.25}$$

satisfying $(7)-(9)$ relative to some fixed antichains $A_0, A_1 \subseteq 2^{<\infty}$ realizing Devlin embedding types, respectively. Using the trees W_l and W_{k-l}, we can build a $B \subseteq X$ such that for every $i = 0, 1$, if $B_i = X_i \cap B$, then $B_i^\wedge$ Em A_i, and such that $B_0^\wedge$ and $B_1^\wedge$ are placed with respect to the $|\cdot|$-ordering in any way we wish. Clearly, there exist $\binom{2k-2}{2l-1}$ possible ways to place $B_0^\wedge$ inside $B^\wedge$, and so this finishes the proof. □

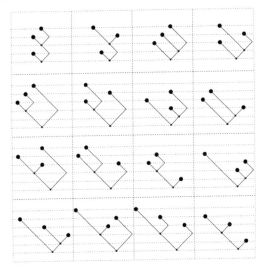

Figure 6.6 Devlin embedding types for $k = 3$.

Combining the two lemmas, we get the following

Theorem 6.22 (First Devlin Theorem) *For every positive integer k there is a coloring of $\mathbb{Q}^{[k]}$ with t_k colors, none of which can be avoided by going to an order-isomorphic copy of $\mathbb{Q}$.*

The sequence $t_1 = 1$, $t_2 = 2$, $t_3 = 16$, $t_4 = 272,\ldots$ is a well-studied sequence of numbers known as *odd tangent numbers* (because $t_k = T_{2k-1}$, where $\tan(z) = \sum_{n=0}^{\infty}(T_n/n!)z^n$). Their crucial role in the partition calculus of the countable dense linear ordering is seen from Theorem 6.22 and the following companion result.

Theorem 6.23 (Second Devlin Theorem) *For every positive integer k and every finite coloring at $\mathbb{Q}^{[k]}$, there exists a subset $Y \subseteq \mathbb{Q}$ order-isomorphic to $\mathbb{Q}$ such that $Y^{[k]}$ uses at most t_k colors.*

Proof. We way assume that we have a finite coloring

$$c : X^{[k]} \rightarrow \{0, \ldots, l-1\}, \tag{6.26}$$

where $X = \{s^\frown 01 : s \in S\}$ and $S \subseteq 2^{<\infty}$ are as in the conclusion of Lemma 6.20. For every finite antichain $A \subseteq 2^{<\infty}$ of size k realizing a Devlin type, we define

$$c_A : [A^\wedge]_{\mathrm{Em}}|S \rightarrow \{0, \ldots, l-1\} \tag{6.27}$$

by

$$c_A(B^\wedge) = c(\{s^\frown 01 : s \in B\}). \tag{6.28}$$

By successive applications of Theorem 6.13, we get a strongly embedded subtree U of S such that for every $A \subseteq 2^{<\infty}$ of size k realizing one of the

Devlin embedding types, we have that c_A is constant on $[A^\wedge]_{\text{Em}} \restriction U$. Pick an antichain $Z \subseteq U$ such that $(Z, <_{\text{lex}})$ is a dense linear ordering. Let

$$Y = \{s ^\smallfrown 01 : s \in Z\} (\subseteq X). \tag{6.29}$$

Then $(Y, <_{\text{lex}})$ is also a dense linear ordering. For $y \in Y$, let y^* be the element of Z obtained by removing the last two digits of y. Then for every $B \in Y^{[k]}$,

$$B^* = \{y^* : y \in B\} \tag{6.30}$$

realizes the same Devlin embedding type as B. It follows that the restriction of c to $Y^{[k]}$ depends only on the embedding type of the $\wedge$-closure of a given k-tuple, and this in view of Lemmas 6.20 and 6.21 gives us that the range of $c \restriction Y^{[k]}$ in $\{0, .., l-1\}$ has size at most t_k. This finishes the proof. $\qquad \square$

Corollary 6.24 (Galvin) *For every finite coloring of the set $\mathbb{Q}^{[2]}$ of all unordered pairs of rationals there is an $X \subseteq \mathbb{Q}$ order-isomorphic to $\mathbb{Q}$ such that $X^{[2]}$ uses at most two colors.*

These results about the countable dense ordering hint toward a general Ramsey theory of ultrahomogeneous structures yet to be developed. Let us exemplify this with the case of, say, *the random graph*, the unique universal countable ultrahomogeneous graph $\mathcal{R}$ characterized by the property that for every pair A and B of disjoint finite sets of vertices of $\mathcal{R}$ there is a vertex of $\mathcal{R}$ connected to every vertex from A and to none from B. Not surprisingly, the binary tree $2^{<\infty}$ (and its subsets) allows a natural realization of $\mathcal{R}$. To see this, define the edge relation E_R on $2^{<\infty}$ as follows:

$$\{s, t\} \in E_R \quad \text{iff} \quad |s| \neq |t|, \quad \text{and} \quad t(|s|) = 0 \quad \text{or} \quad s(|t|) = 0,$$

depending whether $|s| < |t|$ or $|t| < |s|$, respectively. Then, clearly, for any $V \subseteq 2^{<\infty}$ that is dense in $2^{<\infty}$, i.e., has the property

$$(\forall s \in 2^{<\infty})(\exists t \in V) \ \ s \sqsubseteq t$$

and which has at most one node of a given length realizes the random graph, or in other words, (V, E_R) is isomorphic to the random graph. In fact, one needs much less from $V \subseteq 2^{<\infty}$ to realize the random graph. It suffices to assume that V is a dense subset of some strong subtree S of $2^{<\infty}$ besides of course the assumption of not having two different nodes of the same length. Hence, Milliken's theorem (Corollary 6.14) has the following immediate consequence.

Theorem 6.25 *For every finite graph G there is a positive integer $t_{\mathcal{R}}(G)$ with the following two properties:*

(a) *There is a coloring of the set $\binom{\mathcal{R}}{G}$ of all isomorphic copies of G inside the random graph $\mathcal{R}$ into $t_{\mathcal{R}}(G)$ colors such that for every subgraph R' of $\mathcal{R}$ isomorphic to $\mathcal{R}$, the set $\binom{R'}{G}$ uses all the colors.*

(b) *For every finite coloring of $\binom{\mathcal{R}}{G}$ there is a subgraph R' of $\mathcal{R}$ isomorphic to $\mathcal{R}$ such that $\binom{R'}{G}$ uses at most $t_{\mathcal{R}}(G)$ many colors.*

Proof. Since $\mathcal{R}$ is universal, we may assume that it actually contains the graph $(2^{<\infty}, E_R)$ defined above as an induced subgraph. So, to prove the existence of the number $t_{\mathcal{R}}(G)$, we may consider the colorings of the family of all induced subgraphs of $(2^{<\infty}, E_R)$ isomorphic to our given graph G rather than $\binom{\mathcal{R}}{G}$. By Corollary 6.14, for any such finite coloring there will be a strong subtree S of $2^{<\infty}$ such that the color of the induced subgraph G' of (S, E_R) isomorphic to G will depend only on the embedding type of its vertex set. Thus, S reduces the number of colors to not more than the number of different embedding types of k-element subsets of $2^{<\infty}$, where k is the cardinality of the vertex set of G. $\qquad\square$

Remark 6.26 The number $t_{E_m}(|G|)$ of all different embedding types of $|G|$-element subsets of $2^{<\infty}$ is, as in the case of the countable dense linear ordering, too rough an upper bound to $t_{\mathcal{R}}(G)$. So, we need an analog of the Devlin embedding type in this context that would give us the exact values of $t_{\mathcal{R}}(G)$ for every finite graph G. Some results in this direction are already known. For example, it is known that $t_{\mathcal{R}}(\mathcal{K}_2) = t_{\mathcal{R}}(\overline{\mathcal{K}}_2) = 2$. However, obtaining general results in this area seems to require a substantial extension of Milliken's idea.

6.4 A RAMSEY SPACE OF INCREASING SEQUENCES OF RATIONALS

In this section we prove an infinite-dimensional analog of the result from the previous section. To state this result let, $\mathbb{Q} = (2^{<\infty}, <_{\text{lex}})$, where the $<_{\text{lex}}$ is again the usual lexicographical ordering of binary sequences defined by letting $s <_{\text{lex}} t$ if either $s \sqsubset t$ or else s and t are incomparable nodes of the tree $(2^{<\infty}, \sqsubseteq)$ and $s(k) < t(k)$ for k the minimal integer i such that $s(i)$ and $t(i)$ are both defined and different. Let $\mathbb{Q}^{[\infty]}$ denote the collection of all infinite increasing *rapidly converging* sequences of elements of $\mathbb{Q}$, i.e., infinite sequences $(s_n)_n$ of elements of $2^{<\infty}$ such that

(a) $s_n <_{\text{lex}} s_{n+1}$,

(b) $|s_{n+1} \wedge s_n| < |s_n| < |s_{n+2} \wedge s_{n+1}|$.

We view $\mathbb{Q}^{[\infty]}$ as a topological space with the topology induced from $\mathbb{Q}^\infty$.

Theorem 6.27 *For every finite Souslin-measurable coloring of $\mathbb{Q}^{[\infty]}$ there is an $H \subseteq \mathbb{Q}$ order-isomorphic to $\mathbb{Q}$ such that $H^{[\infty]}$ is monochromatic.*

This theorem can be easily deduced from Theorem 6.13 in a manner similar to the way that Corollary 6.14 was deduced from that theorem. However, it is also possible to build a Ramsey space that uses considerably less than the full Halpern-Läuchli theorem as its basic principle A.4. For convenience, we work with the complete binary tree $2^{<\infty}$ with the usual tree ordering $\sqsubseteq$ of end-extension as well as the lexicographical ordering $<_{\text{lex}}$.

Let $\mathcal{C}_\infty$ be the collection of all infinite *combs* inside the complete binary tree $2^{<\infty}$, i.e., infinite antichains $A \subseteq 2^{<\infty}$ such that

(1) $|s| \neq |t|$ for all $s, t \in A^\wedge$ with $s \neq t$,

(2) $|s| < |t \wedge u|$ for every $s, t, u \in A$ such that $|s| < |t| < |u|$,

(3) $s <_{\text{lex}} t$ whenever $s, t \in A$ and $|s| < |t|$,

(4) $t(|s|) = 0$ for all $s, t \in A$ such that $|s| < |t|$.

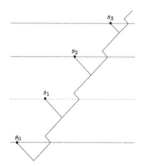

Figure 6.7 An example of a comb $\{s_0, s_1, s_2, \dots\}$.

We approximate elements of $\mathcal{C}_\infty$ as follows. Given a comb

$$A = \{s_0, s_1, s_2, \dots\}$$

enumerated increasingly according to the length function $t \mapsto |t|$, let

$$\bar{A} = A^\wedge \cup \{s_{k+1} \restriction |s_k| : k \in \mathbb{N}\},$$

and let

$$r_n(A) = \bar{A} \restriction n = \bigcup_{k<n} \bar{A}(k). \tag{6.31}$$

Thus $r_0(A) = \emptyset$ for all A, $r_1(A)$ is the root of $A^\wedge$,

$$r_2(A) = \{\text{root}(A^\wedge), s_0, s_1 \restriction |s_0|\},$$
$$r_3(A) = \{\text{root}(A^\wedge), s_0, s_1 \restriction |s_0|, s_1 \wedge s_2\}, \text{ etc.}$$

Let $\mathcal{S}_\infty = \mathcal{S}_\infty(2^{<\infty})$ be the collection at all infinite rooted strong subtrees T of $2^{<\infty}$ approximated as in Section 6.1 above,

$$r_n(T) = T \restriction n = \bigcup_{k<n} T(k). \tag{6.32}$$

We let the inclusion be the ordering on $\mathcal{S}_\infty$ as well as the relation between members of $\mathcal{C}_\infty$ and $\mathcal{S}_\infty$. The inclusion order on $\mathcal{S}_\infty$ is finitized as in the case of Milliken's space, while the inclusion as the relation between members of $\mathcal{C}_\infty$ and $\mathcal{S}_\infty$ is finitized by the following relation $\subseteq_{\mathrm{fin}}$ between the finite approximations $\mathcal{C}_{<\infty}$ to $\mathcal{C}_\infty$ and finite approximations $\mathcal{S}_{<\infty}$ to $\mathcal{S}_\infty$:

$$a \subseteq_{\mathrm{fin}} u \text{ iff } a = u = \emptyset \text{ or } a \subseteq u \text{ and } a\,(\max) \subseteq u\,(\max), \qquad (6.33)$$

where $a\,(\max)$ is the maximal level of a and $u\,(\max)$ is the maximal level of the finite strong subtree u. Checking that the finitizations satisfy $A.2$ and $A.3$ of Section 4.2 is straightforward. It remains to check $A.4$.

Lemma 6.28 *Suppose $[a, T]$ is a nonempty basic set and that a has length l in $\mathcal{C}_{<\infty}$. Let $\mathcal{O}$ be a given subset of $\mathcal{C}_{l+1}$. Then there is an $S \in [\mathrm{depth}_T(a), T]$ such that $r_{l+1}[a, S]$ is included either in $\mathcal{O}$ or its complement.*

Proof. If l is even, then the possible end-extensions of a of length $l+1$ are simply equal to a together with some singletons from the tree that dominate the lexicographically maximal node t_0 at the maximal level $a\,(\max)$ of the finite tree a. Hence, in this case the set $\mathcal{O}$ is simply a coloring of the strong subtree $T\,[\sqsupseteq t_0]$ of $2^{<\infty}$. So by the 1-dimensional version of the strong subtree reformulation of the Halpern-Läuchli theorem, we get a strong subtree $S\,[\sqsupseteq t_0]$ of $T\,[\sqsupseteq t_0]$ such that $S\,[\sqsupseteq t_0]$ is either included or is disjoint from the set

$$\{t \in T : t_0{}^\frown(0) \sqsubseteq t \text{ and } a \cup \{t\} \in Q\}. \qquad (6.34)$$

Let $\mathcal{S}$ be the element of $[\mathrm{depth}_T(a), T]$ whose cone above t_0 is equal to the just selected subtree $S\,[\sqsupseteq t_0]$.

If l is odd, then the top level $a\,(\max)$ is a singleton $\{t_0\}$ and the possible end-extensions of a of length $l+1$ are equal to the union of a and two nodes of the same height, one inside the strong subtree

$$T_0 = \{t \in T : t_0{}^\frown(0) \sqsubseteq t\} \qquad (6.35)$$

and the other inside the strong subtree

$$T_1 = \{t \in T : t_0{}^\frown(1) \sqsubseteq t\}. \qquad (6.36)$$

Hence, the set $\mathcal{O} \subseteq \mathcal{C}_{l+1}$ can be identified with a coloring of the level product $T_0 \otimes T_1$ of the trees T_0 and T_1. By the 2-dimensional version of the strong subtree reformulation of the Halpern-Läuchli theorem, we get two infinite strong subtrees $S_i \subseteq T_i$ ($i = 0, 1$) with the same level set L such that $S_0 \otimes S_1$ is monochromatic with respect to the coloring given by $\mathcal{O}$. Let $S \in [\mathrm{depth}_T(a), T]$ be formed by letting its cone above t_0 be equal to the union of S_0 and S_1, while

$$S\,[\sqsupseteq t] = T\,[\sqsupseteq t] \restriction L \qquad (6.37)$$

for every $t \in T\,(\mathrm{depth}_T(a))$ such that $t \neq t_0$. Then S satisfies the conclusion of the lemma. $\qquad \square$

The abstract infinite-dimensional Ramsey theorem now gives us the following.

Theorem 6.29 $(\mathcal{C}_\infty, \mathcal{S}_\infty, \subseteq, \subseteq, r, r)$ *is a Ramsey space.*

To see that Theorem 6.27 is an immediate consequence of this result, we argue as follows. The hypothesis gives us a finite Souslin-measurable coloring

$$c : \mathbb{Q}^{[\infty]} \to \{0, 1, \dots, l-1\}. \tag{6.38}$$

Recall that we take $\mathbb{Q} = (2^{<\infty}, <_{\text{lex}})$, so that we have the inclusion $\mathcal{C}_\infty \subseteq \mathbb{Q}^{[\infty]}$. The topology of $\mathcal{C}_\infty$ induced from $\mathbb{Q}^{[\infty]}$ $(\subseteq \mathbb{Q}^{\mathbb{N}})$ is clearly weaker than the topology of $\mathcal{C}_\infty$ induced from $\mathcal{C}_{<\infty}^{\mathbb{N}}$. It follows that every color of the induced coloring

$$c : \mathcal{C}_\infty \to \{0, 1, \dots, l-1\} \tag{6.39}$$

is $\mathcal{S}_\infty$-Ramsey. By Theorem 6.29 there is a strong subtree T of $2^{<\infty}$ such that c is constant of $[\emptyset, T]$ = the set of all elements of $\mathcal{C}_\infty$ included in T. Take some $\wedge$-closed subtree $S \subseteq T$ order-isomorphic to the complete binary tree $2^{<\infty}$, which takes at most one node from a given level of T and such that $t(|s|) = 0$ for all $s, t \in S$ with $|s| < |t|$ and $s \not\subseteq t$. Then every rapidly increasing sequence of elements of S is an infinite comb of $2^{<\infty}$, i.e.,

$$S^{[\infty]} \subseteq \mathcal{C}_\infty(T).$$

If follows that c is monochromatic on $S^{[\infty]}$. Since $(S, <_{\text{lex}})$ contains an order-isomorphic copy of $\mathbb{Q}$, this finishes the proof of Theorem 6.27.

Corollary 6.30 (Galvin) *For every finite partition of the set of all unordered pairs of rationals, there is a set of rationals X order-isomorphic to the rationals, so that the set of all ordered pairs of elements of X avoids all but two pieces of the partition.*

Proof. The given coloring of unordered pairs of elements of $\mathbb{Q} = (2^{<\infty}, <_{\text{lex}})$ induces continuous colorings on the sets $\mathcal{C}_\infty^\uparrow$ and $\mathcal{C}_\infty^\downarrow$ of all infinite lexicographically increasing and decreasing combs of $2^{<\infty}$, respectively, by simply letting the color of a comb be equal to the color of its first two elements. Now the rest of the proof follows the steps of our deduction of Theorem 6.27 from Theorem 6.29. The point again is that any strong subtree T of $2^{<\infty}$ contains a perfect subtree S with the property that any unordered pair of elements of S form a beginning of either an increasing or a decreasing comb of S. $\square$

6.5 CONTINUOUS COLORINGS ON $\mathbb{Q}^{[k]}$

The way we have been obtaining order-isomorphic copies of $\mathbb{Q}$ that would simplify given colorings of $\mathbb{Q}^{[k]}$ in the previous two sections gives no information about whether similar reductions are possible with topological copies at $\mathbb{Q}$. The *Sierpinski coloring* of $\mathbb{Q}^{[2]}$ obtained by comparing the usual ordering of $\mathbb{Q}$ with a well ordering shows that the number of colors cannot be reduced to less than 2 not only in order-isomorphic but also topological copies of $\mathbb{Q}$.

The following result shows that there is actually a more fundamental difference between the order theoretic and topological partition calculus on finite powers of $\mathbb{Q}$.

Theorem 6.31 (Baumgartner) *There is a coloring $c : \mathbb{Q}^{[2]} \to \mathbb{N}$ that takes all the values from $\mathbb{N}$ on any set of the form $H^{[2]}$ for $H \subseteq \mathbb{Q}$ homeomorphic to $\mathbb{Q}$.*

A close examination shows that neither Sierpinski's nor Baumgartner's colorings are continuous, so it remains to examine the possibility of a partition calculus for continuous colorings of the symmetric powers $\mathbb{Q}^{[k]}$.

Theorem 6.32 *For every finite continuous coloring*

$$c : \mathbb{Q}^{[2]} \to \{0, 1, \dots, l-1\}, \tag{6.40}$$

there is an $H \subseteq \mathbb{Q}$ homeomorphic to $\mathbb{Q}$ such that c is monochromatic on $H^{[2]}$.

Proof. It is convenient to take $\mathbb{Q}$ to be equal to the family FIN of all finite nonempty subsets of $\mathbb{N}$ with the topology of pointwise convergence, i.e., identify sets with their characteristic functions and take the topology on FIN as a subspace of the Cantor cube. Recall that for FIN we have Hindman's theorem as a basic pigeon hole principle which can be stepped up to all higher dimensions. More precisely, by Corollary 5.26, for every positive integer d, every finite coloring of $\text{FIN}^{[d]}$ is monochromatic on $[P]^{[d]}$ for some *infinite* block sequence P of elements of FIN. Recall that here $\text{FIN}^{[d]}$ denotes the collection of all *block sequences* of length d, i.e., sequences $(x_i)_0^{d-1}$ of elements of FIN such that

$$\max(x_i) < \min(x_j) \text{ whenever } i < j. \tag{6.41}$$

The given continuous coloring

$$c : \{\{u, v\} : u, v \in \text{FIN}, u \neq v\} \to \{0, 1, \dots, l-1\}$$

of the set of all *unordered* pairs of elements of FIN induces the coloring

$$c^* : \text{FIN}^{[3]} \to \{0, 1, \dots, l\} \tag{6.42}$$

of block sequences of elements of FIN of length 3, as follows:

$$c^*(x, y, z) = c(x \cup y, x \cup z). \tag{6.43}$$

By Corollary 5.26, there is an infinite block sequence $A = (a_n)_n$ such that c^* is constant on $[A]^{[3]}$. Recall that $[A]$ denotes the combinatorial subspace of FIN generated by A:

$$[A] = \{a_{n_0} \cup \dots \cup a_{n_p} : p \in \mathbb{N} \text{ and } n_0 < \dots < n_p\}. \tag{6.44}$$

We use the continuity of c to select an infinite subsequence $B = (b_k)_k$ of $A = (a_n)_n$ as follows. Let $b_0 = a_0$ and $b_1 = a_1$. Suppose $b_i = a_{n_i}$ $(i \leq k)$ have been selected. By continuity of c, for every $x \neq y$ in $[(b_i)_{i \leq k}]$, there is an $n = n(x, y) \in \mathbb{N}$ such that

$$c(x, y) = c(s, t) \tag{6.45}$$

for every pair s, t of elements of FIN such that

(a) s end-extends x and t end-extends y.

(b) $\min(s \setminus x), \min(t \setminus y) \geq n$.

Let $b_{k+1} = a_{n_{k+1}}$, where n_{k+1} is an integer such that $n_{k+1} > n_i$ for all $i \leq k$ and

$$\min(a_{n_{k+1}}) > n(x, y) \text{ for all } x, y \in [(b_i)_{i \leq k}]. \tag{6.46}$$

Let P denote the set of positive prime numbers. For a finite increasing sequence σ of elements of P, we choose an element b_σ of $[B]$ recursively as follows:

(c) $b_\emptyset = b_0$,

(d) $b_{\langle p_0, \ldots, p_i \rangle} = b_{\langle p_0 \rangle} \cup b_{\langle p_0 p_1 \rangle} \cup \cdots \cup b_{\langle p_0 \ldots p_i \rangle}$.

Let $H = \{b_\sigma : \sigma \in P^{[<\infty]}\}$. Then H is a subset of $[B]$ homeomorphic to $\mathbb{Q}$ that has the property that

$$(b_\sigma \setminus b_{\sigma \restriction \triangle(\sigma, \tau)}) \cap (b_\tau \setminus b_{\tau \restriction \triangle(\sigma, t)}) = \emptyset \tag{6.47}$$

for every $\sigma, \tau \in P^{[<\infty]}$, where

$$\triangle(\sigma, \tau) = \min\{i : \sigma(i) \neq \tau(i)\}. \tag{6.48}$$

We claim that our original coloring c is constant on the set unordered pairs of elements of H and that in fact the constant value is equal to the constant value of c^* on $[A]^{[3]}$. So, consider two distinct elements b_σ and b_τ of H. If σ is an initial part of τ, then by the choice of the subsequence $B = (b_k = a_{n_k})_k$ we conclude that

$$c(b_\sigma, b_\tau) = c\left(b_{\sigma \frown p}, b_\tau\right), \tag{6.49}$$

where p is any prime above the primes appearing in τ. So, it suffices to consider only the case in which σ and τ do not extend each other. Let $\bar{\sigma}$ be the maximal initial part of σ such that the interval $[\min b_{\bar{\sigma}}, \max b_{\bar{\sigma}}]$ contains no points from $b_\tau \setminus b_\sigma$. Similarly, let $\bar{\tau}$ be the maximal initial part of τ such that $[\min b_{\bar{\tau}}, \max b_{\bar{\tau}}]$ contains no points from $b_\sigma \setminus b_\tau$. By our assumption both $\bar{\sigma}$ and $\bar{\tau}$ strictly end-extend $\varrho = \sigma \restriction \triangle(\sigma, \tau) = \tau \restriction \triangle(\sigma, \tau)$, the maximal common initial part of σ and τ. Thus

$$y = b_{\bar{\sigma}} \setminus b_\varrho \text{ and } z = b_\tau \setminus b_\varrho \tag{6.50}$$

are both nonempty and either $\max(y) < \min(z)$ or $\max(z) < \min(y)$. Assuming the first alternative, we have that $\{b_\varrho, b_{\bar{\sigma}} \setminus b_\varrho, b_{\bar{\tau}} \setminus b_\varrho\} \in [A]^{[3]}$, and therefore,

$$c(b_{\bar{\sigma}}, b_{\bar{\tau}}) = c^*(b_\varrho, b_{\bar{\sigma}} \setminus b_\varrho, b_{\bar{\tau}} \setminus b_\varrho) = l_0, \tag{6.51}$$

where l_0 is the constant value of c^* on $[A]^{[3]}$. Finally, note that by the choice of $B = (a_{n_k})_k$, we have that $c(b_\sigma, b_\tau) = c(b_{\bar{\sigma}}, b_{\bar{\tau}}) = l_0$. This finishes the proof. $\qquad \square$

The following result gives a definite restriction to extending Theorem 6.32 to higher dimensions.

Theorem 6.33 *There is a continuous map* $c : \mathbb{Q}^{[3]} \to \mathbb{N}$ *that takes all the values from* $\mathbb{N}$ *on any set of the form* $H^{[3]}$ *for* $H \subseteq \mathbb{Q}$ *homeomorphic to* $\mathbb{Q}$.

Since the coloring that witnesses this is of independent interest, we give some details. Instead of working with the family FIN of finite nonempty subsets of $\mathbb{N}$ as above, it will be more convenient now to take $\mathbb{Q}$ to be equal to the family $\mathbb{N}^{[<\infty]}$ of *all* finite subsets of $\mathbb{N}$ with the topology of pointwise convergence.[3] For $s, t \in \mathbb{Q}$ one defines an equivalence relation $\sim_{st}$ on $s \triangle t = (s \setminus t) \cup (t \setminus s)$ as follows:

$$i \sim_{st} j \text{ iff } [i, j] \cap (s \setminus t) = \emptyset \text{ or } [i, j] \cap (t \setminus s) = \emptyset, \tag{6.52}$$

where $[i, j]$ denotes the closed interval of integers determined by $\min\{i, j\}$ and $\max\{i, j\}$. Let

$$\text{osc}(s, t) = |s \triangle t / \sim_{st}| . \tag{6.53}$$

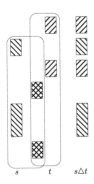

Figure 6.8 $\text{osc}(s, t) = |s \triangle t / \sim_{st}| = 4.$

Clearly, $\text{osc}(t, t) = 0$ and $\text{osc}(s, t) = \text{osc}(t, s)$, so we shall also use the "symmetric" notation $\text{osc}(\{s, t\})$ for the same number. The three-dimensional version

$$\text{osc} : \mathbb{Q}^{[3]} \to \mathbb{N} \tag{6.54}$$

is defined on the basis of $\text{osc} : \mathbb{Q}^{[<3]} \to \mathbb{N}$ as follows:

$$\text{osc}(s, t, u) = \text{osc}(\{s \cap n, t \cap n, u \cap n\}), \tag{6.55}$$

where $n = \triangle(s, t, u)$ and where (see Figure 6.5)

$$\triangle(s, t, u) = \max\{\min(s \triangle t), \min(s \triangle u), \min(t \triangle u)\}. \tag{6.56}$$

Note that the set $\{s \cap n, t \cap n, u \cap n\}$ has indeed at most two elements, so $\text{osc}(\{s \cap n, t \cap n, u \cap n\})$ has an unambiguous meaning. Note also that

$$\text{osc} : \mathbb{Q}^{[3]} \to \mathbb{N}$$

---

[3]Thus, $\mathbb{N}^{[<\infty]} = \{\emptyset\} \cup \text{FIN}.$

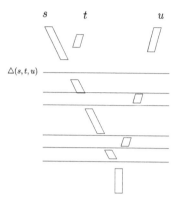

Figure 6.9 $\mathrm{osc}(s, t, u) = 5$

is a continuous map, while the two-dimensional version $\mathrm{osc} : \mathbb{Q}^{[2]} \to \mathbb{N}$ is not.

In order to express some properties of the oscillation mapping, we need the notion of the Cantor-Bendixson derivative $\partial : \mathcal{P}(\mathbb{Q}) \to \mathcal{P}(\mathbb{Q})$, defined as follows:

$$\partial (X) = \left\{ q \in X : q \in \overline{X \setminus \{q\}} \right\}. \tag{6.57}$$

We shall also need the finite iterates of ∂ defined as follows:

$$\partial^1 (X) = \partial(X) \text{ and } \partial^{k+1} (X) = \partial \left(\partial^k (X) \right). \tag{6.58}$$

The following lemma shows that $\mathrm{osc} : \mathbb{Q}^{[3]} \to \mathbb{N}$ is a continuous coloring witnessing the conclusion of Theorem 6.33.

Lemma 6.34 *Suppose that $\partial^k (X) \neq \emptyset$ for some positive integer k and some $X \subseteq \mathbb{Q}$. Then, on $X^{[3]}$, the oscillation mapping takes all the values from the interval $\{1, 2, \ldots, 2k - 1\}$.*

Proof. The proof is by induction on k. Let us first check the case $k = 1$, i.e, when there is an $x \in \partial X$. Since x is a finite set, by the definition of the topology on $\mathbb{Q}$, we can find two elements s and t of X such that x is an initial part of both s and t and

$$s \setminus x < t \setminus x. \tag{6.59}$$

(Recall that for subsets $a, b \subseteq \mathbb{N}$, the inequality $a < b$ is a shorthand for the inequality $\max (a) < \min (b)$.) It follows that $\max \triangle (x, s, t) = \min (t \setminus x)$ and therefore

$$\mathrm{osc} (x, s, t) = \mathrm{osc} (x, s) = 1. \tag{6.60}$$

This establishes the initial case of the conclusion of the lemma. Assume now that $k > 0$ and that the conclusion of the lemma is true for smaller integers. So, it suffices to show that the values $2k - 1$ and $2k - 2$ are taken by osc on $X^{[3]}$. Fix $x \in \partial^k(X)$. Then, recursively on $1 \leq i \leq k - 1$ we can choose $s_i, t_i \in \partial^{k-i}(X)$ end-extending x such that the following holds for $1 \leq i \leq k - 2$:

(1) s_{i+1} properly end-extends s_i,

(2) t_{i+1} properly end-extends t_i,

(3) $s_i \setminus s_{i-1} < t_i \setminus t_{i-1} < s_{i+1} \setminus s_i$, where $s_0 = t_0 = x$.

Then, in particular $s_{k-1}, t_{k-1} \in \partial X$. So, we can find s, t and u in X such that

(4) s properly end-extends s_{k-1},

(5) t and u properly end-extend t_{k-1},

(6) $t_{k-1} < s \setminus s_{k-1} < t \setminus t_{k-1} < u \setminus t_{k-1}$.

It follows that

$$\triangle (s, t, u) = \triangle (s_{k-1}, t, u) = \min (t \setminus t_{k-1}), \qquad (6.61)$$

and therefore,

$$\mathrm{osc} (s, t, u) = \mathrm{osc} (s, t_{k-1}) = 2k - 1, \qquad (6.62)$$

$$\mathrm{osc} (s_{k-1}, t, u) = \mathrm{osc} (s_{k-1}, t_{k-1}) = 2k - 2. \qquad (6.63)$$

This finishes the proof.   □

Note that the proof of Lemma 6.34 gives us the following information about the two-dimensional oscillation mapping, which is sufficient to give us the conclusion of Theorem 6.31.

Lemma 6.35 *Suppose $\partial^k (X) \neq \emptyset$ for some positive integer k and $X \subseteq \mathbb{Q}$. Then on $X^{[2]}$ the oscillation mapping takes all the values from the interval $\{1, 2, \ldots, 2k - 1\}$.*

It is clear that our definition of the two-dimensional oscillation mapping makes sense for pairs of arbitrary subsets of $\mathbb{N}$ rather than just the finite sets. So we can define

$$\mathrm{osc} : \mathcal{P} (\mathbb{N})^{[3]} \to \mathbb{N} \qquad (6.64)$$

on the basis of $\mathrm{osc} : \mathcal{P} (\mathbb{N})^{[2]} \to \mathbb{N} \cup \{\infty\}$ in exactly the same manner is in the case of finite sets,

$$\mathrm{osc} (x, y, z) = \mathrm{osc} (\{x \cap n, y \cap n, z \cap n\}), \qquad (6.65)$$

where

$$n = \max \{\min (x \triangle y), \min (y \triangle z), \min (x \triangle z)\}. \qquad (6.66)$$

Then the proof of Lemma 6.34 gives us the following fact, where of course we take $\mathcal{P} (\mathbb{N})$ with the topology of pointwise convergence.

Lemma 6.36 *Suppose that $\partial^k (\overline{X} \cap \mathbb{Q}) \neq \emptyset$ for some $X \subseteq P(\mathbb{N})$ and some positive integer k. Then the oscillation mapping on $X^{[3]}$ takes all the values from the interval $\{1, 2, \ldots, 2k - 1\}$.*

Corollary 6.37 *There is a continuous map $c : (\mathbb{R} \setminus \mathbb{Q})^{[3]} \to \mathbb{N}$ that takes all the values from $\mathbb{N}$ on any set of the form $X^{[3]}$ for X a closed but not σ-compact subset of $\mathbb{R} \setminus \mathbb{Q}$.*

With almost no extra work one can replace the range $\mathbb{N}$ of the mapping c in this corollary by the Baire space $\mathbb{N}^{\mathbb{N}}$. This is done by simply iterating the oscillation mapping on $P(\mathbb{N})^{[3]}$ as follows:

$$\mathrm{osc}^0 (x, y, z) = \mathrm{osc} (x, y, z)$$
$$\mathrm{osc}^{k+1} (x, y, z) = \mathrm{osc} (x \setminus n, y \setminus n, z \setminus n),$$

where $n = \triangle^k (x, y, z)$ and where $\triangle^k : P(\mathbb{N})^{[3]} \to \mathbb{N}$ is defined recursively on k as follows:

$$\triangle^1 (x, y, z) = \max \{\min (x \triangle y), \min (y \triangle z), \min (x \triangle z)\}, \qquad (6.67)$$

$$\triangle^{k+1} (x, y, z) = \triangle \left(x \setminus \triangle^k (x, y, z), y \setminus \triangle^k (x, y, z), z \setminus \triangle^k (x, y, z) \right). \tag{6.68}$$

The corresponding continuous map

$$\mathrm{osc}^\infty : P(\mathbb{N})^{[3]} \to \mathbb{N}^{\mathbb{N}} \tag{6.69}$$

defined by

$$\mathrm{osc}^\infty (x, y, t) = \left(\mathrm{osc}^k (x, y, z) \right)_{k=0}^\infty \tag{6.70}$$

takes all the values from $\mathbb{N}^{\mathbb{N}}$ on any set of the form $X^{[3]}$ for any closed subset X of $P(\mathbb{N}) \setminus \mathbb{Q}$ such that $\partial^k (\overline{X} \cap \mathbb{Q}) \neq \emptyset$ for all k. Since $P(\mathbb{N}) \setminus \mathbb{Q}$ is homeomorphic to the irrationals, we have the following version of Corollary 6.37.

Theorem 6.38 *There is a continuous map $c : (\mathbb{R} \setminus \mathbb{Q})^{[3]} \to \mathbb{R} \setminus \mathbb{Q}$ that takes all the values from $\mathbb{R} \setminus \mathbb{Q}$ on any set of the form $X^{[3]}$ for X a closed but not σ-compact subset of $\mathbb{R} \setminus \mathbb{Q}$.*

6.6 SOME PERFECT SET THEOREMS

In this section we continue the list of applications of Milliken's space of strong subtrees, or more precisely of the following result which appears above as Theorem 6.13.

Theorem 6.39 *Suppose U is a nonempty downward closed finitely branching subtree of $\mathbb{N}^{<\infty}$ with no terminal nodes and let A be an arbitrary subset of U. Then for every finite Souslin-measurable coloring of the class $[A]_{\mathrm{Em}}$ of all subsets of U realizing the same embedding type as A, there is a strong subtree T of U such that the restriction $[A]_{\mathrm{Em}}|T$ is monochromatic.*

The common feature of the list of corollaries in this section is that they are all about the partition calculus of perfect sets of reals. We start with the following basic reduction principle.

Theorem 6.40 (Mycielski) *Suppose $M_n \subseteq \mathbb{R}^{k_n}(n = 0, 1, \ldots)$ is a sequence of sets such that[4] $\lambda_{k_n}(M_n) = 0$ for all n, or suppose that M_n is a meager subset of $\mathbb{R}^{k_n}$ for all n. Then there is a perfect set $P \subseteq \mathbb{R}$ such that $P^{[k_n]} \cap M_n = \emptyset$ for all n.*

Here, for an integer $k \geq 1$ and $P \subseteq \mathbb{R}$, the set $P^{[k]}$ of all k-element subsets of P is identified with the following subset of $\mathbb{R}^k$:

$$\{(x_0, \ldots, x_{k+1}) \in \mathbb{R}^k : x_0 < \ldots < x_{k-1} \text{ and } \{x_0, \ldots, x_{k-1}\} \subseteq P\}. \quad (6.71)$$

We give a sketch of the proof of the case of a single G_δ subset $M \subseteq \mathbb{R}^k$ of k-dimensional Lebesgue measure-zero. The general case is an obvious variation of this argument. (The category case, in fact, is considerably easier to prove.) Recall the notion of Lebesgue density in dimension k:[5]

$$d_k(x, H) = \lim_{\delta \to 0^+} \frac{\lambda_k(B_\delta(x) \cap H)}{\lambda_k(B_\delta(x))}. \quad (6.72)$$

The Lebesgue density theorem says that for every measurable set[6] $H \subseteq \mathbb{R}^k$,

$$\lambda_k(\{x \in \mathbb{R}^k : d_k(x, H) \neq \chi_H(x)\}) = 0.$$

A subset H of $\mathbb{R}^k$ is *regular* if it is Borel and if $d_k(x, H) = 1$ for all $x \in H$. Clearly, $H = \mathbb{R}^k \setminus M$ is a regular subset of $\mathbb{R}^k$. The following is an immediate consequence of the Lebesgue density theorem and the inner regularity of the k-dimensional Lebesgue measure λ_k.

Lemma 6.41 *Suppose $Q \subseteq H \subseteq \mathbb{R}^{[k]}$ are such that Q is finite and H is regular. Then there is a regular set F such that $Q \subseteq F \subseteq \overline{F} \subseteq H$.*

Lemma 6.42 *Suppose F is a regular subset of $\mathbb{R}^{[k]}$, that $K = \{z_1, \ldots, z_n\}$, is a finite subset of $\mathbb{R}$ enumerated increasingly such that $K^{[k]} \subseteq F$. Let*

$$X = \{(x_0, x_1, \ldots, x_n) \in \mathbb{R}^{n+1} : x_0 \neq x_1 \text{ and } \{x_0, x_1, \ldots, x_n\}^{[k]} \subseteq F\}. \quad (6.73)$$

Then $\lambda_{n+1}(B_\delta(z_1, z_1, \ldots, z_n) \cap X) > 0$ for all $\delta > 0$.

Proof. First of all note that if we let $z_0 = z_1$, and $z = (z_0, z_1, \ldots, z_n)$ then for every $i_0, \ldots, i_{k-1} \leq n$,

$$(z_{i_0}, \ldots, z_{i_{k-1}}) \in F \text{ implies } d_{n+1}(z, F_{n+1}(i_0, \ldots, i_{k-1})) = 1, \quad (6.74)$$

where

$$F_{n+1}(i_0, \ldots, i_{k-1}) = \{(x_0, \ldots, x_n) \in \mathbb{R}^{n+1} : (x_{i_0}, \ldots, x_{i_{k-1}}) \in F\}. \quad (6.75)$$

---

[4]Here, λ_{k_n} denotes the k_n-dimensional Lebesgue measure on $\mathbb{R}^{k_n}$.

[5]Here, $B_\delta(x)$ is the ball of radius δ and center x with respect to the usual Euclidean metric of $\mathbb{R}^k$.

[6]In other words, the set of all $x \in \mathbf{R}$, where the Lebesgue density disagrees with the characteristic function of H has measure 0.

If follows that for every $i_0, \ldots, i_{k-1} \le n$ such that $(z_{i_0}, \ldots, z_{i_{k-1}}) \in F$ we have that

$$\lim_{\delta \to 0+} \frac{\lambda_{n+1} \left(B_\delta(z) \setminus F_{n+1}(i_0, \ldots, i_{k-1}) \right)}{\lambda_{n+1} \left(B_\delta(z) \right)} = 0. \tag{6.76}$$

Since there are only finitely many choices of $i_0, \ldots, i_{k-1} \le n$, for all sufficiently small $\delta > 0$,

$$\lambda_{n+1} \left(B_\delta(z) \cap \left(\bigcap_{i_0, \ldots, i_{k-1} \le n} F_{n+1}(i_0, \ldots, i_{k-1}) \right) \right) > 0, \tag{6.77}$$

which was to be proved. $\qquad\square$

We are now ready to start the proof of Theorem 6.40 by constructing a perfect set $P \subseteq \mathbb{R}$ such that $P^{[k]} \cap M = \emptyset$. Recall that $H = \mathbb{R}^k \setminus M$ is regular, so by Lemma 6.41 we can choose regular set F such that $\overline{F} \subseteq H$. Recursively on $\sigma \in 2^{<\infty}$ we construct a Cantor scheme

$$q_\sigma \in \text{int}\,(I_\sigma) \subseteq I_\sigma \;\; (\sigma \in 2^{<\infty}, |\sigma| \ge k) \tag{6.78}$$

of closed intervals and points of $\mathbb{R}$ such that

(1) $I_\sigma \cap I_\tau = \emptyset$ whenever $\sigma, \tau \in 2^k$, $\sigma \neq \tau$,

(2) $I_{\sigma^\frown 0}, I_{\sigma^\frown 1} \subseteq I_\sigma$, $\;\; I_{\sigma^\frown 0} \cap I_{\sigma^\frown 1} = \emptyset$,

(3) $\text{diam}\,(I_\sigma) \le 2^{-|\sigma|}$,

(4) $\{q_\sigma : \sigma \in 2^n\}^{[k]} \subseteq F$ for all $n \ge k$.

To see the recursive step from n to $n+1$, one starts with $K_n = \{q_\sigma : \sigma \in 2^n\}$ such that $K_n^{[k]} \subseteq F$ and successively applies Lemma 6.42 finitely many times and obtains $q_{\sigma^\frown 0}, q_{\sigma^\frown 1} \in I_\sigma$ $(\sigma \in 2^n)$ such that if we let

$$K_{n+1} = \{q_{\sigma^\frown i} : \sigma \in 2^n, i < 2\}$$

then $K_{n+1}^{[k]} \subseteq F$. For each $\sigma \in 2^n$ we pick closed intervals $I_{\sigma^\frown 0}\, I_{\sigma^- 1}$ included in I_σ of diameter $\le 2^{n+1}$ such that $q_{\sigma^\frown i} \in \text{int}\,(I_{\sigma^\frown 1})$ for $i = 0, 1$. This completes the inductive step. Let

$$P = \bigcup_{x \in 2^{\mathbb{N}}} \bigcap_{n=k}^{\infty} I_{x \restriction n}.$$

Then P is a perfect subset of $\mathbb{R}$ such that $P^{[k]} \subseteq \overline{F} \subseteq H$, as required. This finishes the proof of Theorem 6.40.

Corollary 6.43 *Every perfect set of real numbers contains a perfect subset that is algebraically independent.*

Proof. This follows from Theorem 6.40 and the following simple fact. $\qquad\square$

Lemma 6.44 *If P is a perfect set of real numbers and $f(v_0, \ldots, v_{k-1})$ a nonzero polynomial then*

$$M_f = \{(x_0, \ldots x_{k-1}) \in P^k : f(x_0, \ldots, x_{k-1}) = 0\} \tag{6.79}$$

is nowhere dense in P^k.

Similar arguments deduce the following corollary from the measure-zero case of Mycielski's theorem.

Corollary 6.45 *Let S be a given measure-zero set of irrational numbers. Then there is a perfect set P such that the field generated by P is disjoint from S.*

Let us now give an application of Mycielski's theorem to a different area.

Corollary 6.46 *Suppose X is a Polish space and that $\Phi : 2^{\mathbb{N}} \times X \to \mathbb{R}$ is a Borel function such that the corresponding sequence of functions $f_a = \Phi(a, .)$ $(a \in 2^{\mathbb{N}})$ is uniformly bounded and such that the set*

$$\{a \in 2^{\mathbb{N}} : f_a(x) \neq 0\}$$

is at most countable for every $x \in X$. Then there is a perfect set $P \subseteq 2^{\mathbb{N}}$ such that the sequence $(f_a)_{a \in P}$ is 1-unconditional[7] in the Banach space $\ell_\infty(X)$.

Proof. By Mycielski's theorem it suffices to show that for each positive integer n, the set

$$U_n = \{(a_0, \ldots, a_{n-1}) \in (2^{\mathbb{N}})^n : (f_{a_i})_{i<n} \text{ is 1-unconditional in } \ell_\infty(X)\}$$

is comeager in $(2^{\mathbb{N}})^n$. Otherwise, since by our assumption about Φ this set has the property of Baire, we can find an $F \subseteq \{0, 1, \ldots, n-1\}$, an $\varepsilon > 0$, a sequence λ_i $(i < n)$ of rationals, and a sequence of pairwise disjoint perfect subsets $P_0, \ldots, P_{n-1}$ of $2^{\mathbb{N}}$ such that for every choice $a_i \in P_i$ $(i < n)$, we have

$$\left\| \sum_{i \in F} \lambda_i f_{a_i} \right\|_\infty \geq (1 + \varepsilon) \cdot \left\| \sum_{i<n} \lambda_i f_{a_i} \right\|_\infty.$$

Fix $a_i \in P_i$ for $i \in F$. Pick an $x \in X$ such that

$$\left\| \sum_{i \in F} \lambda_i f_{a_i} \right\|_\infty < (1 + \varepsilon) \cdot \left| \sum_{i \in F} \lambda_i f_{a_i}(x) \right|.$$

By our assumption, the set $D = \{a \in 2^{\mathbb{N}} : f_a(x) \neq 0\}$ is countable, so we can find $a_i \in P_i \setminus D$ for each $i < n$ such that $i \notin F$. It follows that

$$\left\| \sum_{i \in F} \lambda_i f_{a_i} \right\|_\infty < (1 + \varepsilon) \cdot \left| \sum_{i \in F} \lambda_i f_{a_i}(x) \right| =$$

---

[7]Meaning that for every pair E and F of nonempty finite subsets of the index set P with $E \subseteq F$ and every choice $(\lambda_i)_{i \in F}$ of scalars, we have $\left\| \sum_{i \in E} \lambda_i x_i \right\| \leq \left\| \sum_{i \in F} \lambda_i x_i \right\|$ (see Appendix 9.4).

$$= (1 + \varepsilon) \cdot |\sum_{i<n} \lambda_i f_{a_i}(x)| \le (1 + \varepsilon) \cdot \|\sum_{i<n} \lambda_i f_{a_i}\|_\infty,$$

a contradiction.

□

We are now ready to start our list of the perfect-set corollaries of Theorem 6.2. We start with the following finite-dimensional result, which reveals the interesting phenomenon that in some situations, when it is not possible to have a purely Ramsey theoretic result, one still would like to know if there is a maximally complex coloring witnessing this.

Corollary 6.47 (Blass) *For every positive integer k and every finite Baire- or Lebesgue-measurable coloring of the set $\mathbb{R}^{[k]}$ of all k-element sets of reals, there is a perfect set $P \subseteq \mathbb{R}$ such that $P^{[k]}$ meets at most $(k-1)!$ colors.*

Proof. By Mycielski's theorem, we find a perfect set $P \subseteq \mathbb{R}$ such that the restriction of each color to $P^{[k]}$ is a relatively open subset of $P^{[k]}$. In other words, we may restrict ourselves to proving Blass's theorem for a continuous coloring of $(2^{\mathbb{N}})^{[k]}$. Going still to a thinner perfect set we may assume that the color of any given k-element subset X of $2^{\mathbb{N}}$ is determined at the level

$$\Delta(X) = \max\{\Delta(x,y) : x \ne y \in X\} + 1, \tag{6.80}$$

where $\Delta(x,y) = \min\{n : x(n) \ne y(n)\}$. In other words, for every $X, Y \in (2^{\mathbb{N}})^{[k]}$ if $\Delta(X) = \Delta(Y) = n$ and if $X \upharpoonright n = Y \upharpoonright n$, then X and Y have the same color.

Let $c : (2^{\mathbb{N}})^{[k]} \to \{0, \dots, l-1\}$ be the given coloring. Define

$$c^* : (2^{<\infty})^{[k]} \to \{0, \dots, l\} \tag{6.81}$$

by letting $c^*(\{s_0, \dots, s_{k-1}\}) = l$ if either $\{s_0, \dots, s_{k-1}\}$ is not an antichain or else c is not monochromatic on $[s_0] \times \dots \times [s_{k-1}]$; in all other cases put $c^*(\{s_0, \dots, s_{k-1}\}) = c(X)$ for some (all) $X \in [s_0] \times \dots \times [s_{k-1}]$. By Corollary 6.14 there is a strong subtree T of $2^{<\infty}$ such that c^* is monochromatic on any class of the form $[A]_{\text{Em}} \upharpoonright T$ where A is some k-element subset of $2^{<\infty}$. Choose some $\wedge$–closed perfect subtree S of T such that

(*i*) $(\forall n) |S \cap T(n)| \le 1$,

(*ii*) $(\forall s, t \in S) (|s| < |t| \,\&\, s \not\subseteq t \Rightarrow t(|s|) = 0)$.

Note that the embedding type of a k-element antichain $A = \{s_0, \dots, s_{k-1}\}$ of S enumerated according to the lexicographical ordering of $2^{<\infty}$ is uniquely determined by the way the distances

$$\Delta(s_i, s_{i+1}) \, (i < k-1) \tag{6.82}$$

are ordered according to the natural ordering of $\mathbb{N}$. It follows that a k-element antichain of S realizes one of exactly $(k-1)!$ different embedding types in $2^{<\infty}$. Since for every such antichain $A = \{s_0, \dots, s_{k-1}\} \subseteq S$, the color

$c^*(A)$ is equal to the constant value of c on $[s_0] \times \ldots \times [s_{k-1}]$, this shows that c is monochromatic in $P^{[k]}$ where P is the perfect subset of $2^{\mathbb{N}}$ formed by taking the unions of all infinite chains of S. □

The $(k-1)!$ in Blass's theorem is only a possible upper bound on how much a given measurable coloring can be reduced. To see that this is in fact optimal, with every permutation $\pi \in \mathcal{S}_{k-1}$ we associate the following open subset of $\mathbb{R}^{[k]}$:

$$\mathcal{C}_\pi = \{\{x_0, \ldots, x_{k-1}\}_< : (\forall i, j < k-1)$$
$$(x_{i+1} - x_i < x_{j+1} - x_j \leftrightarrow \pi(i) < \pi(j))\}.$$

Then $P^{[k]} \cap \mathcal{C}_\pi \neq \phi$ for all perfect $P \subseteq \mathbb{R}$ and $\pi \in \mathcal{S}_{k-1}$. Note that this example also shows that in any infinite-dimensional perfect set theorem of this sort we must restrict ourselves to colorings of increasing (or decreasing) infinite sequences of $\mathbb{R}$. In fact, this example shows that we must also restrict ourselves to sequences $(x_n) \subseteq \mathbb{R}$ that are, moreover, *rapidly converging* in the following sense.

Definition 6.48 *An increasing sequence $(x_n) \subseteq \mathbb{R}$ is* rapidly converging *if the sequence of consecutive distances is monotone and it conveges to 0, i.e.,*

$$x_{n+2} - x_{n+1} < x_{n+1} - x_n \text{ for all } n \text{ and } x_{n+1} - x_n \longrightarrow_n 0. \qquad (6.83)$$

Let $\mathbb{R}^{[\infty]}$ denote the collection of all increasing rapidly converging sequences of the reals. Of course, this definition makes perfect sense for many other spaces, such us for example the Baire space $\mathbb{N}^{\mathbb{N}}$ or the Cantor space $2^{\mathbb{N}}$, which beside the natural metrics have the natural (lexicographical) orderings.

It turns out that the partition theorem for embedding types (Theorem 6.13) has perfect set properties of this sort as immediate corollaries. The following is an example of this kind of corollaries of Theorem 6.13.

Corollary 6.49 (Louveau-Shelah-Velickovic) *For every finite Souslin-measurable coloring of the set $\mathbb{R}^{[\infty]}$ of all increasing rapidly converging infinite sequences of the reals, there is a perfect set $P \subseteq \mathbb{R}$ such that $P^{[\infty]}$ is monochromatic.*

Proof. Without loss of generality, we prove this result for the space $(2^{\mathbb{N}})^{[\infty]}$ of increasing rapidly converging sequences of the Cantor space $2^{\mathbb{N}}$. Let $A = \{1^{(n)}0^{(m)} : n, m \in \mathbb{N}\}$. An element B of the embedding class $[A]_{\text{Em}}$ determines a lexicographically strictly increasing sequence, which is actually rapidly convergent, i.e., an element of $(2^{\mathbb{N}})^{[\infty]}$. This gives us a coloring of $[A]_{\text{Em}}$. By Theorem 6.39, there is a strong subtree T of $2^{<\infty}$ such that $[A]_{\text{Em}} \upharpoonright T$ is monochromatic. Let S be a $\wedge$-closed subtree of T satisfying (i) and (ii) from the proof of Blass' theorem and let P be the perfect subset of $2^{\mathbb{N}}$ obtained by taking the unions of all infinite chains of S. Then *every* member of $P^{[\infty]}$ is determined by some $B \in [A]_{\text{Em}} \upharpoonright S$, and so $P^{[\infty]}$ is monochromatic. □

For a permutation π of $\mathbb{N}$, let $\mathbb{R}^{[\pi]}$ be the collection of all strictly increasing converging sequences $(x_n) \subseteq \mathbb{R}$ such that

(a) $x_{n+1} - x_n \neq x_{m+1} - x_m$ whenever $m \neq n$,

(b) $x_{n+1} - x_n < x_{m+1} - x_m$ iff $\pi(m) < \pi(n)$.

Thus, $\mathbb{R}^{[\pi]} = \mathbb{R}^{[\infty]}$ for $\pi = id$. Then the partition theorem for embedding types has the following consequence, which incorporates both Corollary 6.47 and Corollary 6.49.

Corollary 6.50 *For every permutation π of $\mathbb{N}$ and every finite Souslin-measurable coloring of $\mathbb{R}^{[\pi]}$, there is a perfect set $P \subseteq \mathbb{R}$ such that $P^{[\pi]}$ is monochromatic.*

The following variation on Blass's theorem can now be easily deduced from these results.

Corollary 6.51 (Parametrized Blass Theorem) *For every positive integer k and every finite Souslin-measurable coloring of the product $\mathbb{R}^{[k]} \times \mathbb{N}^{[\infty]}$ that is invariant under finite changes on the second coordinate, there is a perfect $P \subseteq \mathbb{R}$ and an infinite $M \subseteq \mathbb{N}$ such that the product $P^{[k]} \times M^{[\infty]}$ meets at most $(k-1)!$ colors.*

Proof. Clearly, we may replace $\mathbb{R}$ by the Cantor space $2^{\mathbb{N}}$ of all infinite binary sequences and assume that we have a Souslin-measurable l-coloring of the form

$$c : \left(2^{\mathbb{N}}\right)^{[k]} \times \mathbb{N}^{[\infty]} \to \{0, \dots, l-1\}. \tag{6.84}$$

For $\pi \in S_{k-1}$, let $\pi^* \in S_{\infty}$ be determined as follows: $\pi^* \upharpoonright k-1 = \pi$ and $\pi^*(n) = n$ for $n \geq k-1$. For $\pi \in S_{k-1}$ define

$$c_{\pi}^* : (2^{\mathbb{N}})^{[\pi^*]} \to \{0, \dots, n-1\} \tag{6.85}$$

by

$$c_{\pi}^* \left((x_n)\right) = c \left((x_n)_{n<k}, \{\Delta(x_n, x_{n+1}) : n \geq k\}\right). \tag{6.86}$$

By Corollary 6.50 we get a perfect set $P \subseteq 2^{\mathbb{N}}$ such that c_{π}^* is monochromatic on $P^{[\pi^*]}$ for all $\pi \in S_{k-1}$. Let s_0 be the stem of P, let

$$P_0 = \{t \in P : t \subseteq s_0^\frown(0) \text{ or } s_0^\frown(0) \subseteq t\}, \tag{6.87}$$

and let (z_n) be an arbitrary rapidly increasing sequence of elements of P that extend $s_0^\frown(1)$. Let

$$M = \{\Delta(z_n, z_{n+1}) : n \in \mathbb{N}\}. \tag{6.88}$$

Then c takes at most $(k-1)!$ values on $P_0^{[k]} \times M^{[\infty]}$. $\qquad \square$

The following example shows that the restriction that the coloring is invariant under finite changes on the second coordinate in Corollary 6.51 is essential.

Example 6.6.1 *There is a continuous coloring $c : (2^{\mathbb{N}})^{[2]} \times \mathbb{N}^{[\infty]} \to \mathbb{N}$ that takes all the values from $\mathbb{N}$ on any set of the form $P^{[2]} \times M^{[\infty]}$, where P is a perfect subset of $2^{\mathbb{N}}$ and M is an infinite subset of $\mathbb{N}$. One such coloring c is defined by*

$$c(\{x, y\}, Z) = \max\{k : 2^k \text{ divides } |Z \cap \{0, 1, \dots, \Delta(x, y)\}|\}. \qquad (6.89)$$

We have seen our next corollary in a different context above, but we shall now easily deduce it also from the parametrized version of Blass's theorem.

Corollary 6.52 (Harrington) *For every positive integer k and every bounded sequence (f_n) of Baire- or Lebsgue-measurable maps from $\mathbb{R}^k$ into $\mathbb{R}$, there is a subsequence (f_{n_i}) and a perfect set $P \subseteq \mathbb{R}$ such that (f_{n_i}) is uniformly convergent on P^k.*

Proof. First of all, applying Mycielski's theorem and going to a perfect subset of $\mathbb{R}$, we may assume that the functions f_n are in fact continuous. As before, working by induction on k and using the fact that $\mathbb{R}^k$ is naturally decomposed into a finite union of subsets isomorphic to $\mathbb{R}^{[l]}$ $(l \leq k)$, it suffices to prove only the symmetric version of the theorem, or in other words, assume that the mappings are defined on the set $\mathbb{R}^{[k]}$ of all increasing k-tuples. The sequence (f_n) of Borel functions leads to coloring

$$c : \mathbb{R}^{[k]} \times \mathbb{N}^{[\infty]} \to \{0, 1, 2\}$$

defined as follows:

$$c(x, M) = \begin{cases} 0 \text{ if } (f_n(x))_{n \in M} \text{ is eventually nondecreasing} \\ 1 \text{ if } (f_n(x))_{n \in M} \text{ is eventually nonincreasing} \\ 2 \text{ if } (f_n(x))_{n \in M} \text{ is none of the above.} \end{cases} \qquad (6.90)$$

By Corollary 6.51 there is a perfect set $P \subseteq \mathbb{R}$ and an infinite $M \subseteq \mathbb{N}$ such that c on $P^{[k]} \times M^{[\infty]}$ depends only on the relationship between the consecutive distances of k-tuples $x_0 < x_1 < \cdots < x_{k-1}$ of elements of P. If follows that $P^{[k]}$ can be split into $(k-1)!$ relatively clopen subset on which the convergence is monotone. By Mycielski's theorem, we may assume that the pointwise limit of $(f_n)_{n \in M}$ is a continuous function on $P^{[k]}$. So by Dini's theorem we can conclude that the convergence of $(f_n)_{n \in M}$ on $P^{[k]}$ is in fact uniform. This finishes the proof. $\qquad \square$

6.7 ANALYTIC IDEALS AND POINTS IN COMPACT SETS OF THE FIRST BAIRE CLASS

In this section we apply the Ramsey theory of perfect trees to the analysis of characters of points in separable compact sets of Baire Class 1 functions on some Polish space X. This eventually reduces to studying a special kind of ideals on $\mathbb{N}$. Recall that an *ideal* on $\mathbb{N}$ is simply a family of subsets of $\mathbb{N}$ closed under taking finite unions and subsets. The descriptive properties of ideals like for example F_σ, *Borel*, or *analytic* refer to the topology they inherit from the power set of $\mathbb{N}$ when this set is naturally identified with the Cantor space $2^\mathbb{N}$. When working with ideals on $\mathbb{N}$, one implicitly assumes that they are all *proper*, i.e., not equal to the power set of $\mathbb{N}$ and *nonprincipal* in the sense that they contain all singletons. Ideals on $\mathbb{N}$ are typically related to various notions of smallness for subsets of $\mathbb{N}$. Consider, for example, the ideal FIN of all finite subsets of $\mathbb{N}$ or the ideal $\mathcal{Z}_0$ of all subsets M of $\mathbb{N}$ having asymptotic density 0. Ideals on $\mathbb{N}$ can also be viewed as points in topological spaces via the identification

$$\mathcal{I}_K(x,(x_n)) = \{M \subseteq \mathbb{N} : x \notin \overline{\{x_n : n \in M\}}\} \tag{6.91}$$

for some Hausdorff space K, some point $x \in K$ and a sequence $(x_n) \subseteq K\backslash\{x\}$ accumulating to x. It is in this context that one realizes that many of the natural examples of ideals on $\mathbb{N}$ are in fact *analytic*, i.e., equal to continuous images of the irrationals when considered as topological spaces in the topology of pointwise convergence. Thus in particular, every Borel ideal on $\mathbb{N}$ is analytic although of course not vice versa. The following lemma explains the place of analytic ideals in this representation.

Lemma 6.53 *An ideal $\mathcal{I}$ on $\mathbb{N}$ is analytic if and only if it is represented by a sequence (f_n) of continuous functions on a Polish space X accumulating pointwise to a continuous function f on X.*

Proof. To see the direct implication, note that $\mathcal{I}$ allows the representation $\mathcal{I} = \mathcal{I}_K(\bar{0}, (\pi_n))$ with K being the Tychonov cube $2^\mathcal{I}$. Pick a continuous map $g : \mathbb{N}^\mathbb{N} \to 2^\mathbb{N}$ whose range is equal to $\mathcal{I}$, and for $n \in \mathbb{N}$ define $f_n = \pi_n \circ g$ and let $f = \bar{0} \circ g$. Then all these functions are continuous functions from $\mathbb{N}^\mathbb{N}$ into $\{0,1\}$. Note also that $\mathcal{I} = \mathcal{I}_{K_0}(f, (f_n))$, where K_0 is the Tychonov cube $2^{\mathbb{N}^\mathbb{N}}$.

To see the reverse implication, we may assume that the accumulation point f is actually equal to the constantly zero function $\bar{0}$ on X. For $\epsilon > 0$ and a finite set $S \subseteq X$, let

$$M(S, \epsilon) = \{n \in \mathbb{N} : (\exists s \in S)\, |f_n(s)| \geq \epsilon\}.$$

Note that $(S, \epsilon) \mapsto M(S, \epsilon)$ is a Borel map from the product $X^{[<\infty]} \times \mathbb{R}_+$ into $2^\mathbb{N}$ when the set $X^{[<\infty]}$ of finite subsets of X is equipped with the Vietoris topology. It follows that $\{M(S, \epsilon) : S \in X^{[<\infty]}, \epsilon > 0\}$ is an analytic subset of $2^\mathbb{N}$ that generates the given ideal $\mathcal{I}_{\mathbb{R}^X}(\bar{0}, (f_n))$. $\qquad\square$

Many structural results about analytic ideals on $\mathbb{N}$ are obtained by analyzing a notion of *cofinal similarity* among analytic ideals on $\mathbb{N}$. Thus, we consider ideals on $\mathbb{N}$ as directed sets under the inclusion ordering and study their cofinal structure. When ideals are viewed as points in separable topological spaces, the theory of cofinal similarity between ideals on $\mathbb{N}$ is transferred to a particularly fine theory about characters of points in such spaces. So let us recall the basic notions of this so-called *Tukey theory* of cofinal types in this context. Given a pair of ideals $\mathcal{I}$ and $\mathcal{J}$ on $\mathbb{N}$, we say that $\mathcal{I}$ is *cofinally finer* than $\mathcal{J}$ if there is a mapping $f : \mathcal{I} \to \mathcal{J}$ such that (cf1) $a \subseteq b \to f(a) \subseteq f(b)$,

(cf2) $(\forall a \in \mathcal{J})(\exists b \in \mathcal{I}) \, f(b) \supseteq a$.

This is equivalent to saying that $\mathcal{J}$ is *Tukey-reducible* to $\mathcal{I}$ in the sense that there is a *Tukey map* $g : \mathcal{J} \to \mathcal{I}$, i.e., a map with the following property:

(tk) $(\forall \mathcal{X} \subseteq \mathcal{J})[\bigcup \mathcal{X} \notin \mathcal{J} \Longrightarrow \bigcup g''\mathcal{X} \notin \mathcal{I}]$.

For, given $f : \mathcal{I} \to \mathcal{J}$ satisfying (cf1) and (cf2), we can define $g : \mathcal{J} \to \mathcal{I}$ satisfying (tk) by letting $g(a)$ be any $b \in \mathcal{I}$ such that $f(b) \supseteq a$. Conversely, given $g : \mathcal{J} \to \mathcal{I}$ satisfying (tk), we can define $f : \mathcal{I} \to \mathcal{J}$ satisfying (cf1) and (cf2) by letting $f(b) = \bigcup\{a \in \mathcal{J} : g(a) \subseteq b\}$. We use the notation $\mathcal{J} \leq_T \mathcal{I}$ whenever there is a Tukey map from $\mathcal{J}$ into $\mathcal{I}$, and we write $\mathcal{I} \equiv_T \mathcal{J}$ whenever $\mathcal{J} \leq_T \mathcal{I}$ and $\mathcal{I} \leq_T \mathcal{J}$.

It turns out that the class of analytic ideals on $\mathbb{N}$ has a maximal element $\mathcal{I}_{\max}$ relative to $\leq_T$. The best way to visualize $\mathcal{I}_{\max}$ is to look at its copy on the index set $2^{<\mathbb{N}}$ rather than $\mathbb{N}$. Thus we let $\mathcal{I}_{\max}$ be the ideal of subsets of $2^{<\mathbb{N}}$ generated by the family of all infinite branches of the complete binary tree $(2^{<\mathbb{N}}, \sqsubseteq)$. It turns out that $\mathcal{I}_{\max}$ is represented by a separable compact set of Baire Class 1 functions defined on the Cantor set $2^{\mathbb{N}}$. To see this, let

$$\hat{A}(2^{\mathbb{N}}) = \bar{0} \cup \{\delta_x : x \in 2^{\mathbb{N}}\} \cup \{\delta_s : s \in 2^{<\mathbb{N}}\},$$

where for $x \in 2^{\mathbb{N}}$, the δ_x denotes the characteristic function of the singleton $\{x\}$, and where for $s \in 2^{<\mathbb{N}}$, the δ_s denotes the characteristic function of the basic clopen set of all $x \in 2^{\mathbb{N}}$ that end-extend s. Referring to the notation introduced at the beginning of this section, with index set $\mathbb{N}$ replaced by $2^{<\mathbb{N}}$, we have the following equality

$$\mathcal{I}_{\max} = \mathcal{I}_K(\bar{0}, (\delta_s)_{s \in 2^{<\mathbb{N}}}) \quad \text{with} \quad K = \hat{A}(2^{\mathbb{N}}).$$

As indicated by its name, this ideal has the following property.

Lemma 6.54 FIN $\leq_T \mathcal{I} \leq_T \mathcal{I}_{\max}$ *for every proper (analytic) ideal $\mathcal{I}$ on $\mathbb{N}$.*

Proof. To verify the second inequality, choose a homeomorphic embedding $f : \mathcal{I} \to 2^{\mathbb{N}}$, for example the one obtained by identifying a member of $\mathcal{I}$ with its characteristic function. Now note that f is a Tukey map. □

The problem of characterizing analytic ideals $\mathcal{I}$ on $\mathbb{N}$ such that $\mathcal{I} \equiv_T \mathcal{I}_{\max}$ naturally arises. The purpose of this section is to prove the following result, which is given below in a more precise form as Theorem 6.64 after we develop some theory of analytic gaps, a theory which is very closely related to the Ramsey theory of perfect trees developed so far in this chapter.

Theorem 6.55 *Let $\mathcal{I}$ be a proper analytic ideal on $\mathbb{N}$ represented by a compact set of Baire Class 1 functions defined on some Polish space X. Then either $\mathcal{I} \equiv_T$ FIN or $\mathcal{I} \equiv_T \mathcal{I}_{\max}$.*

Note that the representation $\mathcal{I} = \mathcal{I}_K(x, (x_n))$ indicates that part of the structure of an analytic ideal $\mathcal{I}$ will likely be hidden inside its orthogonal

$$\mathcal{I}^{\perp} = \{N \subseteq \mathbb{N} : N \cap M \text{ is finite for all } M \in \mathcal{I}\}.$$

To see this, note that, modulo indexing, $\mathcal{I}^{\perp}$ is simply the set of all subsequences of (x_n) that converge to x. In nontrivial cases the pair $(\mathcal{I}, \mathcal{I}^{\perp})$ forms a *gap* with one of its sides an analytic family of subsets of $\mathbb{N}$. So in order to take advantage of this representation, we recall some of the theory of *analytic gaps*. Two subsets M and N of $\mathbb{N}$ are *orthogonal* if their intersection is finite. Let $M \perp N$ denote this fact. Two families $\mathcal{M}$ and $\mathcal{N}$ of subsets of $\mathbb{N}$ are orthogonal if $M \perp N$ for all $M \in \mathcal{M}$ and $N \in \mathcal{N}$. We use $\mathcal{M} \perp \mathcal{N}$ to denote this fact and we use $\mathcal{M}^{\perp}$ to denote *the orthogonal* of $\mathcal{M}$, the set $\{N \subseteq \mathbb{N} : N \perp M \text{ for all } M \in \mathcal{M}\}$. Given a pair $\mathcal{M}$ and $\mathcal{N}$ of families of subsets of $\mathbb{N}$, we say that $\mathcal{M}$ is *countably generated* in $\mathcal{N}$ if there is a countable sequence (N_k) of elements of $\mathcal{N}$ such that every element of $\mathcal{M}$ is almost included in one of the N_k. The following example gives a typical situation when this happens.

Example 6.7.1 *Let $\mathcal{M}$ be a countable family of infinite pairwise disjoint subsets of $\mathbb{N}$ that covers $\mathbb{N}$ and let $\mathcal{N} = \mathcal{M}^{\perp}$. Then $(\mathcal{M}, \mathcal{N})$ forms an analytic gap in which $\mathcal{M}$ is countably generated in $\mathcal{N}^{\perp}$.*

This example motivates the following definition.

Definition 6.56 *Let $\mathcal{N}$ be a given family of subsets of $\mathbb{N}$. A subtree T of the tree $(\mathbb{N}^{[<\infty]}, \sqsubseteq)$ of all finite subsets of $\mathbb{N}$ ordered by the relation $\sqsubseteq$ of end-extension is an $\mathcal{N}$-tree if it satisfies the following conditions:*

(a) $\emptyset \in T$,

(b) $t \in T$ *implies that $\{n \in \mathbb{N} : n > \max(t) \text{ and } t \cup \{n\} \in T\}$ is an infinite set included in a member of $\mathcal{N}$.*

The following result characterizes one kind of analytic gap.

Theorem 6.57 (Analytic Gap Theorem) *Suppose $\mathcal{M}$ and $\mathcal{N}$ are two orthogonal families of subsets of $\mathbb{N}$ and that $\mathcal{M}$ is analytic and closed under taking subsets. Then either*

(1) $\mathcal{M}$ *is countably generated in $\mathcal{N}^{\perp}$, or else*

(2) *there is an $\mathcal{N}$-tree all of whose branches are members of $\mathcal{M}$.*

In order to get alternative (2) in a symmetric form, we need to weaken condition (1) of separation. Thus, we shall say that two orthogonal families $\mathcal{M}$ and $\mathcal{N}$ of subsets of $\mathbb{N}$ are *countably separated* if there is a countable sequence (C_k) of subsets of $\mathbb{N}$ such that for every pair $(M, N) \in \mathcal{M} \times \mathcal{N}$ there is a k such that $M \subseteq^* C_k$ and $C_k \perp N$. Note that the two families of Example 6.7.1 are countably separated, so we need a different example that would lead us to the second alternative of the analytic gap dichotomy.

Definition 6.58 *Two orthogonal families* $\mathcal{M} = \{M_i : i \in I\}$ *and* $\mathcal{N} = \{N_i : i \in I\}$ *of subsets of* $\mathbb{N}$ *indexed by the same set* I *form a* biorthogonal gap *if for all* $i, j \in I$,
$$(M_i \cap N_j) \cup (M_j \cap N_i) = \emptyset \text{ iff } i = j.$$
When the index set I *is a perfect set of reals and the map* $i \mapsto (M_i, N_i)$ *is continuous and one-to-one, then we say that* $\mathcal{M} = \{M_i : i \in I\}$ *and* $\mathcal{N} = \{N_i : i \in I\}$ *form a* perfect biorthogonal gap.

Lemma 6.59 *Two orthogonal families* $\mathcal{M} = \{M_i : i \in I\}$ *and* $\mathcal{N} = \{N_i : i \in I\}$ *of subsets of* $\mathbb{N}$ *indexed by an uncountable set* I *and forming a biorthogonal gap cannot be countably separated.*

Proof. To see this, note that for every $C \subseteq \mathbb{N}$, the set
$$\{i \in I : M_i \subseteq^* C \text{ and } C \perp N_i\}$$
is at most countable. $\square$

Example 6.7.2 (Perfect Biorthogonal Gap) *This is a gap of subsets of the set* $2^{<\mathbb{N}}$ *of all finite binary sequences rather than of subsets of* $\mathbb{N}$. *We view* $2^{<\mathbb{N}}$ *as a tree ordered by the relation* $\sqsubseteq$ *of end-extension, and we view* $2^{\mathbb{N}}$ *as the set of all branches of the complete binary tree* $2^{<\mathbb{N}}$. *For* $x \in 2^{\mathbb{N}}$, *let*
$$M_x = \{\sigma \in 2^{<\mathbb{N}} : \sigma^\frown 0 \sqsubset x\} \quad \text{and} \quad N_x = \{\sigma \in 2^{<\mathbb{N}} : \sigma^\frown 1 \sqsubset x\}.$$
Then $\mathcal{M} = \{M_x : x \in 2^{\mathbb{N}}\}$ *and* $\mathcal{N} = \{N_x : x \in 2^{\mathbb{N}}\}$ *form a perfect biorthogonal gap of subsets of the complete binary tree* $2^{<\mathbb{N}}$.

The following result shows that this is a typical example of a noncountably separated analytic gap.

Theorem 6.60 (Perfect Biorthogonal Gap Theorem) *Suppose* $\mathcal{M}$ *and* $\mathcal{N}$ *are two orthogonal analytic families of infinite subsets of* $\mathbb{N}$ *closed under subsets and finite changes of their elements. Then either*

(1) $\mathcal{M}$ *and* $\mathcal{N}$ *can be separated by countably many sets, or else*

(2) *there is a biorthogonal gap* $\{M_x : x \in 2^{\mathbb{N}}\} \subseteq \mathcal{M}$ *and* $\{N_x : x \in 2^{\mathbb{N}}\} \subseteq \mathcal{N}$ *indexed continuously by the Cantor space.* $\square$

Here is a typical application of these two gap theorems, which shows that the conclusion of the Perfect Biorthogonal Gap Theorem can sometimes be extended to families of larger descriptive complexities, provided one adds some structural conditions.

Lemma 6.61 *Suppose that an analytic ideal* $\mathcal{I}$ *on* $\mathbb{N}$ *is represented as*
$$\mathcal{I} = \mathcal{I}_K(f, (f_n)),$$
where K *is some compact set of Baire Class 1 functions.[8] Then either* $\mathcal{I}$ *is countably generated or there is a perfect biorthogonal gap* $\{M_x : x \in 2^{\mathbb{N}}\} \subseteq \mathcal{I}$ *and* $\{N_x : x \in 2^{\mathbb{N}}\} \subseteq \mathcal{I}^\perp$ *such that* $(f_n)_{n \in M_x}$ *is a convergent sequence in* K *for all* $x \in 2^{\mathbb{N}}$.

---

[8]Thus, K is simply a collections of Baire Class 1 functions on $\mathbb{R}$ that is compact when equipped with the topology of pointwise convergence on $\mathbb{R}$.

Proof. Beside the assumption that the ideal $\mathcal{I}$ is analytic, we use only the following two additional properties, to be established in the next chapter when we develop some local Ramsey theory (see Corollary 7.52 and Theorem 7.53):

$$(\mathcal{I}^{\perp})^{\perp} = \mathcal{I}. \tag{6.92}$$

$$(\forall\{a_n : n \in \mathbb{N}\} \subseteq \mathcal{I}^{\perp} \cap \mathbb{N}^{[\infty]})(\exists b \in \mathcal{I}^{\perp})(\exists^{\infty} n) \ b \cap a_n \neq \emptyset. \tag{6.93}$$

Applying the Analytic Gap Theorem to the pair $(\mathcal{I}, \mathcal{I}^{\perp})$ and property (6.92) of $\mathcal{I}$, if the ideal $\mathcal{I}$ is not countably generated, we can find an $\mathcal{I}^{\perp}$-tree T all of whose branches are members of $\mathcal{I}$. Refining T, we may assume that

$$\{n \in \mathbb{N} : s \cup \{n\} \in T\} \cap \{n \in \mathbb{N} : t \cup \{n\} \in T\} = \emptyset \text{ for } s \neq t \in T. \tag{6.94}$$

It follows that a node $t \neq \emptyset$ in T is uniquely determined by its maximal element $\max(t)$. This means that we can transfer the restriction $\mathcal{I} \upharpoonright (\bigcup T)$ to the tree T itself, and since $T \setminus \{\emptyset\}$ is naturally isomorphic to the collection FIN of all nonempty finite subsets of $\mathbb{N}$ (considered a tree under the ordering $\sqsubseteq$ of end-extension), we may actually assume that $\mathcal{I}$ lives on the index set FIN. More precisely, we may assume that $\mathcal{I} = \mathcal{I}_K(f, (f_t)_{t \in \text{FIN}})$, for some compact set K of Baire Class 1 functions such that

(i) Every branch of the tree (FIN, $\sqsubseteq$) belongs to $\mathcal{I}$.

(ii) The set $\{t \cup \{n\} : n \in \mathbb{N}, n > \max(t)\}$ of all immediate successors of a given node $t \in \{\emptyset\} \cup \text{FIN}$ belongs to $\mathcal{I}^{\perp}$.

Color the space $\text{FIN}^{[\infty]}$ of all infinite sequences $B = (b_n)$ of finite subsets of $\mathbb{N}$ according to whether the sequence $(f_{b_{\restriction n} \cup \{\min(b_{n+1})\}})$ is convergent in K or not, where for an infinite block sequence $Z = (z_n)$ of elements of FIN and $n \in \mathbb{N}$, we let $z_{\restriction n} = \bigcup_{k=0}^{n} z_k$. Clearly, this is a Souslin-measurable coloring, so by Corollary 5.24, there is an $X \in \text{FIN}^{[\infty]}$ such that the set $[X]^{[\infty]}$ of all infinite block-subsequences of X is monochromatic. By Rosenthal Dichotomy Theorem 5.56, it must be that $(f_{b_{\restriction n} \cup \{\min(b_{n+1})\}})$ is convergent in K for all $B \in [X]^{[\infty]}$. Note that, given an infinite block subsequence Y of X, applying property (6.93) to the sequence $(\{y_n \cup \{\min(y_k)\} : k \in \mathbb{N}, k > n\})$ of elements of $\mathcal{I}^{\perp}$, we can find an infinite block-subsequence B of Y such that $(f_{b_{\restriction 2n} \cup \{\min(b_{2n+2})\}})$ converges to f, as well as an infinite block subsequence C of Y that has this property relative to the odd indexes. Applying Corollary 5.24 to the two respective Souslin-measurable colorings and observing that we can take union of two elements of $\mathcal{I}^{\perp}$ and remain in $\mathcal{I}^{\perp}$, we can find an infinite block subsequence Y of X such that the sequence $(f_{b_{\restriction n} \cup \{\min(b_{n+2})\}})$ converges to f for *all* $B \in [Y]^{[\infty]}$. By taking the limit of an appropriate Cantor scheme of finite increasing subsequences of $Y = (y_n)$, we arrive at a continuous one-to-one mapping $\sigma \mapsto B^{\sigma} = (b_n^{\sigma})$ from the Cantor set $2^{\mathbb{N}}$ into the set of all increasing subsequences of Y such that for every $\sigma <_{\text{lex}} \tau$ in $2^{\mathbb{N}}$, if $n = \min\{k : \sigma(k) \neq \tau(k)\}$, then

$$\langle b_l^{\sigma} : l < n \rangle = \langle b_l^{\tau} : l < n \rangle \tag{6.95}$$

and for some $k \in \mathbb{N}$ larger than all j for which y_j appears in the finite subsequence $\langle b_l^\sigma : l < n \rangle = \langle b_l^\tau : l < n \rangle$ of Y, we have

$$b_n^\sigma = y_{k+1} \text{ and } b_n^\tau = y_k \text{ and } b_{n+1}^\tau = y_{k+1}. \tag{6.96}$$

For $\sigma \in 2^{\mathbb{N}}$, set

$$M_\sigma = \{b_{\upharpoonright n}^\sigma \cup \{\min(b_{n+1})\} : n \in \mathbb{N}\} \text{ and } N_\sigma = \{b_{\upharpoonright n}^\sigma \cup \{\min(b_{n+2})\} : n \in \mathbb{N}\}.$$

Clearly, $\sigma \mapsto (M_\sigma, N_\sigma)$ is a continuous map from the Cantor set into the power set of FIN and $M_\sigma \cap N_\sigma = \emptyset$ for all $\sigma \in 2^{\mathbb{N}}$. To check that (M_σ, N_σ) $(\sigma \in 2^{\mathbb{N}})$ forms a perfect biorthogonal gap, consider $\sigma <_{\mathrm{lex}} \tau$ in $2^{\mathbb{N}}$ such that $n = \min\{k : \sigma(k) \neq \tau(k)\} > 0$. From Equation (6.95), we learn that $b_{\upharpoonright n-1}^\sigma = b_{\upharpoonright n-1}^\tau = s$, while from Equation (6.96), we learn that

$$s \cup \{\min(b_n^\sigma)\} = s \cup \{\min(b_{n+1}^\tau)\} \in M_\sigma \cap N_\tau, \tag{6.97}$$

as required. Note also that by our initial choice of the block sequence X, we know that $(f_t)_{t \in M_\sigma}$ is a converging sequence in K, for all $\sigma \in 2^{\mathbb{N}}$. This finishes the proof. □

Before proceeding further, let us give applications of Lemma 6.61 to the unconditional basic sequence problem and the separable quotient problem in Banach spaces.

Theorem 6.62 *Suppose X is a separable Banach space that contains no ℓ_1 but whose dual is not separable. Then its double dual X^{**} contains a normalized 1-unconditional sequence $(f_a)_{a \in 2^{\mathbb{N}}}$ indexed by the Cantor space $2^{\mathbb{N}}$.*

Proof. By Rosenthal's ℓ_1-theorem and by Rosenthal's dichotomy the double dual ball $K = B_{X^{**}}$ can be viewed as a compact set of Baire Class 1 functions defined on the dual ball B_{X^*} equipped with its weak*-topology. By our assumption that X^* is not separable, the point 0^{**} is not G_δ in K. By Lemma 6.61 we get a perfect family

$$f_a = \lim_{n \in M_a} f_n \quad (a \in 2^{\mathbb{N}})$$

of functionals satisfying the hypothesis of Corollary 6.46, so an application of this result gives us an 1-unconditional perfect subsequence f_a $(a \in 2^{\mathbb{N}})$. □

Recall that a Banach space E is said to be *analytically representable* if it is isomorphic to a subspace F of ℓ_∞ such that F as a topological subspace of $\mathbb{R}^{\mathbb{N}}$ is analytic, i.e., a continuous image of some Polish space. Equivalently, the dual ball B_{E^*} of E has a countable subset D that norms E such that E equipped with the topology τ_D that makes all the functionals from D continuous is an analytic space. While this class includes all separable Banach spaces, its nonseparable part seems to avoid all the standard pathologies seen in the class of nonseparable Banach spaces.

Theorem 6.63 *Every analytically representable Banach space E has a quotient[9] with a Schauder basis that can be taken unconditional if E is not separable.*

---

[9] All Banach spaces in question are assumed to be infinite-dimensional.

Proof. The separable case of this result is of course the well-known theorem of Johnson and Rosenthal (see Theorem 9.60). We shall reduce the rest of this result to Lemma 6.61 via Collorary 6.46, but this time we also use some well-known results from Banach-space theory which all could be found in textbooks of this area such as, for example, [64] and [42] or in Appendix 9.4 to this book. Applying Rosenthal's dichotomy to the dual ball B_{E^*}, viewed as a set of functions on the analytic space (E, τ_D), we conclude that either B_{E^*} is a pointwise compact set of Baire Class 1 functions on (E, τ_D), or else B_{E^*} equipped with the weak* topology contains a copy of $\beta\mathbb{N}$. In the latter case by Talagrand's theorem (see Theorem 9.63), E has quotient isomorphic to ℓ_∞, which is more than we need here. So we are left with the case that $K = B_{E^*}$ is a compact set of Baire Class 1 functions. If E is not separable, then 0^* is not a G_δ point of K, so, as above, we get 1-unconditional sequence in E^* of length continuum, so in particular 1-unconditional infinite sequence $(f_n) \subseteq E^*$. Now the conclusion follows from Theorem 9.60.

$\square$

Let us now go back to the main goal of this section and discuss the problem of characterizing analytic ideals $\mathcal{I}$ on $\mathbb{N}$ that are Tukey-equivalent to the ideal $\mathcal{I}_{\max}$ realizing the maximal Tukey type. We have the following partial information about this problem, in which for an infinite subset M of $\mathbb{N}$, we let $\mathcal{I}_{\max}^M$ be the isomorphic copy of $\mathcal{I}_{\max}$ induced by a bijection between $2^{<\mathbb{N}}$ and M.

Theorem 6.64 *Suppose that a proper analytic ideal $\mathcal{I}$ on $\mathbb{N}$ is represented by a compact set of Baire Class 1 functions defined on some Polish space X. Then either $\mathcal{I}$ is countably generated, or there is an infinite subset M of $\mathbb{N}$ such that $\mathcal{I} \upharpoonright M = \mathcal{I}_{\max}^M$.*

Proof. Let $\mathcal{I} = \mathcal{I}_K(\bar{0}, (f_n))$ for some sequence (f_n) of continuous real functions defined on some Polish space X whose closure K relative to the topology of pointwise convergence[10] on X consists only of Baire Class 1 functions defined on X. Let

$$\mathcal{M} = \{M \in \mathbb{N}^{[\infty]} : (f_n)_{n \in M} \text{ converges to } f_M \neq \bar{0}\},$$

$$\mathcal{N} = \{N \in \mathbb{N}^{[\infty]} : (f_n)_{n \in N} \text{ converges to } \bar{0}\}.$$

Then $\mathcal{M} \subseteq \mathcal{I}$ and $\mathcal{N} \subseteq \mathcal{I}^\perp$, where we recall that $\mathcal{I}^\perp$ denotes the *orthogonal* of $\mathcal{I}$, the family of all subsets X of $\mathbb{N}$ such that $X \cap M$ is finite for all $M \in \mathcal{I}$. Thus in particular, $\mathcal{M}$ and $\mathcal{N}$ are two orthogonal families of infinite subsets of $\mathbb{N}$. Suppose that $\mathcal{I}$ is not countably generated. By Lemma 6.61, there is a sequence (M_x, N_x) $(x \in 2^{\mathbb{N}})$ of elements of $\mathcal{M} \times \mathcal{N}$ forming a perfect biorthogonal gap, or in other words, a sequence with the following three properties:

(1) $x \mapsto (M_x, N_x)$ is continuous,

---

[10]In other words, we consider $\{f_n : n \in \mathbb{N}\}$ a subset of $\mathbb{R}^X$ and let K be its closure in the Tychonov topology of this power of the real line $\mathbb{R}$.

(2) $M_x \cap N_x = \emptyset$ for all x,

(3) $(M_x \cap N_y) \cup (N_x \cap M_y) \neq \emptyset$ whenever $x \neq y$.

For $x \in 2^{\mathbb{N}}$, we let f_x denote the pointwise limit f_{M_x} of the sequence $(f_n)_{n \in M_x}$. Recall that $(2^{\mathbb{N}})^{[\infty]}$ denotes the collection of all infinite *increasing* rapidly converging sequences of elements of the Cantor space $2^{\mathbb{N}}$. Similarly, we let $(2^{\mathbb{N}})^{[-\infty]}$ denote the collection of all infinite *decreasing* rapidly converging sequences of elements of $2^{\mathbb{N}}$. Consider the following two subsets of $(2^{\mathbb{N}})^{[\infty]}$ and $(2^{\mathbb{N}})^{[-\infty]}$, respectively:

$$\mathfrak{X}^+ = \{(x_n) \in (2^{\mathbb{N}})^{[\infty]} : (f_{x_n}) \text{ accumulates to } \bar{0}\},$$

$$\mathfrak{X}^- = \{(x_n) \in (2^{\mathbb{N}})^{[-\infty]} : (f_{x_n}) \text{ accumulates to } \bar{0}\}.$$

We claim that there is a perfect subset P of $2^{\mathbb{N}}$ such that $P^{[\infty]} \subseteq \mathfrak{X}^+$ and $P^{[-\infty]} \subseteq \mathfrak{X}^-$. Since $\mathfrak{X}^+$ and $\mathfrak{X}^-$ are Souslin-measurable, by Corollary 6.49 it suffices to show that there is no perfect set $Q \subseteq 2^{\mathbb{N}}$ such that $Q^{[\infty]} \cap \mathfrak{X}^+ = \emptyset$ or $Q^{[-\infty]} \cap \mathfrak{X}^- = \emptyset$. By symmetry, it suffices to show that

$$Q^{[\infty]} \cap \mathfrak{X}^+ \neq \emptyset \text{ for all perfect } Q \subseteq 2^{\mathbb{N}}. \tag{6.98}$$

Before showing this let us first establish the following fact:

$$\{f_x : x \in R\} \text{ accumulates to } \bar{0} \text{ for all uncountable } R \subseteq Q. \tag{6.99}$$

Otherwise, we can find a basic open neighborhood

$$B(Z; \epsilon) = \{g \in K : (\forall z \in Z) \, |g(z)| < \epsilon\}$$

of $\bar{0}$ given by some $\epsilon > 0$ and some finite set $Z \subseteq X$ whose closure misses $\{f_x : x \in R\}$. Then

$$A = \{n \in \mathbb{N} : f_n \notin B(Z; \epsilon)\}$$

is a member of $\mathcal{I}$ with the property that

$$M_x \subseteq^* A \text{ and } N_x \cap A =^* \emptyset \text{ for all } x \in R. \tag{6.100}$$

Choose $n \in \mathbb{N}$ and uncountable $R_0 \subseteq R$ such that

$$M_x \subseteq A \cup \{0, 1, \ldots, n\} \text{ and } N_x \cap A \subseteq \{0, 1, \ldots, n\} \text{ for all } x \in R_0. \tag{6.101}$$

Moreover, shrinking R_0, we may assume that for some $s, t \subseteq \{0, 1, \ldots, n\}$,

$$M_x \cap \{0, 1, \ldots, n\} = s \text{ and } N_x \cap \{0, 1, \ldots, n\} = t \text{ for all } x \in R_0. \tag{6.102}$$

It follows that $s \cap t = \emptyset$ (as $M_x \cap N_x = \emptyset$) and thus that

$$(M_x \cap N_y) \cup (N_x \cap M_y) = \emptyset \text{ for all } x, y \in R_0,$$

contradicting the condition (3). Having established Equation (6.99), we are ready to start with the proof of Equation (6.98). Let $q = \max(Q)$ and for $n \in \mathbb{N}$, let

$$X_n = \{f_x : x \in Q \text{ and } \triangle(x, q) \geq n\}.$$

Then by 6.99, $\bar{0} \in \overline{X_n}$ for all n, so we can choose an ultrafilter $\mathcal{U}$ on K extending the neighborhood filter of $\bar{0}$ such that $X_n \in \mathcal{U}$ for all n. By Corollary 7.55, there is a decreasing sequence (Z_n) of elements of $\mathcal{U}$ which converges to $\bar{0}$. We may assume that $Z_n \subseteq X_n$ for all n. Then for each $n \in \mathbb{N}$ we can choose a point $x_n \in Q$ such that $\triangle(x_n, q) \geq n$ and $f_{x_n} \in Z_n$. Then (x_n) converges to q, while (f_{x_n}) converges to $\bar{0}$. Going to a subsequence, we may assume that (x_n) is rapidly converging and increasing. Then $(x_n) \in Q^{[\infty]} \cap \mathfrak{X}^+$ and this finishes the proof that Equation (6.98) holds.

Fix a perfect subset P of $2^{\mathbb{N}}$ such that $P^{[\infty]} \subseteq \mathfrak{X}^+$ and $P^{[-\infty]} \subseteq \mathfrak{X}^-$. It follows that

$$\{f_x : x \in R\} \quad \text{accumulates to} \quad \bar{0} \quad \text{for every infinite} \quad R \subseteq P. \tag{6.103}$$

Fix a natural homeomorphism $\varphi : 2^{\mathbb{N}} \to P$ and choose a Cantor scheme (n_σ) $(\sigma \in 2^{<\mathbb{N}})$ of nonegaitive integers such that

(4) $n_\sigma \neq n_\tau$ whenever $\sigma \neq \tau$,

(5) $n_\sigma < n_\tau$ whenever $\sigma \sqsubset \tau$,

(6) $\{n_{(x \restriction k)} : k \in \mathbb{N}\} \subseteq M_{\varphi(x)}$ for all $x \in 2^{\mathbb{N}}$.

Recall now Theorem 6.27, which discusses about colorings of the space $\mathbb{Q}^{[\infty]}$ of rapidly increasing sequences of the rationals $\mathbb{Q} = (2^{<\mathbb{N}}, <_{\text{lex}})$. Of course we also have at our disposal the space $\mathbb{Q}^{[-\infty]}$ of rapidly decreasing sequences of the rationals. Consider the following two subsets of $(2^{<\mathbb{N}})^{[\infty]}$ and $(2^{<\mathbb{N}})^{[-\infty]}$, respectively,

$$\mathfrak{A}^+ = \{(\sigma_k) \in (2^{<\mathbb{N}})^{[\infty]} : (f_{n_{\sigma_k}}) \text{ accumulates to } \bar{0}\},$$

$$\mathfrak{A}^- = \{(\sigma_k) \in (2^{<\mathbb{N}})^{[-\infty]} : (f_{n_{\sigma_k}}) \text{ accumulates to } \bar{0}\}.$$

We claim that there is a strong subtree T of $2^{\mathbb{N}}$ of height ω such that $T^{[\infty]} \subseteq \mathfrak{A}^+$ and $T^{[-\infty]} \subseteq \mathfrak{A}^-$. By Theorem 6.27 or by Theorem 6.29 and by symmetry, it suffices to show that

$$T^{[\infty]} \cap \mathfrak{A}^+ \neq \emptyset \text{ for every strong subtree } T \subseteq 2^{<\mathbb{N}} \text{ of infinite height.} \tag{6.104}$$

To see this, choose an infinite increasing rapidly converging sequence $(x_n) \in 2^{\mathbb{N}}$ such that for all n,

$$C_n = \{x_n \restriction k : k > \triangle(x_n, x_{n+1})\} \cap T$$

is an infinite chain of T. For $n \in \mathbb{N}$, let $B_n = \{n_\sigma : \sigma \in C_n\}$. Then B_n is an infinite subset of $M_{\varphi(x_n)}$ for all n. By Equation (6.103),

$$\{f_{n_\sigma} : (\exists n \geq m) \ \sigma \in B_n\} \quad \text{accumulates to} \quad \bar{0} \quad \text{for all} \quad m \in \mathbb{N}. \tag{6.105}$$

By Corollaries 7.52 and 7.56, there is an infinite $B \subseteq \bigcup_{n=0}^{\infty} B_n$ such that $B \cap B_n$ is finite for all n and such that the sequence $(f_\sigma)_{\sigma \in B}$ converges pointwise to $\bar{0}$. Refining further, we may assume that there is an enumeration (σ_n) of B that forms an increasing rapidly converging sequence of elements of T. Since clearly (σ_n) belongs also to $\mathfrak{A}^+$, this establishes the validity of Equation (6.104). By Theorem 6.27 or Theorem 6.29, we can fix a strong subtree T of $2^{<[\infty]}$ of infinite height such that $T^{[\infty]} \subseteq \mathfrak{A}^+$ and $T^{[-\infty]} \subseteq \mathfrak{A}^-$. Let Q be the set of unions of all infinite chains of T, i.e., the set of all $x \in 2^{\mathbb{N}}$ for which the set $C_x = \{\sigma \in T : \sigma \sqsubset x\}$ is infinite. For $x \in Q$, let $\bar{M}_x = \{n_\sigma : \sigma \in C_x\}$. Then $\bar{M}_x$ is an infinite subset of $M_{\varphi(x)}$ for all $x \in Q$. Let

$$\bar{M} = \bigcup_{x \in Q} \bar{M}_x.$$

Let $\bar{\mathcal{I}}_{\max}$ be the ideal on the index set generated by the perfect almost disjoint family $(\bar{M}_x)$ $(x \in Q)$. Note that the bijection $\sigma \mapsto n_\sigma$ between T and $\bar{M}$ moves $\bar{\mathcal{I}}_{\max}$ as defined on the tree T rather than the complete binary tree $2^{<\mathbb{N}}$ to a realization $\bar{\mathcal{I}}_{\max}^{\bar{M}}$ of the ideal $\mathcal{I}_{\max}$ on the set $\bar{M}$. Note also that Equation (6.104) and the fact that for all $x \in Q$ the sequence $(f_{n_\sigma})_{\sigma \in \bar{M}_x}$ converges to $f_{\varphi(x)} \neq \bar{0}$ show that

$$\mathcal{I} \cap \mathcal{P}(\bar{M}) = \bar{\mathcal{I}}_{\max}^{\bar{M}}.$$

This finishes the proof.   □

The following reformulation of Theorem 6.64 is also worth mentioning.

Theorem 6.65 *Suppose K is a separable compact set of Baire class-1 functions defined on some Polish space X. Let D be a countable dense subset of K, and let f be a point of K that is not G_δ in K. Then there is a homeomorphic embedding*

$$\varphi : \hat{A}(2^{\mathbb{N}}) \to K$$

such that $\varphi(\bar{0}) = f$ and $\varphi[\{\delta_s : s \in 2^{<\mathbb{N}}\}] \subseteq D$.   □

Remark 6.66 Note that this shows that the 1-unconditional sequence of Theorem 6.62 can be taken to be weak*-null. This would make the proof of Theorem 6.63 even more direct, thus avoiding some of the cases (see [42]; Statement 5.10 on p. 205).

We finish this section with a discussion of the following problem about the Tukey-maximal analytic ideal $\mathcal{I}_{\max}$, where for two ideals $\mathcal{I}$ and $\mathcal{J}$, we define their ideal sum by,

$$\mathcal{I} \oplus \mathcal{J} = \{a \cup b : a \in \mathcal{I}_{\text{even}},\ b \in \mathcal{J}_{\text{odd}}\}, \tag{6.106}$$

where we let $\mathcal{I}_{\text{even}}$ be the isomorphic copy of $\mathcal{I}$ on the set $2\mathbb{N}$ of even integers, while $\mathcal{J}_{\text{odd}}$ is the isomorphic copy of $\mathcal{J}$ on the set $2\mathbb{N} + 1$ of odd integers.

Question 6.67 *Suppose that $\mathcal{I}$ and $\mathcal{J}$ are analytic ideals on $\mathbb{N}$ and that $\mathcal{I}_{\max} \leq_T \mathcal{I} \oplus \mathcal{J}$. Is it true that then either $\mathcal{I}_{\max} \leq_T \mathcal{I}$, or else $\mathcal{I}_{\max} \leq_T \mathcal{J}$?*

One route toward this question might be through the variation $\mathcal{J} \leq_{ST} \mathcal{I}$ of the relation $\mathcal{J} \leq_T \mathcal{I}$ of Tukey-reducibility defined to hold whenever there is a *Souslin-measurable* map $g : \mathcal{J} \to \mathcal{I}$ satisfying (tk), or equivalently, a Souslin-measurable map $f : \mathcal{I} \to \mathcal{J}$ satisfying (cf1) and (cf2). This is shown above in the case of unrestricted Tukey-reducibility, but to see that the equivalence remains true in the Souslin-measurable realm, consider a *Souslin-Tukey* map $g : \mathcal{J} \to \mathcal{I}$ witnessing $\mathcal{J} \leq_{ST} \mathcal{I}$, or in other words, a Souslin-measurable map satisfying (tk), and define $f : \mathcal{I} \to \mathcal{J}$ by

$$f(a) = \bigcup \{ b \in \mathcal{J} : g(b) \subseteq a \}. \tag{6.107}$$

Then f is a Souslin-measurable map satisfying (cf1) and (cf2). Conversely, suppose $f : \mathcal{I} \to \mathcal{J}$ is a Souslin-measurable map satisfying (cf1) and (cf2) and define $g : \mathcal{J} \to \mathcal{I}$ by letting

$$g(b) = \bigcap \{ a \in \mathcal{I} : b \subseteq f(a) \}. \tag{6.108}$$

Then g is a Souslin-measurable map satisfying (tk). The variation $\leq_{ST}$ is called the *Souslin-Tukey reducibility* and the corresponding equivalence relation $\equiv_{ST}$, the *Souslin-Tukey equivalence*. The exact relationship between the two quasi-orderings $\leq_T$ and $\leq_{ST}$ is not yet clear although all known results establishing a Tukey reducibility also establish the stronger Souslin-Tukey reducibility. So in particular, the following problem remains open.

Question 6.68 *Suppose $\mathcal{I}$ and $\mathcal{J}$ are analytic ideals on $\mathbb{N}$ such that the relation $\mathcal{I} \leq_T \mathcal{J}$ holds. Can one conclude then that $\mathcal{I} \leq_{ST} \mathcal{J}$ also holds? In particular, is this true when $\mathcal{I} = \mathcal{I}_{\max}$?*

It is known, however, that in many important situations the answer to this question is affirmative:

Theorem 6.69 *If $\mathcal{I}$ and $\mathcal{J}$ are analytic P-ideals[11] on $\mathbb{N}$, then $\mathcal{I} \leq_T \mathcal{J}$ is equivalent to $\mathcal{I} \leq_{ST} \mathcal{J}$.*

The following result shows that if this particular case of Question 6.68 has an affirmative answer, so does also our initial Question 6.67.

Theorem 6.70 *Suppose that $\mathcal{I}$ and $\mathcal{J}$ are analytic ideals on $\mathbb{N}$ and that $\mathcal{I}_{\max} \equiv_{ST} \mathcal{I} \oplus \mathcal{J}$. Then either $\mathcal{I}_{\max} \equiv_{ST} \mathcal{I}$ or $\mathcal{I}_{\max} \equiv_{ST} \mathcal{J}$.*

Proof. Let $P \subseteq \mathcal{I}_{\max}$ be a fixed perfect almost disjoint family that generates $\mathcal{I}_{\max}$. Let $g : \mathcal{I}_{\max} \to \mathcal{I} \oplus \mathcal{J}$ be a fixed Souslin-Tukey map witnessing $\mathcal{I}_{\max} \leq \mathcal{I} \oplus \mathcal{J}$. Define a coloring $P^{[\infty]} = C_0 \cup C_1$ as follows

$$(x_n) \in C_0 \text{ iff } \left(\bigcup_{n=0}^{\infty} g(x_n) \right) \cap 2\mathbb{N} \notin \mathcal{I}_{\text{even}}. \tag{6.109}$$

---

[11] Recall that an ideal $\mathcal{I}$ is a *P-ideal* if for every sequence $(A_n)_{n=0}^{\infty} \subseteq \mathcal{I}$, there is $B \in \mathcal{I}$ such that $A_n \setminus B$ is finite for all n.

Clearly, this is a Souslin-measurable coloring, so Corollary 6.49 applies to it, and we have the following two cases to consider.

Case 1. There is perfect $Q \subseteq P$ such that $Q^{[\infty]} \subseteq C_0$. Pick two disjoint perfect subsets Q_0 and Q_1 of Q and a homeomorphism $\varphi : Q_0 \to Q_1$ such that $x <_{\text{lex}} y$ implies $\varphi(y) <_{\text{lex}} \varphi(x)$. Let $R = \{x \cup \varphi(x) : x \in Q_0\}$. Then R is also a perfect almost disjoint family of infinite subsets of $\mathbb{N}$, so the ideal $\mathcal{I}_R$ is Souslin-Tukey equivalent to $\mathcal{I}_{\text{max}}$. Then for every $b \in \mathcal{I}_{\text{even}}$, the set

$$Q_0(b) = \{x \in Q_0 : g(x) \cup g(\varphi(x)) \subseteq b\} \qquad (6.110)$$

is finite. This follows from the fact that every infinite sequence $(x_n) \subseteq Q_0$ contains an infinite subsequence (x_{n_k}) such that either $(x_{n_k}) \in Q^{[\infty]}$ or $(\varphi(x_{n_k})) \in Q^{[\infty]}$. So

$$h(b) = \{0, 1, \dots |Q_0(b)|\} \cup \bigcup \{x \cup \varphi(x) : x \in Q_0(b)\} \qquad (6.111)$$

defines a Souslin-measurable map $h : \mathcal{I}_{\text{even}} \to \mathcal{I}_R$ that is monotone and that has a cofinal range in $\mathcal{I}_R$; i.e., it satisfies conditions (cf 1) and (cf 2) above. This shows that $\mathcal{I}_R \leq_{ST} \mathcal{I}_{\text{even}}$.

Case 2. There is a perfect $Q \subseteq P$ such that $Q^{[\infty]} \cap C_0 = \emptyset$. The argument from the previous case gives us a perfect almost disjoint family S such that $\mathcal{I}_S \leq_{ST} \mathcal{I}_{\text{odd}}$. This finishes the proof. $\qquad \square$

NOTES TO CHAPTER SIX

Milliken's space of strong subtrees was developed in his paper [78]. We have seen above that on the basis of this space, most of the known perfect-set theorems could naturally be proved. The result of Stern appears in his paper [103]. Devlin's two theorems appear in his thesis [19]. Devlin's work was motivated by a correction to an earlier version of the finite-dimensional Ramsey theorem for the rationals due to Laver, who was the first to show the existence of Ramsey degrees in this case. The corresponding results about the random graph can be found in the papers of Pouzet and Sauer [87] and Laflamme-Sauer-Vuksanovic [60]. Galvin's lemma appears originally in his announcement [33]. It is a substantial extension of an earlier generalization of Ramsey's theorem due to Nash-Williams [82] although in some literature the two results are identified. Baumgartner proved his Theorem 6.31 in his paper [7] using a coloring that gives a less precise result than the one we presented, provided it is stated in terms of the Cantor-Bendixson derivative. Similarly, Theorem 6.38 appears in a paper of Velickovic and Woodin [111], with a coloring that is not continuous. Corollary 6.17 appears in a paper of Stern [103], who uses a metamathematical argument in its proof. The two theorems of Mycielski appear in his papers [80] and [81]. Corollary 6.46 to Mycielski's theorem appears in the paper of Argyros-Dodos-Kanellopoulos [1]. The result of Blass appears in his paper [9], and the result of Louveau-Shelah-Velickovic appears in their paper [70]. It should also be mentioned that Blass's theorem was conjectured by Galvin [34] long before who proved it for dimensions $k = 2$ and 3. The measure case of Blass' theorem in dimension $k = 2$ was more recently reproved by Laczkovich [59] and more results in this direction can be found in papers [105], [62], and [53]. A first attempt toward a parametrized version of Blass's theorem appears in the paper of Pawlikowski [85]. Krawczyk's theorem appears in his paper [57]. Tukey-reducibility between directed sets has been introduced by Tukey in [110] and was motivated by the theory of Moore-Smith convergence. That this notion can express rough classification results in various areas of mathematics was realized much latter. That Tukey-reducibility is equivalent to Tukey-Souslin-reducibility in a rather natural class of "basic" directed orders is a result of Solecki-Todorcevic [101], where the reader in particular can find a proof of Theorem 6.69. The Analytic Gap Theorem and the Biorthogonal Gap Theorem appear in the paper [106] of the author, where the reader can find their proofs and a more complete treatment of the subject of gaps in the quotient algebra $\mathcal{P}(\mathbb{N})/\text{FIN}$. Lemma 6.61 and Theorem 6.64 appear in the paper [107] of the author, where we the reader can find a more extensive analysis of analytic ideals realizable by compact sets of Baire Class 1 functions. The applications of this result given above as Theorems 6.62 and 6.63 were recently found by Argyros-Dodos-Kanellopoulos [1]. Regarding the part of the proof of Theorem 6.63, we note that the fact that if the dual E^* of a given Banach space has an infinite unconditional sequence, then E has a quotient with an unconditional basis was first pointed out to us by Rosenthal [94]. Indeed, this follows rather directly from the results due to Johnson-Rosenthal [51]. We also note that Lemma 6.61 is one of the interesting instances in which we have the conclusion of the Biorthogonal Gap Theorem for the case in which the two orthogonal families are of a bigger complexity. We shall encounter this again in the next chapter when treating some problems about sequential convergence in the context of countable spaces with analytic topologies.

Chapter Seven

Local Ramsey Theory

7.1 LOCAL ELLENTUCK THEORY

The local Ellentuck theory in its most fundamental form deals with Ramsey spaces of the form

$$(\mathbb{N}^{[\infty]}, \mathcal{H}, \subseteq, r), \tag{7.1}$$

where $\mathbb{N}^{[\infty]}$ is the collection of all infinite subsets of $\mathbb{N}$, $\emptyset \neq \mathcal{H} \subseteq \mathbb{N}^{[\infty]}$ and where $r = (r_n)$ is the standard sequence of restriction maps:

$$r_n(A) = \text{ the first } n \text{ members of } A. \tag{7.2}$$

Thus, the extremal case $\mathcal{H} = \mathbb{N}^{[\infty]}$ is the original Ellentuck space, but as we shall soon see there are many other interesting choices for $\mathcal{H}$. The need to develop such a theory comes from the fact that sometimes when considering colorings of $\mathbb{N}^{[\infty]}$ one needs a monochromatic cube $M^{[\infty]}$ of a special form, not just an arbitrary cube given to us by the Ellentuck theory. Here is a typical situation in which such a need arises. Suppose K is a given topological space, $x \in K$, and $(x_n) \subseteq K \setminus \{x\}$ is a sequence of points *accumulating* to x, and suppose that we wish to find a subsequence (x_{n_k}) *converging* to x. This is equivalent to analyzing the following coloring of $\mathbb{N}^{[\infty]}$:

$$\mathcal{C}_0 = \{M \in \mathbb{N}^{[\infty]} : x \in \overline{\{x_n : n \in M\}}\},$$

$$\mathcal{C}_1 = \{M \in \mathbb{N}^{[\infty]} : x \notin \overline{\{x_n : n \in M\}}\},$$

Note that $(x_n)_{n \in M}$ converges to x if and only if $M^{[\infty]} \subseteq \mathcal{C}_0$. The local Ellentuck theory gives us $M \in \mathcal{H}$ such that $M^{[\infty]}$ is monochromatic if the coloring is of a reasonable complexity. The point is that the new theory gives us a chance to get the alternative $M^{[\infty]} \subseteq \mathcal{C}_0$ by appropriately choosing the coideal $\mathcal{H}$.

The first necessary assumption on $\mathcal{H} \subseteq \mathbb{N}^{[\infty]}$ is that it must be a *coideal*, i.e., it must have the following two properties:

(a) $A \supseteq B$ and $B \in \mathcal{H} \rightarrow A \in \mathcal{H}$,

(b) $A \cup B \in \mathcal{H} \rightarrow A \in \mathcal{H} \vee B \in \mathcal{H}$.

Thus $\mathcal{H}$ is a coideal on $\mathbb{N}$ iff its complement $\mathcal{I} = P(\mathbb{N})\backslash\mathcal{H}$ is an ideal on $\mathbb{N}$ containing all finite subsets of $\mathbb{N}$. The prototype example of a coideal on $\mathbb{N}$ is of course the collection $\mathbb{N}^{[\infty]}$ of all infinite subsets of $\mathbb{N}$, but we shall see that there exist many other interesting examples.

We fix from now on a coideal $\mathcal{H}$ on $\mathbb{N}$, and we consider the corresponding $\mathcal{H}$-*basic* subsets of $\mathbb{N}^{[\infty]}$ defined as follows:

$$[a, M] = \{A \in \mathbb{N}^{[\infty]} : a \text{ is an initial segment of } A \text{ and } A \subseteq a \cup M\}, \quad (7.3)$$

where $a \in \mathbb{N}^{[<\infty]}$ [1] and where $M \in \mathcal{H}$. Note that all basic sets are nonempty and closed in the metrizable topology of $\mathbb{N}^{[\infty]}$. Having defined $\mathcal{H}$-basic sets, one immediately has the corresponding Baire-category notions defined as follows.

Definition 7.1 *A subset $\mathcal{X} \subseteq \mathbb{N}^{[\infty]}$ is $\mathcal{H}$-Baire if for every $a \in \mathbb{N}^{[<\infty]}$ and $M \in \mathcal{H}$, there exist $b \in \mathbb{N}^{[<\infty]}$ and $N \in \mathcal{H}$ such that $[b, N] \subseteq [a, M]$ and $[b, N] \subseteq \mathcal{X}$ or $[b, N] \subseteq \mathcal{X}^c$. If every $\mathcal{H}$-basic set can be refined to an $\mathcal{H}$-basic subset avoiding $\mathcal{X}$, then $\mathcal{X}$ is called $\mathcal{H}$-meager.*

Of course, we also have the corresponding stronger Ramsey notions.

Definition 7.2 *A subset $\mathcal{X} \subseteq \mathbb{N}^{[\infty]}$ is $\mathcal{H}$-Ramsey if for every $a \in \mathbb{N}^{[<\infty]}$ and $M \in \mathcal{H}$, there is an $N \in [a, M] \cap \mathcal{H}$ such that $[a, N] \subseteq \mathcal{X}$ or $[a, N] \subseteq \mathcal{X}^c$. If for every $[a, M]$ with $a \in \mathbb{N}^{[<\infty]}$ and $M \in \mathcal{H}$ we can find an $N \in [a, M] \cap \mathcal{H}$ such that $[a, N] \subseteq \mathcal{X}^c$, then the set $\mathcal{X}$ is called $\mathcal{H}$-Ramsey null.*

The purpose of this section is to describe properties of coideals $\mathcal{H}$ on $\mathbb{N}$ that guarantee rich fields of $\mathcal{H}$-Ramsey sets. In particular, we wish to find minimal restrictions on such $\mathcal{H}$ that guarantee the analog of the Ellentuck theorem for the triple $(\mathbb{N}^{[\infty]}, \mathcal{H}, \subseteq)$.

Notation. For $M \in \mathbb{N}^{[\infty]}, a \in \mathbb{N}^{[<\infty]}$ and $n \in \mathbb{N}$, we set

$$\begin{aligned}
\mathcal{H}|M &= \{N \in \mathcal{H} : N \subseteq M\}, \\
M/n &= M\backslash\{0, 1, ..., n\}, \\
M/a &= M\backslash\{0, 1, ..., \max(a)\}.
\end{aligned} \quad (7.4)$$

Definition 7.3 *A coideal $\mathcal{H}$ on $\mathbb{N}$ is selective if for every decreasing sequence (A_n) of elements of $\mathcal{H}$, there is a $B \in \mathcal{H}$ such that $B/n \subseteq A_n$ for all $n \in B$. Such a set B will be called a diagonalization of (A_n).*

The following prototype example for this notion should of course be mentioned first.

Example 7.1.1 *The coideal $\mathcal{H} = \mathbb{N}^{[\infty]}$ of all infinite subsets of $\mathbb{N}$ is selective.*

The following example is, however, much more subtle.

---

[1]Recall that $\mathbb{N}^{[<\infty]}$ denotes the collection of finite subsets of $\mathbb{N}$ including $\emptyset$.

Example 7.1.2 (Mathias). *Let $\mathcal{A}$ be an infinite family of infinite subsets of $\mathbb{N}$ such that $A \cap B$ is finite for every pair A, B of distinct elements of $\mathcal{A}$. Let $\mathcal{H}$ be the collection of all subsets of $\mathbb{N}$ that cannot be covered up to a finite set by finitely many members of $\mathcal{A}$. Then $\mathcal{H}$ is a selective coideal on $\mathbb{N}$.*

Proof. To see this, let (A_n) be a given decreasing sequence of elements of $\mathcal{H}$. We need to show that one of its diagonalizations belongs to $\mathcal{H}$. We split the discussion into the following two cases.

Case 1. There is an infinite sequence (C_k) of distinct members of $\mathcal{A}$ such that $C_k \cap A_n$ is infinite for all k and n. Construct a strictly increasing sequence $(n_j) \subseteq \mathbb{N}$ as follows:

$$n_0 = \min A_0 \text{ and } n_{i+1} = \min(A_{n_i} \cap (C_k/n_i)), \qquad (7.5)$$

where $k = \max\{\ell : 2^\ell \text{ divides } i\}$. Let $B = \{n_i : i \in \mathbb{N}\}$. Then B belongs to $\mathcal{H}$, since it has infinite intersection with infinitely many members of $\mathcal{A}$. On the other hand, $B/n_i \subseteq A_{n_i}$ for all i, as required.

Case 2. There is no $(C_k) \subseteq \mathcal{A}$ as in Case 1. Choose an arbitrary diagonalization B_0 of (A_n). If $B_0 \in \mathcal{H}$, we are done. Otherwise, there is a finite $\mathcal{A}_0 \subseteq \mathcal{A}$ such that $B_0 \subseteq^* \bigcup \mathcal{A}_0$. Note that $(A_n \backslash \cup \mathcal{A}_0) \in \mathcal{H}$ for all n. Choose an infinite diagonalization B_1 of $(A_n \backslash \cup \mathcal{A}_0)$. If $B_1 \in \mathcal{H}$, we are done. Otherwise, there is a finite $\mathcal{A}_1 \subseteq \mathcal{A}$ such that $B_1 \subseteq^* \bigcup \mathcal{A}_1$, and so on. This process must stop at some finite stage or else we enter into Case 1. $\qquad \square$

The following reformulation of the notion of selectivity is quite useful.

Lemma 7.4 *A coideal $\mathcal{H}$ on $\mathbb{N}$ is selective iff it has the following two properties:*

(p) *For every decreasing sequence $(A_n) \subseteq \mathcal{H}$, there is $B \in \mathcal{H}$ such that $B \subseteq^* A_n$ for all n.*

(q) *For every $A \in \mathcal{H}$ and every disjoint partition $A = \bigcup_{k=0}^{\infty} F_k$ with F_k finite for all k, there is $B \in \mathcal{H}|A$ such that $|B \cap F_k| \leq 1$ for all k.*

Proof. It is clear that every selective coideal $\mathcal{H}$ satisfies (p) so let us check that it satisfies (q). For $n \in \mathbb{N}$, let

$$A_n = \bigcup \{F_k : k \in \mathbb{N}, F_k \cap \{0, ..., n\} = \emptyset\}. \qquad (7.6)$$

Choose $B \in \mathcal{H}|A$ such that $B/n \subseteq A_n$ for all $n \in B$. Then $|B \cap F_k| \leq 1$ for all k.

Suppose now that $\mathcal{H}$ is a coideal on $\mathbb{N}$ satisfying (p) and (q). Let (A_n) be a given infinite decreasing sequence of members of $\mathcal{H}$. By (p) we choose $A \in \mathcal{H}$ such that $A \subseteq^* A_n$ for all n. Choose a strictly increasing sequence $(n_k) \subseteq \mathbb{N}$ such that $n_0 = 0$ and

$$A \backslash A_{n_i} \subseteq \{0, ..., n_{i+1}\}. \tag{7.7}$$

By (q), there is $B \in \mathcal{H}|A$ such that $|B \cap [n_k, n_{k+1})| \leq 1$ for all k. Splitting B into its intersection with the union of even- and the union of odd-numbered intervals $[n_k, n_{k+1})$, letting the coideal $\mathcal{H}$ decide for one of these subsets of B, and then changing B with the chosen set, we may assume that, say, $B \cap [n_{2k}, n_{2k+1}) = \emptyset$ for all k. It follows that $B/n \subseteq A_n$ for all $n \in B$, as required. This finishes the proof. □

The following fact relates the second of the two restrictions on a given coideal $\mathcal{H}$ on $\mathbb{N}$ to the Ramsey theoretic notions associated with $\mathcal{H}$ introduced above in Definition 7.2.

Lemma 7.5 *If a coideal $\mathcal{H}$ on $\mathbb{N}$ fails to satisfy the property* (q)*, then there is an $\mathcal{H}$-Baire set that is not $\mathcal{H}$-Ramsey.*

Proof. Choose a partition $N = \bigcup_{k=0}^{\infty} F_k$ of an $N \in \mathbb{N}$ with all F_k finite for which there is no $M \in \mathcal{H}|N$ such that $|M \cap F_k| \leq 1$ for all k. Let[2]

$$\mathcal{X} = \{X \in \mathbb{N}^{[\infty]} : (\exists k)\ X(2) \subseteq F_k\}.$$

Clearly, $\mathcal{X}$ is a clopen subset of $\mathbb{N}^{[\infty]}$ and so in particular is $\mathcal{H}$-Baire. On the other hand, note that there is no $M \in \mathcal{H}|N$ such that $M^{[\infty]} \subseteq \mathcal{X}$, or $M^{[\infty]} \cap \mathcal{X} = \emptyset$. So $\mathcal{X}$ is not $\mathcal{H}$-Ramsey. □

This shows that requirement (q) is in some sense necessary if we are to have that the triple $(\mathbb{N}^{[\infty]}, \mathcal{H}, \subseteq)$ forms a Ramsey space. However, as we shall see now, property (p) can be further weakened without affecting the Ramsey theoretic properties of a given coideal $\mathcal{H}$ on $\mathbb{N}$. To introduce this weakening, we need the following notion.

Definition 7.6 *A set $\mathcal{D} \subseteq \mathcal{H}$ is* dense-open *in $\mathcal{H}$ if*

(1) $(\forall M \in \mathcal{H})(\exists N \in \mathcal{D})\ N \subseteq M$,

(2) $(\forall M \in \mathcal{H})(\forall N \in \mathcal{D})(M \subseteq N \rightarrow M \in \mathcal{D})$.

Definition 7.7 *A coideal $\mathcal{H}$ on $\mathbb{N}$ has the* weak (p) *property or simply, property* (p^w)*, if for every sequence $(\mathcal{D}_n)$ of dense-open subsets of $\mathcal{H}$ and every $N \in \mathcal{H}$, there is an $M \in \mathcal{H}|N$ such that for all n there is an $X \in \mathcal{D}_n$ such that $M \subseteq^* X$.*

Note that if $\mathcal{H}$ is a maximal coideal on $\mathbb{N}$, or equivalently, a nonprincipal ultrafilter on $\mathbb{N}$, then the weak (p) property is in fact equivalent to the property (p). In this case $\mathcal{H}$ becomes what is usually called a *P-point ultrafilter*, and this classical notion is the reason behind our use of the letter "p" for this variety of properties. The following fact gives one of the reasons we would consider this property at all.

---

[2]Recall that $X(2)$ was our notation for the set formed by the first two elements of X according to the natural increasing enumeration of X.

Lemma 7.8 *If a coideal $\mathcal{H}$ on $\mathbb{N}$ fails to satisfy the weak property* (p), *then the ideal of $\mathcal{H}$-Ramsey null sets is not a σ-ideal.*

Proof. Choose a sequence $(\mathcal{D}_n)$ of dense-open subsets of $\mathcal{H}$ and $M \in \mathcal{H}$ such that there is no $N \in \mathcal{H}|N$ such that for all n we can find $X \in \mathcal{D}_n$ with the property that $N \subseteq^* X$. For a given integer n, set

$$\mathcal{X}_n = \{Y \in \mathbb{N}^{[\infty]} : (\forall X \in \mathcal{D}_n)\ Y \not\subseteq^* X\}.$$

Then $\mathcal{X}_n$ is an $\mathcal{H}$-Ramsey null set for each n, but there is no $N \in \mathcal{H}|M$ such that $N^{[\infty]} \cap \bigcup_{n=0}^{\infty} \mathcal{X}_n = \emptyset$. So, in particular, the union $\bigcup_{n=0}^{\infty} \mathcal{X}_n$ is not $\mathcal{H}$-Ramsey null. □

We now come to the notion needed for the development of something that we call *the local Ellentuck theory* and a notion that is in some sense optimal for this purpose.

Definition 7.9 *A coideal $\mathcal{H}$ on $\mathbb{N}$ is* semiselective *if it has the following two properties:*

(p^w) *For every sequence $(\mathcal{D}_n)$ of dense-open subsets of $\mathcal{H}$ and every $N \in \mathcal{H}$ there is an $M \in \mathcal{H}|N$ such that for all n, there is an $X \in \mathcal{D}_n$ such that $M \subseteq^* X$.*

(q) *For every $N \in \mathcal{H}$ and every disjoint partition $N = \bigcup_{k=0}^{\infty} F_k$ with F_k finite for all k, there is an $M \in \mathcal{H}|N$ such that $|M \cap F_k| \leq 1$ for all k.*

Lemma 7.10 *A coideal $\mathcal{H}$ on $\mathbb{N}$ is semiselective if and only if for every sequence $(\mathcal{D}_n)$ of dense-open subsets of $\mathcal{H}$ and every $N \in \mathcal{H}$, there is $M \in \mathcal{H}|N$ such that $M/n \in \mathcal{D}_n$ for all $n \in M$.*[3]

Proof. To prove the direct implication, consider a sequence $(\mathcal{D}_n)$ of dense-open subsets of $\mathcal{H}$ and $N \in \mathcal{H}$. Using the property p^w of $\mathcal{H}$, find $M_0 \in \mathcal{H}|N$ such that for all n we can fix an $X_n \in \mathcal{H}$ such that $M \subseteq^* X_n$. For a given n, let i_n be the minimal integer $i \geq n$ such that $M \subseteq (X_n)/i$. Choose a strictly increasing sequence (m_k) of nonnegative integers such that $m_0 = 0$, and such that m_{k+1} is the minimal integer $m > m_k$ such that $m > i_n$ for all $n \leq m_k$. By property (q) of $\mathcal{H}$, there is an $M_1 \in \mathcal{H}|M_0$ such that

$$|M_1 \cap [n_k, n_{k+1})| \leq 1 \text{ for all } k. \tag{7.8}$$

Splitting M_1 into its intersection with the union of even and odd numbered intervals $[m_k, m_{k+1})$ and letting $\mathcal{H}$ decide for one of these two sets, we may further assume that for, example,

$$M_1 \cap [n_{2k}, n_{2k+1}) = \emptyset \text{ for all } k. \tag{7.9}$$

It follows that $(M_1)/m \in \mathcal{D}_m$ for all $m \in M_1$, as required.

For the converse implication, it remains to prove only that any coideal $\mathcal{H}$ that can diagonalize any given sequence of its dense-open subsets has

---

[3]When this happens we say that M *diagonalizes* the given sequence $(\mathcal{D}_n)$.

property (q) since property (p^w) is immediate. So let $N = \bigcup_{k=0}^{\infty} F_k$ be a given partition of an $N \in \mathcal{H}$ into pairwise disjoint finite sets F_k. Note that we can actually assume that there is a strictly increasing sequence (m_k) of integers such that $m_0 = 0$ and such that $F_k = [m_k, m_{k+1}) \cap N$ for all k. For a given $n \in \mathbb{N}$, let $\mathcal{D}_n$ be the collection of all $X \in \mathcal{H}$ such that either $X \cap N = \emptyset$, or else $|X \cap [m_k, m_{k+1})| \leq 1$ for the unique k such that $m_k \leq n < n_{k+1}$. Clearly, $\mathcal{D}_n$ is dense-open in $\mathcal{H}$ for all n. By our hypothesis, there is an $M \in \mathcal{H}|N$ such that $M/n \in \mathcal{D}_n$ for all $n \in M$. It follows that $|M \cap [m_k, m_{k+1})| \leq 1$ for all k, as required. $\qquad\square$

Note also the following immediate variation on the previous lemma, which, however, is the way use the semiselectivity of a given coideal in the arguments below.

Lemma 7.11 *A coideal $\mathcal{H}$ on $\mathbb{N}$ is semiselective if for every sequence $(\mathcal{D}_b)$ ($b \in \mathbb{N}^{[<\infty]}$) of dense-open subsets of $\mathcal{H}$ and every $N \in \mathcal{H}$, there is an $M \in \mathcal{H}|N$ such that $M/b \in \mathcal{D}_b$ for all $b \in M^{<\infty}$.*[4]

It turns out that the semiselectivity requirement on a given coideal $\mathcal{H}$ is all that is needed in order to guarantee the existence of monochromatic cubes of the form $M^{[k]}$ with $M \in \mathcal{H}$ for arbitrary finite colorings of finite symmetric powers $\mathbb{N}^{[k]}$ of $\mathbb{N}$. In fact, there always seem to be simple and natural ways to transfer even proofs into the new context. The following is an illustration of how one transfers a standard fact about the coideal $\mathcal{H} = \mathbb{N}^{[\infty]}$ to the context of an arbitrary semiselective ideal $\mathcal{H}$ on $\mathbb{N}$.

Lemma 7.12 (Semiselective Galvin Lemma) *Suppose $\mathcal{H}$ is a semiselective coideal on $\mathbb{N}$ and $\mathcal{F}$ is a family of nonempty finite subsets of $\mathbb{N}$. Then for every $N \in \mathcal{H}$, there is an $M \in \mathcal{H}|N$ such that either*

(1) M contains no member of $\mathcal{F}$, or

(2) every infinite subset of M has an initial segment in $\mathcal{F}$.

Proof. To save on notation, we assume that the initial set N is equal to $\mathbb{N}$ and in what follows we use the letters $M, N, ...$ for members of $\mathcal{H}$ and $a, b, ...$ for finite subsets of $\mathbb{N}$.

Definition 7.13 *Let us say that M accepts a if every $X \in [a, M]$ has an initial segment in $\mathcal{F}$. If there is no $N \in \mathcal{H}|M$ accepting a we say that M rejects a. Let us say that M decides a if M either accepts a or it rejects a.*

Note the following properties which follow immediately from the definitions and the assumption that we are working with a coideal $\mathcal{H}$.

(i) If M accepts (rejects) a, then every $N \in \mathcal{H}|M$ accepts (rejects) a.

---

[4]When this happens we say that M *diagonalizes* the sequence $(\mathcal{D}_b)$ ($b \in \mathbb{N}^{[<\infty]}$).

(*ii*) For every a and M, there is an $N \in \mathcal{H}|M$ such that N decides a.

(*iii*) If M accepts a, then M accepts $a \cup \{n\}$ for every $n \in M/a$.

(*iv*) If M rejects a, then $\{n \in M/a : M \text{ accepts } a \cup \{n\}\} \notin \mathcal{H}$.

Claim 7.13.1 *There is an $M \in \mathcal{H}$ that decides all of its finite subsets.*

Proof. For a finite set a, put

$$\mathcal{D}_a = \{N \in \mathcal{H} : N \text{ decides } a\}. \tag{7.10}$$

By (*i*) and (*ii*), we have that $\mathcal{D}_a (a \in \mathbb{N}^{[<\infty]})$ is a sequence of dense-open subsets of $\mathcal{H}$, so by semiselectivity of $\mathcal{H}$, there is an $M \in \mathcal{H}$ such that $M/a \in \mathcal{D}_a$ for all finite $a \subseteq M$. This M decides all of its finite subsets. $\square$

Fix $M \in \mathcal{H}$ satisfying the conclusion of the claim. If M accepts $\emptyset$, conclusion (2) of Lemma 7.12 holds. So let us consider the case that M rejects $\emptyset$. For a finite $a \subseteq M$, let $\mathcal{C}_a = \mathcal{H}$ if M accepts $\mathcal{H}$; otherwise, let

$$\mathcal{C}_a = \{N \in \mathcal{H}|M : (\forall n \in N/a) \ N \text{ rejects } a \cup \{n\}\}. \tag{7.11}$$

By (*i*) and (*iv*), we conclude that $\mathcal{C}_a \ (a \in M^{[<\infty]})$ is a sequence of dense-open subsets of $\mathcal{H}|M$, so by applying the semiselectivity of $\mathcal{H}$, we find $N \in \mathcal{H}|M$ such that

$$N/a \in \mathcal{C}_a \text{ for all finite } a \subseteq N. \tag{7.12}$$

Using this property of N, one shows by induction on the size of a that N rejects all finite subsets $a \subseteq N$. Thus in particular, N cannot contain a set from the family $\mathcal{F}$. This establishes alternative (1) of Lemma 7.12 and finishes the proof. $\square$

Corollary 7.14 *If $\mathcal{H}$ is a semiselective coideal in $\mathbb{N}$, then every $\mathcal{H}$-Baire subset of $\mathbb{N}^{[\infty]}$ is $\mathcal{H}$-Ramsey.*

Proof. Consider an $\mathcal{H}$-Baire set $\mathcal{X} \subseteq \mathbb{N}^{[\infty]}$ and a basic set $[a, M]$, where $a \in \mathbb{N}^{[<\infty]}$ and $M \in \mathcal{H}$. We again save on notation by assuming that $a = \emptyset$. For a finite set $b \subseteq \mathbb{N}$, let $\mathcal{D}_b$ be the collection of all $N \in \mathcal{H}|M$ such that either $[b, N] \subseteq \mathcal{X}$, or $[b, N] \subseteq \mathcal{X}^c$, or

$$(\forall P \in \mathcal{H}|N)([b, P] \not\subseteq \mathcal{X} \ \& \ [b, P] \not\subseteq \mathcal{X}^c). \tag{7.13}$$

Clearly, each $\mathcal{D}_b$ is dense-open in $\mathcal{H}$, so we can find $N \in \mathcal{H}|M$ such that

$$N/b \in \mathcal{D}_b \text{ for all finite } b \subseteq \mathbb{N}. \tag{7.14}$$

Let $\mathcal{F}_0$ be the collection of all $b \in N^{[<\infty]}$ such that $[b, N] \subseteq \mathcal{X}$, let $\mathcal{F}_1$ be the collection of all $b \in N^{[<\infty]}$ such that $[b, N] \subseteq \mathcal{X}^c$, and let

$$\mathcal{F}_2 = N^{[<\infty]} \backslash (\mathcal{F}_0 \cup \mathcal{F}_1). \tag{7.15}$$

By the semiselective Galvin lemma, there are $P \in \mathcal{H}|N$ and $i \leq 2$ such that every infinite subset of P has an initial segment in $\mathcal{F}_i$. If $i = 0$, this gives that $[\emptyset, P] \subseteq \mathcal{X}$ and if $i = 1$ this gives that $[\emptyset, P] \subseteq \mathcal{X}^c$. The case $i = 2$ can easily be eliminated using the assumption that $\mathcal{X}$ is $\mathcal{H}$-Baire. This finishes the proof. $\qquad\square$

Corollary 7.15 *If $\mathcal{H}$ is a semiselective coideal on $\mathbb{N}$, then every $\mathcal{H}$-meager subset of $\mathbb{N}^{[\infty]}$ is $\mathcal{H}$-Ramsey null.*

We now come to a particularly important property of a given semiselective coideal $\mathcal{H}$ which, as we have seen above in Lemma 7.8, is equivalent to the weak (p) property of $\mathcal{H}$.

Lemma 7.16 *If $\mathcal{H}$ is a semiselective coideal, then the ideal of $\mathcal{H}$-Ramsey null subsets of $\mathbb{N}^{[\infty]}$ is σ-additive.*

Proof. Let $(\mathcal{X}_n)$ be a given sequence of $\mathcal{H}$-Ramsey null subsets of $\mathbb{N}^{[\infty]}$ and let $[a, M]$ be a given $\mathcal{H}$-basic set. Since the case $a = \emptyset$ is general enough, we make the assumption that a is empty. For a finite set $b \subseteq \mathbb{N}$, let

$$\mathcal{D}_b = \{N \in \mathcal{H}|M : (\forall n \leq \max(b))[b, N] \cap \mathcal{X}_n = \emptyset\}. \qquad (7.16)$$

Then $\mathcal{D}_b$ ($b \in \mathbb{N}^{[<\infty]}$) is a sequence of dense-open subsets of $\mathcal{H}$, so we can find $N \in \mathcal{H}|M$ such that $N/b \in \mathcal{D}_b$ for all finite $b \subseteq N$. It follows that $[\emptyset, N] \cap \mathcal{X}_n = \emptyset$ for all n, as required. $\qquad\square$

Corollary 7.17 *For every semiselective coideal $\mathcal{H}$ on $\mathbb{N}$, the fields of $\mathcal{H}$-Baire and $\mathcal{H}$-Ramsey sets are σ-fields on $\mathbb{N}^{[\infty]}$ that coincide. Moreover, the ideals of $\mathcal{H}$-meager and $\mathcal{H}$-Ramsey null sets are σ-ideals on $\mathbb{N}^{[\infty]}$ that also coincide.*

The following lemma gives us a considerably deeper closure property of these fields of sets, one that will allow us to claim that the triple $(\mathbb{N}^{[\infty]}, \mathcal{H}, \subseteq)$ forms a local Ramsey space.

Lemma 7.18 *If $\mathcal{H}$ is a semiselective coideal, then the field of $\mathcal{H}$-Ramsey subsets of $\mathbb{N}^{[\infty]}$ is closed under the Souslin operation.*

Proof. Suppose we are given a Souslin scheme $\mathcal{X}_s$ ($s \in \mathbb{N}^{[<\infty]}$) of $\mathcal{H}$-Ramsey subsets of $\mathbb{N}^{[\infty]}$ indexed by finite subsets of $\mathbb{N}$ rather than finite sequences of elements of $\mathbb{N}$. We have to show that

$$\mathcal{X} = \bigcup_{A \in \mathbb{N}^{[\infty]}} \bigcap_{n \in \mathbb{N}} \mathcal{X}_{r_n(A)} \qquad (7.17)$$

is $\mathcal{H}$-Ramsey. So let $[a, M]$ be a given $\mathcal{H}$-basic set. We assume $a = \emptyset$, since this case shows all the difficulties. We may also assume that $\mathcal{X}_t \subseteq \mathcal{X}_s$ whenever t end-extends s. For a given $s \in \mathbb{N}^{[<\infty]}$, set

$$\mathcal{X}_s^* = \bigcup_{A \in [s, \mathbb{N}]} \bigcap_{n \in \mathbb{N}} \mathcal{X}_{r_n(A)}. \tag{7.18}$$

Then $\mathcal{X}_s^* \subseteq \mathcal{X}_s$, and

$$\mathcal{X}_s^* = \bigcup_{n \in \mathbb{N}, n > \max(s)} \mathcal{X}_{s \cup \{n\}}^*. \tag{7.19}$$

Note also that $\mathcal{X}_\emptyset^* = \mathcal{X}$.

For a finite subset $b \subseteq \mathbb{N}$, let $\mathcal{D}_b$ be the collection of all $N \in \mathcal{H}|M$ such that for all $s \subseteq \{0, 1, \ldots, \max(b)\}$ either

(1) $[b, N] \subseteq \mathcal{X}_s^*$, or

(2) $[b, N] \cap \mathcal{X}_s^* = \emptyset$, or

(3) $(\forall P \in \mathcal{H}|N)([b, N] \not\subseteq \mathcal{X}_s^* \ \& \ [b, N] \cap \mathcal{X}_s^* \neq \emptyset)$.

Then $\mathcal{D}_b$ $(b \in \mathbb{N}^{[<\infty]})$ is a sequence of dense-open subsets of $\mathcal{H}$ so we can find $N \in \mathcal{H}|M$ such that $N/b \in \mathcal{D}_b$ for all finite $b \subseteq \mathbb{N}$.

For $s \in \mathbb{N}^{[<\infty]}$, set

$$\Phi(\mathcal{X}_s^*) = [\emptyset, N] \setminus \bigcup \{[b, N] : \max(b) \geq \max(s) \& [b, N] \cap \mathcal{X}_s^* = \emptyset\}.$$

Then $\Phi(\mathcal{X}_s^*)$ is $\mathcal{H}$-Ramsey and $\mathcal{X}_s^* \cap [\emptyset, N] \subseteq \Phi(\mathcal{X}_s^*)$.

Claim 7.18.1 *For every $s \in \mathbb{N}^{[<\infty]}$, every $(\mathcal{H}|N)$-Baire set Z included in the difference $\Phi(\mathcal{X}_s^*) \setminus (\mathcal{X}_s^* \cap [\emptyset, N])$ is $(\mathcal{H}|N)$-meager.*

Proof. If Z is not $\mathcal{H}$-meager, being $\mathcal{H}$-Baire, it would contain a basic subset of the form $[b, P]$ for some finite $b \subseteq \mathbb{N}$ and $P \in \mathcal{H}|N$. Shrinking the basic set, we may assume that $\max(b) \geq \max(s)$. It follows, in particular, that $[b, P] \cap \mathcal{X}_s^* = \emptyset$. Since $N/b \in \mathcal{D}_b$, we have the alternatives (1), (2), and (3) about b, N, and $\mathcal{X}_s^*$. The existence of $[b, P]$, which is disjoint from $\mathcal{X}_s^*$, rules out (1) and (3), so we are left with (2). But this means that $[b, N]$ has been subtracted from $\Phi(\mathcal{X}_s^*)$, contradicting the fact that it includes the set $[b, P]$ included in $\Phi(\mathcal{X}_s^*)$. The completes the proof of the claim. □

Back to the proof of Lemma 7.18. For $s \in \mathbb{N}^{[<\infty]}$, set $\Psi(\mathcal{X}_s^*) = \mathcal{X}_s \cap \Phi(\mathcal{X}_s^*)$. Then $\Psi(\mathcal{X}_s^*)$ is still an $\mathcal{H}$-Ramsey superset of $\mathcal{X}_s^* \cap [\emptyset, N]$ satisfying the conclusion of Claim 7.18.1, and therefore the $\mathcal{H}|N$-Baire subset

$$\mathcal{M}_s = \Psi(\mathcal{X}_s^*) \setminus \bigcup_{n \in \mathbb{N}, n > \max(s)} \Psi(\mathcal{X}_{s \cup \{n\}}^*) \tag{7.20}$$

of the difference $\Psi(\mathcal{X}_s^*) \setminus (\mathcal{X}_s^* \cap [\emptyset, N])$ must be $\mathcal{H}|N$-meager and therefore $\mathcal{H}|N$-Ramsey null. By (the local version of) Lemma 7.16, we can find a $P \in \mathcal{H}|N$ such that

$$[\emptyset, P] \cap \mathcal{M}_s = \emptyset \text{ for all } s \in \mathbb{N}^{[<\infty]}. \tag{7.21}$$

As in the proof of Lemma 4.39, one has that

$$[\emptyset, P] \cap \Psi(\mathcal{X}_\emptyset^*) = [\emptyset, P] \cap \mathcal{X}_\emptyset^*. \tag{7.22}$$

If follows that $[\emptyset, P] \cap \mathcal{X}$ is $\mathcal{H}|P$-Ramsey, so we can find $R \in \mathcal{H}|P$ such that $[\emptyset, R] \subseteq \mathcal{X}$ or $[\emptyset, R] \cap \mathcal{X} = \emptyset$, as required. This finishes the proof. □

Combining the lemmas proved so far we have the following result which gives us the local version of the original topological Ramsey theorem of Ellentuck.

Theorem 7.19 (Local Ellentuck Theorem) *The following two properties are equivalent for every coideal $\mathcal{H}$ on $\mathbb{N}$:*

(1) $\mathcal{H}$ is semiselective.

(2) The ideals of $\mathcal{H}$-meager and $\mathcal{H}$-Ramsey null subsets of $\mathbb{N}^{[\infty]}$ coincide and are σ-additive, and the fields of $\mathcal{H}$-Baire and $\mathcal{H}$-Ramsey subsets of $\mathbb{N}^{[\infty]}$ coincide and are closed under the Souslin operation.

Corollary 7.20 *For every semiselective coideal $\mathcal{H}$ on $\mathbb{N}$ and every finite metrically Souslin-measurable coloring of the space $\mathbb{N}^{[\infty]}$ of all infinite subsets of $\mathbb{N}$, there is an $M \in \mathcal{H}$ such that $M^{[\infty]}$ is monochromatic.*

Proof. This follows from the immediate fact that for every semiselective coideal $\mathcal{H}$ on $\mathbb{N}$, every basic open subset of the metrizable product topology of $\mathbb{N}^{[\infty]}$ is $\mathcal{H}$-Baire. □

Clearly, every selective coideal is semiselective, so we also have the following form of the Local Ellentuck Theorem which suffices in most of the known applications of the local theory.

Theorem 7.21 (Selective Ellentuck Theorem) *Let $\mathcal{H}$ be a selective coideal on $\mathbb{N}$. The ideals of $\mathcal{H}$-meager and $\mathcal{H}$-Ramsey null subsets of $\mathbb{N}^{[\infty]}$ coincide and are σ-additive. The fields of $\mathcal{H}$-Baire and $\mathcal{H}$-Ramsey subsets of $\mathbb{N}^{[\infty]}$ coincide and are closed under the Souslin operation.*

Corollary 7.22 (Ellentuck) *$(\mathbb{N}^{[\infty]}, \subseteq, r)$ is a topological Ramsey space.*

Proof. The coideal $\mathbb{N}^{[\infty]}$ of all infinite subsets of $\mathbb{N}$ is selective, and the corresponding family of basic sets form a topology whose Baire-category notions coincide with the Ramsey theoretic notions relative to this coideal. So the conclusion follows from Theorem 7.21. □

Corollary 7.23 (Mathias) *For every selective coideal $\mathcal{H}$ on $\mathbb{N}$, and every finite metrically Souslin-measurable coloring of $\mathbb{N}^{[\infty]}$, there is an $M \in \mathcal{H}$ such that $M^{[\infty]}$ is monochromatic.*

Corollary 7.24 (Louveau) *If $\mathcal{U}$ is a selective ultrafilter on $\mathbb{N}$, then the triple $(\mathbb{N}^{[\infty]}, \mathcal{U}, \subseteq)$ is a topological Ramsey space.*

Proof. Being a nonprincipal ultrafilter on $\mathbb{N}$ is the same as being a minimal coideal on $\mathbb{N}$, so the $\mathcal{U}$-basic sets form a topology that extends the usual product topology of $\mathbb{N}^{[\infty]}$. Again, the Baire-category notions relative to this topology coincide with the Ramsey theoretic notions relative to $\mathcal{U}$.   □

Corollary 7.25 (Mathias) *For every selective ultrafilter $\mathcal{U}$ on $\mathbb{N}$ and every finite metrically Souslin-measurable coloring of $\mathbb{N}^{[\infty]}$, there is an $M \in \mathcal{U}$ such that $M^{[\infty]}$ is monochromatic.*

We finish this section with an application of this corollary and Example 7.1.2 above.

Theorem 7.26 (Mathias) *Every analytic maximal family of infinite pairwise almost disjoint subsets of $\mathbb{N}$ must be finite.*

Proof. Let $\mathcal{A}$ be an infinite family of infinite pairwise almost disjoint subsets of $\mathbb{N}$ that cannot be extended to a larger family with this property. Let $\mathcal{H}$ be the coideal of subsets of $\mathbb{N}$ that cannot be covered modulo a finite set by finitely many members of $\mathcal{A}$. Then $\mathcal{H}$ is selective. If $\mathcal{A}$ were analytic, then $\mathcal{H}$ would be coanalytic, so Corollary 7.23 would give us an $M \in \mathcal{H}$ such that $M^{[\infty]} \subseteq \mathcal{H}$ or $M^{[\infty]} \cap \mathcal{H} = \emptyset$. The second case is impossible, and the first case in particular means that $M \cap A$ is finite for all $A \in \mathcal{A}$. This contradicts the fact that $\mathcal{A}$ is maximal and finishes the proof.   □

Remark 7.27 Recall the Abstract Ramsey Theorem from Chapter Four which discusses about a structure of the form

$$(\mathcal{R}, \mathcal{S}, \leq, \leq^o, r, s)$$

and conditions that guarantee the equivalence between Baire and Ramsey properties. The results of this section about the prototype Ramsey space

$$(\mathbb{N}^{[\infty]}, \mathbb{N}^{[\infty]}, \subseteq, \subseteq_{\text{fin}}, r, r)$$

show that the the second factor $\mathbb{N}^{[\infty]}$ can be reduced to a sufficiently rich family $\mathcal{H} \subseteq \mathbb{N}^{[\infty]}$ still guaranteeing the equivalence between Baire and Ramsey properties. We would like to point out that it is more or less straightforward to extend this into the abstract context in the sense that there is a natural notion of a semiselective coideal $\mathcal{H} \subseteq \mathcal{S}$ that guarantees the equivalence between Baire and Ramsey properties in the corresponding reduced structure

$$(\mathcal{R}, \mathcal{H}, \leq, \leq^o, r, s).$$

We leave the details to the interested reader and instead develop in later sections of this chapter a parallel theory of so-called *Ultra-Ramsey Spaces* that is general enough to capture basically all known application of the abstract theory of semiselective coideals.

7.2 TOPOLOGICAL ULTRA-RAMSEY SPACES

In this section we choose the set $\mathbb{N}$ of all nonnegative integers as a convenient countable index set and build on it what one can call an "Ultra-Ramsey Theory". As before, we let $\mathbb{N}^{[<\infty]}$ denote the collection of *all* finite subsets of $\mathbb{N}$ and consider it a tree ordered by end-extension. A *subtree* T of $\mathbb{N}^{[<\infty]}$ is always assumed to be downward closed, and we work most of the time with subtrees that have a *stem*, $st(T)$, the maximal node that is comparable with every other node of T. From now on in this section, a *tree* is always assumed to be a subtree of $\mathbb{N}^{[<\infty]}$.

Let $\vec{\mathcal{U}} = (\mathcal{U}_s : s \in \mathbb{N}^{[<\infty]})$ be a family of nonprincipal ultrafilters on $\mathbb{N}$. If $\vec{\mathcal{U}}$ is a constant sequence $\mathcal{U}_s = \mathcal{U}$, we suppress the arrow.

Definition 7.28 *Given a sequence $\vec{\mathcal{U}} = (\mathcal{U}_s : s \in \mathbb{N}^{[<\infty]})$ of nonprincipal ultrafilters on $\mathbb{N}$, a $\vec{\mathcal{U}}$-tree is a tree T with the property that*

$$\{n \in \mathbb{N} : t \cup \{n\} \in T\} \in \mathcal{U}_t \qquad (7.23)$$

for all $t \in T$.

Definition 7.29 *For two $\vec{\mathcal{U}}$-trees T and T', we write $T' \leq^0 T$ and say that T' is a* pure refinement *of T if $T' \subseteq T$ and $st(T') = st(T)$. For an integer $n \geq 0$, let $T' \leq^n T$ mean that $T' \leq^0 T$ and the first n levels of T' above the common stem are equal to the first n levels of T above the common stem.*

Definition 7.30 *A fusion sequence is an infinite sequence (T_n) of trees with the property that $T_{n+1} \leq^n T_n$ for all n. Note that if (T_n) is a fusion sequence of $\vec{\mathcal{U}}$-trees, then the intersection*

$$T_\infty = \bigcap_{n=0}^{\infty} T_n \qquad (7.24)$$

is also a $\vec{\mathcal{U}}$-tree.

Lemma 7.31 *Suppose that T and T' are two $\vec{\mathcal{U}}$-trees. Then $T \cap T'$ is a $\mathcal{U}$-tree iff $st(T)$ and $st(T')$ are comparable.*

For a tree T and $s \in T$, set

$$T/s = \{t \in T : s \subseteq t\}. \qquad (7.25)$$

Fix for a while a sequence $\vec{\mathcal{U}} = (\mathcal{U}_s : s \in \mathbb{N}^{[<\infty]})$ of nonprincipal ultrafilters on $\mathbb{N}$.

Definition 7.32 *A subset $G \subseteq \mathbb{N}^{[<\infty]}$ is $\vec{\mathcal{U}}$-open if for every $t \in G$ there is a $\vec{\mathcal{U}}$-tree T such that $t = st(T)$ and $T/t \subseteq G$.*

Note that by Lemma 7.31, the collection of all $\vec{\mathcal{U}}$-open subsets of $\mathbb{N}^{[<\infty]}$ forms a zero-dimensional Hausdorff topology on $\mathbb{N}^{[<\infty]}$. The following characterization of $\vec{\mathcal{U}}$-open subsets of $\mathbb{N}^{[<\infty]}$ is quite useful.

Lemma 7.33 *A subset $G \subseteq \mathbb{N}^{[<\infty]}$ is $\vec{\mathcal{U}}$-open iff*

$$G_s = \{n : s \cup \{n\} \in G\} \in \mathcal{U}_s \qquad (7.26)$$

for all $s \in G$.

Proof. The direct implication is an immediate consequence of the definition, so let us prove the converse. Let G be a subset of $\mathbb{N}^{[<\infty]}$ with the property that $G_s \in \mathcal{U}_s$ for all $s \in G$. Let s_0 be a given element of G. We need to produce a $\vec{\mathcal{U}}$-tree T with stem s_0 such that $T/s_0 \subseteq G$. We build the levels $T(1), T(2), \ldots, T(n), \ldots$ of T above the stem as follows. Let $T(1) = G_s$, and provided we have $T(n)$, let

$$T(n+1) = \bigcup_{s \in T(n)} G_s. \qquad (7.27)$$

Clearly, the tree T constructed this way is a $\vec{\mathcal{U}}$-tree with stem s_0 such that $T/s_0 \subseteq G$. This finishes the proof. $\qquad\qquad\square$

The following fact shows that the $\vec{\mathcal{U}}$-topology on $\mathbb{N}^{[<\infty]}$ is a quite unusual example of a topology on a countable index set.

Lemma 7.34 *The $\vec{\mathcal{U}}$-closure of every $\vec{\mathcal{U}}$-open set is $\vec{\mathcal{U}}$-open.*

Proof. Let G be a $\vec{\mathcal{U}}$-open set, and let F be its closure relative the $\vec{\mathcal{U}}$-topology on $\mathbb{N}^{[<\infty]}$. By Lemma 7.33, it suffices to show that

$$F_s = \{n \in \mathbb{N} : s \cup \{n\} \in F\} \in \mathcal{U}_s \qquad (7.28)$$

for all $s \in F$. Suppose $F_s \notin \mathcal{U}_s$ for some $s \in F$. For every $n > \max(s)$ such that $n \notin F_s$, the node $s \cup \{n\}$ does not belong to the closure of G, so we can choose a $\vec{\mathcal{U}}$-tree T_n with stem $s \cup \{n\}$ such that $T_n/s \cup \{n\}$ is disjoint from G. Let T be the $\vec{\mathcal{U}}$-tree with stem s such that

$$T/s = \bigcup\{T_n : n > \max(s), n \notin F_s\}. \qquad (7.29)$$

Since G is $\vec{\mathcal{U}}$-open and since the complement of F_s and therefore the complement of G_s belongs to $\mathcal{U}_s$, we must have that $s \notin G$. It follows that $(T/s) \cap G = \emptyset$, so s is not in the closure of G, a contradiction. $\qquad\square$

We are however more interested in an analogously defined topology on the set $\mathbb{N}^{[\infty]}$ of all infinite subsets of $\mathbb{N}$ rather than on $\mathbb{N}^{[<\infty]}$. To define this topology, let for a given tree T,

$$[T] = \{A \in \mathbb{N}^{[\infty]} : A \cap \{0, \ldots, n-1\} \in T \text{ for all } n \in \mathbb{N}\}. \qquad (7.30)$$

Definition 7.35 *A subset G of $\mathbb{N}^{[\infty]}$ is $\vec{\mathcal{U}}$-open if for every $A \in G$, there is a $\vec{\mathcal{U}}$-tree T such that $A \in [T]$ and $[T] \subseteq G$.*

Note that by Lemma 7.31, the family of all $\vec{\mathcal{U}}$-open subsets of $\mathbb{N}^{[\infty]}$ form a topology on $\mathbb{N}^{[\infty]}$ with basis $[T]$ (T a $\vec{\mathcal{U}}$-tree). Recall that $\mathbb{N}^{[\infty]}$ also has the natural metrizable topology generated by basic open sets of the form

$$[s] = \{A \in \mathbb{N}^{[\infty]} : A \cap \{0, \dots, \max(s)\} = s\} \tag{7.31}$$

for $s \in \mathbb{N}^{[<\infty]}$. Since $[s]$ can also be written as $[T]$ for the $\vec{\mathcal{U}}$-tree T of all nodes of $\mathbb{N}^{[<\infty]}$ comparable to s, this shows that the $\vec{\mathcal{U}}$-topology includes the metric topology of $\mathbb{N}^{[\infty]}$. Since every set of the form $[T]$ is metrically closed, the $\vec{\mathcal{U}}$-topology is zero-dimensional. Combining these observations, we get the following.

Lemma 7.36 *The $\vec{\mathcal{U}}$-topology on $\mathbb{N}^{[\infty]}$ is a zero-dimensional topology satisfying the countable chain condition and extending the metric topology of the set $\mathbb{N}^{[\infty]}$ of all infinite subsets of $\mathbb{N}$.*

Definition 7.37 *A subset $\mathcal{X}$ of $\mathbb{N}^{[\infty]}$ is $\vec{\mathcal{U}}$-Ramsey if for every $\vec{\mathcal{U}}$-tree T there is a pure extension $T' \leq^0 T$ such that $[T'] \subseteq \mathcal{X}$ or $[T'] \subseteq \mathcal{X}^c$.*

This leads us to the following crucial lemma of this section.

Lemma 7.38 *$\vec{\mathcal{U}}$-open sets are $\vec{\mathcal{U}}$-Ramsey.*

Proof. Let $\mathcal{X}$ be a given $\vec{\mathcal{U}}$-open set. Let

$$G = \{s \in \mathbb{N}^{[<\infty]} : \exists \, \vec{\mathcal{U}}\text{-tree } T \quad (\mathrm{st}(T) = s \text{ and } [T] \subseteq \mathcal{X})\}. \tag{7.32}$$

Clearly, G is a $\vec{\mathcal{U}}$-open subset of $\mathbb{N}^{[<\infty]}$. Note that

$$s \notin G \text{ implies } \{n : s \cup \{n\} \in G\} \notin \mathcal{U}_s.$$

It follows that $F = \mathbb{N}^{[<\infty]} \setminus G$ satisfies the criterion of Lemma 7.33 for being open. It follows that F is, in fact, clopen. The lemma will follow if we can show that for every $t \in F$ there is a $\vec{\mathcal{U}}$-tree T with stem t such $[T] \cap \mathcal{X} = \emptyset$. In fact, we claim that if T is any $\vec{\mathcal{U}}$-tree with stem t such that $T/t \subseteq F$ then $[T] \cap \mathcal{X} = \emptyset$. Suppose the contrary and fix an $A \in [T] \cap \mathcal{X}$. Since $\mathcal{X}$ is $\vec{\mathcal{U}}$-open, there is an $n \in A$ above $\max(t)$ and a $\vec{\mathcal{U}}$-tree T' with stem equal to $u = A \cap \{0, \dots, n\}$ such that $A \in [T'] \subseteq \mathcal{X}$. If follows that $T'/u \subseteq G$ and therefore $u \in F \cap G$, a contradiction. $\square$

Definition 7.39 *A subset $\mathcal{X}$ of $\mathbb{N}^{[\infty]}$ is $\vec{\mathcal{U}}$-Ramsey null if for every $\mathcal{U}$-tree T there is a $T' \leq^0 T$ such that $[T'] \cap \mathcal{X} = \emptyset$.*

Lemma 7.40 *$\vec{\mathcal{U}}$-nowhere dense-sets are $\vec{\mathcal{U}}$-Ramsey null.*

Proof. Let $\mathcal{X} \subseteq \mathbb{N}^{[\infty]}$ be a given $\vec{\mathcal{U}}$-nowhere dense set. Let $\overline{\mathcal{X}}$ be its closure in the $\vec{\mathcal{U}}$ topology. Apply the previous Lemma to the given $\vec{\mathcal{U}}$-tree T and the complement of $\overline{\mathcal{X}}$ and get $T' \leq^0 T$ such that $[T'] \subseteq \overline{\mathcal{X}}$ or $[T'] \cap \overline{\mathcal{X}} = \emptyset$. Note that, since $\mathcal{X}$ is nowhere dense, the first alternative is impossible. $\square$

Lemma 7.41 *The $\vec{\mathcal{U}}$-Ramsey null sets form a σ-ideal.*

Proof. Let $(\mathcal{X}_n)$ be a given sequence of $\vec{\mathcal{U}}$-Ramsey null sets and let T be a given $\vec{\mathcal{U}}$-tree. Define a fusion sequence (T_n) of $\vec{\mathcal{U}}$-trees as follows. Let $T_0 = T$ and having defined T_n let L_n be the nth level of T_n above

$$st(T_n) = \ldots = st(T_0) = st(T).$$

For each $t \in L_n$ we apply the fact that $\bigcup_{k=0}^{n} \mathcal{X}_k$ is $\vec{\mathcal{U}}$-Ramsey null and get a $\vec{\mathcal{U}}$-tree $S_t \subseteq T_n$ with stem t such that $[S_t]$ is disjoint from $\bigcup_{k=0}^{n} \mathcal{X}_k$. Let T_{n+1} be the $\vec{\mathcal{U}}$-tree such that $L_k \subseteq T_{n+1}$ for all $k \leq n$ (or in other words $T_{n+1} \leq^n T_n$), and such that $T_{n+1}/t = S_t$ for all $t \in L_n$. Finally, let

$$T_\infty = \bigcap_{n=0}^{\infty} T_n. \qquad (7.33)$$

Then T_∞ is a $\vec{\mathcal{U}}$-tree such that $T_\infty \leq^0 T$ and such that

$$[T_\infty] \cap \left(\bigcup_{n=0}^{\infty} \mathcal{X}_n \right) = \emptyset.$$

This finishes the proof. $\qquad \square$

Combining these results we arrive at the following version of the local Ellentuck theorem.

Theorem 7.42 (Ultra-Ellentuck Theorem) *Every subset of $\mathbb{N}^{[\infty]}$ that has the property of Baire relative to the $\vec{\mathcal{U}}$-topology is $\vec{\mathcal{U}}$-Ramsey and vice versa. Moreover, the ideals of $\vec{\mathcal{U}}$-meager and $\vec{\mathcal{U}}$-Ramsey null subsets of $\mathbb{N}^{[\infty]}$ are σ-ideals that coincide.*

Proof. Let $\mathcal{X} \subseteq \mathbb{N}^{[\infty]}$ be a given $\vec{\mathcal{U}}$-Baire set. Pick a $\vec{\mathcal{U}}$-open set $\mathcal{O}$ and a $\vec{\mathcal{U}}$-meager set $\mathcal{M}$ such that $\mathcal{X} = \mathcal{M} \triangle \mathcal{O}$. Let T be a given $\vec{\mathcal{U}}$-tree. By Lemma 7.41, there is a $\vec{\mathcal{U}}$-tree $T' \leq^0 T$ such that $[T'] \cap \mathcal{M} = \emptyset$. By Lemma 7.38, there is a $\mathcal{U}$-tree $T'' \leq^0 T'$ such that $[T''] \subseteq \mathcal{O}$ or $[T''] \subseteq \mathcal{O}^c$. It follows that $[T''] \subseteq \mathcal{X}$ or $[T''] \subseteq \mathcal{X}^c$. $\qquad \square$

Lemma 7.43 *Suppose $\mathcal{U}_s = \mathcal{U}$ for all $s \in \mathbb{N}^{[<\infty]}$. Let T be a $\mathcal{U}$-tree with stem s. Then there is $A \in \mathbb{N}^{[\infty]}$ such that $[s, A] \subseteq [T]$.*

Proof. We show this only for $s = \emptyset$. For $t \in T$, let

$$A_t = \{n > \max(t) : t \cup \{n\} \in T\}. \qquad (7.34)$$

Define a strictly increasing sequence $(n_i) \subseteq \mathbb{N}$ as follows:

$$n_k = \min \bigcap \{A_t : t \subseteq \{n_0, \ldots, n_{k-1}\}\}. \qquad (7.35)$$

Let $A = \{n_i : i \in \mathbb{N}\}$. Then every infinite subset of A is a branch of T. $\qquad \square$

Corollary 7.44 (Silver) *The field of Ramsey subsets of $\mathbb{N}^{[\infty]}$ is closed under the Souslin operation.*

Proof. First of all, note that this conclusion is equivalent to the fact that metrically analytic subsets of $\mathbb{N}^{[\infty]}$ are Ramsey. To see this, given a Souslin scheme $\mathcal{X}_s$ ($s \in \mathbb{N}^{[<\infty]}$) of Ramsey sets and given a basic set $[n, A]$, we first find a set $B \in [n, A]$ such that for all $s, t \in \mathbb{N}^{[<\infty]}$, if $m = \max\{n, \max(s \cup t)\}$ then $[t, B/m] \subseteq \mathcal{X}_s$ or $[t, B/m] \subseteq \mathcal{X}_s^c$. It follows that $\mathcal{X}_s \cap [n, B]$ is a relatively clopen subset of $[n, B]$ for all $s \in \mathbb{N}^{[<\infty]}$, so $[n, B] \cap (\bigcup_{M \in \mathbb{N}^{[\infty]}} \bigcap_{n=0}^{\infty} \mathcal{X}_{r_n(M)})$ is an analytic subset of $[n, B]$.

Back to the proof of Silver's theorem. So let $[n, A]$ be a given basic set and let $\mathcal{X} \subseteq \mathbb{N}^{[\infty]}$ be a given metrically analytic set. Pick a nonprincipal ultrafilter $\mathcal{U}$ on $\mathbb{N}$ such that $A \in \mathcal{U}$. Then if s is the set formed by taking the first n elements of A, then there is a $\mathcal{U}$-tree T with stem s such that $[T] = [n, A]$. Since the $\mathcal{U}$-topology extends the metric topology, the set $\mathcal{X}$ is $\mathcal{U}$-Baire and therefore $\mathcal{U}$-Ramsey by the Ultra-Ellentuck Theorem. So, pick a $\mathcal{U}$-tree $T' \leq^0 T$ such that $[T'] \subseteq \mathcal{X}$ or $[T'] \cap \mathcal{X} = \emptyset$. Apply the previous lemma and get $B \in [n, A]$ such that $[n, B] \subseteq [T']$. Then $[n, B] \subseteq \mathcal{X}$ or $[n, B] \cap \mathcal{X} = \emptyset$, as required. □

Corollary 7.45 (Galvin-Prikry) *Every metrically Borel subset of $\mathbb{N}^{[\infty]}$ is Ramsey.*

In the next three sections of this chapter, we shall give several applications of the Ultra-Ellentuck Theorem. This will be typically done by replacing $\mathbb{N}$ with some other convenient countable index set S, carefully choosing an ultrafilter $\mathcal{U}$ on S, and then applying the Ultra-Ramsey theory developed above.

7.3 SOME EXAMPLES OF SELECTIVE COIDEALS ON $\mathbb{N}$

Note that every coideal $\mathcal{H}$ on $\mathbb{N}$ has the form

$$\mathcal{H}_K(x, (x_n)) = \{M \subseteq \mathbb{N} : x \in \overline{\{x_n : n \in M\}}\} \tag{7.36}$$

for some regular space K, a point $x \in K$, and a sequence $(x_n) \subseteq K \setminus \{x\}$ accumulating to x. To see this, let $K = \mathbb{N} \cup \{\infty\}$ be topologized by letting the points of $\mathbb{N}$ be isolated and by letting the neighborhoods of ∞ be sets $\{\infty\} \cup G$, where G ranges over the filter of all subsets of $\mathbb{N}$ that meet every set in $\mathcal{H}$. When $\mathcal{H} = \mathcal{H}_K(x, (x_n))$, we say that K *represents* $\mathcal{H}$. There is another rather canonical way to get such a representation of an arbitrary coideal $\mathcal{H}$ on $\mathbb{N}$. To see this, let $\mathcal{I} = \mathcal{P}(\mathbb{N}) \setminus \mathcal{H}$, and for $n \in \mathbb{N}$, let $\pi_n : 2^{\mathcal{I}} \to 2$ be the projection map

$$\pi_n(a) = 1 \text{ iff } n \in a. \tag{7.37}$$

Let K be the closure of (π_n) in the Tychonov cube 2^J and let $\bar{1}$ denote the constantly equal to 1 map. Then $\mathcal{H} = \mathcal{H}_K(\bar{1}, (\pi_n))$. Note that $\bar{1}$ and π_n

are all continuous maps on $\mathcal{I}$ when we view $\mathcal{I}$ as a topological space with the topology induced from the Cantor cube $2^{\mathbb{N}}$. So every coideal $\mathcal{H}$ on $\mathbb{N}$ is represented by a set of continuous maps on some separable metric space. The following reformulation of Lemma 6.53 answers a natural question.

Lemma 7.46 *A coideal $\mathcal{H}$ on $\mathbb{N}$ is coanalytic iff it is represented by a sequence of continuous functions on a Polish space.*

We shall say that a coideal $\mathcal{H}$ on $\mathbb{N}$ is *hereditarily Baire* if $\mathcal{H}|A$ has the property of Baire as a subset of 2^A for all $A \in \mathcal{H}$. Clearly, every coideal $\mathcal{H}$ on $\mathbb{N}$ that is either analytic or coanalytic is hereditarily Baire. Recall that a topological space K is *countably tight* if any subset $A \subseteq K$ accumulating to a point $b \in K$ contains a countable subset $A_0 \subseteq A$ still accumulating to the point $b \in K$.

The following result gives us a rich source of examples of selective coideals on $\mathbb{N}$ and therefore a rich source of potential applications of the Ultra-Ramsey Theory developed above.

Theorem 7.47 *Every hereditarily Baire coideal $\mathcal{H}$ on $\mathbb{N}$ represented by a countably tight compactum is selective.*

Proof. We may assume that $\mathcal{H} = \mathcal{H}_K(x, (x_n))$ for some compact countably tight space K in which the sequence (x_n) is dense. We need to check that $\mathcal{H}$ satisfies properties (p) and (q) of Lemma 7.4.

To check the property (p) let (A_n) be a given decreasing sequence of elements of $\mathcal{H}$. For an infinite set $M \subseteq \mathbb{N}$, let K_M be the set of all accumulation points of $(x_n)_{n \in M}$ in K. Clearly, K_M is nonempty, and the set

$$Y = \bigcup \{K_M : M \text{ infinite } \& (\forall n) M \subseteq^* A_n\} \tag{7.38}$$

accumulates to x. So we can find a sequence (M_k) of infinite subsets such that $M_k \subseteq^* A_n$ for all k and n and such that

$$Y_0 = \bigcup_{k=0}^{\infty} K_{M_K} \tag{7.39}$$

still accumulates to x_0. Choose a set $B \subseteq \mathbb{N}$ such that $M_k \subseteq^* B \subseteq^* A_n$ for all k and n. Then K_B contains x, so B is as required for property (p).

To check the property (q), let $A = \bigcup_{k=0}^{\infty} F_k$ be a given partition of an $A \in \mathcal{H}$ into finite sets. By our assumption $\mathcal{H}|A$ is a Baire-measurable subset of 2^A. Since it is closed under finite changes, it must be meager or comeager. If $\mathcal{H}|A$ were meager, the set $\{A \backslash H : H \in \mathcal{H}|A\}$ would also be meager, so we could find a set $B \subseteq A$ such that neither B nor $A \backslash B$ belongs to $\mathcal{H}$, contradicting the fact that $\mathcal{H}$ is coideal. Hence $\mathcal{H}|A$ is comeager in 2^A, so by a well-known facts about comeager subsets of the Cantor set[5] and the

---

[5] A subset $\mathcal{C}$ of the Cantor set $2^{\mathbb{N}}$ is comeager if and only if it contains a set of the form $\{x \in 2^{\mathbb{N}} : (\exists^{\infty} k)\, x \supseteq s_k\}$ for some sequence (s_k) of finite disjointly supported partial maps from $\mathbb{N}$ into 2 (see Lemma 9.34).

fact that $\mathcal{H}|A$ is closed upwards under inclusion, gluing together some of the finite sets F_k, we may assume that

$$F_M = \bigcup_{k \in M} F_k \in \mathcal{H} \tag{7.40}$$

for all infinite $M \subseteq \mathbb{N}$. Pick a family $\mathcal{M}$ of size continuum of infinite subsets of $\mathbb{N}$ such that $M \cap N$ is finite for $M \neq N \in \mathcal{M}$. Use $\mathcal{M}$ to index a local base $\mathcal{O}_M (M \in \mathcal{M})$ of x in K. Then for each $M \in \mathcal{M}$, we can choose infinite $B_M \subseteq F_M$ such that $|B_M \cap F_k| \leq 1$ for all $k \in M$ and such that the set K_{B_M} of all accumulation points of $(x_n)_{n \in B_M}$ is a subset of $\mathcal{O}_M$. Such a B_M can be found since x is an accumulation point of $(x_n)_{n \in F_M}$. It follows that x is an accumulation point of

$$\bigcup \{K_{B_M} : M \in \mathcal{M}\}. \tag{7.41}$$

Since the compactum K is countably tight, there is a sequence (M_i) of elements of $\mathcal{M}$ such that x is an accumulation point of

$$\bigcup_{i=0}^{\infty} K_{B_{M_i}}. \tag{7.42}$$

Since (M_i) is a sequence of almost disjoint sets, for each i we can choose a tail $\bar{M}_i$ of M_i such that $\bar{M}_i \cap \bar{M}_j = \emptyset$ whenever $i \neq j$. Let $\bar{B}_i = B_{M_i} \cap F_{\bar{M}_i}$. Then $B_{M_i} \setminus \bar{B}_i$ is finite, so $K_{B_{M_i}} = K_{\bar{M}_i}$. Let $B = \bigcup_{i=0}^{\infty} \bar{B}_i$. Then $|B \cap F_k| \leq 1$ for all k and $\bigcup_{i=0}^{\infty} K_{B_{M_i}} \subseteq K_B$. It follows that $(x_n)_{n \in B}$ accumulates to x, or in other words, that B belongs to $\mathcal{H}$. This finishes the proof. $\qquad \square$

Recall that a topological space K is said to be *sequentially compact* if every sequence (x_n) of elements of K contains a convergent subsequence. It is also worth pointing out the following variation on a previous theorem.

Theorem 7.48 *Every coideal on $\mathbb{N}$ represented by a countably tight sequentially compact space is selective.*

Proof. Pick a representation $\mathcal{H} = \mathcal{H}_K(x, (x_n))$, where K is a countably tight sequentially compact space and let (A_n) be a given decreasing sequence of elements of $\mathcal{H}$. Recall that a *diagonalization* of the sequence (A_n) is any infinite set $B \subseteq \mathbb{N}$ such that

$$B \setminus \{0, \ldots, n\} \subseteq A_n \text{ for all } n \in B. \tag{7.43}$$

Let $\mathcal{D}(A_n)$ be the collection of all diagonalizations of the sequence (A_n) and let $\mathcal{CD}(A_n)$ denote the family of all $B \in \mathcal{D}(A_n)$ for which the sequence $(x_n)_{n \in B}$ is convergent in K. For $B \in \mathcal{CD}(A_n)$, let x_B be the limit of $(x_n)_{n \in B}$. Our assumption that K is sequentially compact (and regular) implies that the set

$$\{x_B : B \in \mathcal{CD}(A_n)\} \tag{7.44}$$

accumulates to x. Since K is also assumed to be countably tight, we can find a sequence (B_k) of elements of $\mathcal{CD}(A_n)$ such that $\{x_{B_k} : k \in \mathbb{N}\}$ accumulates to x as well. We may assume that $x_{B_k} \neq x_{B_\ell}$ whenever $k \neq \ell$, which gives us that the B_k are pairwise almost disjoint. Note that, during the course of analyzing Example 7.1.2, we have given a procedure that produces a diagonalization C of (A_n) such that $C \cap B_k$ is infinite for all k. It follows that the set of accumulation points of $(x_n)_{n \in C}$ includes all the points x_{B_k} and therefore our point x. If follows that the diagonalization C belongs to the coideal $\mathcal{H}$, and this finishes the proof. $\qquad\square$

Let us now examine the assumptions in the previous two results. The following consequence of Rosenthal Dichotomy Theorem 5.56 gives us a rich source of coideals satisfying one of these assumptions.

Theorem 7.49 (Rosenthal) *Every compact set K of Baire Class 1 functions defined on some Polish space X is sequentially compact.*

The second assumption is given by the following general topological fact which is of independent interest.

Theorem 7.50 (Rosenthal) *Every compact set K of Baire Class 1 functions defined on some Polish space X is countably tight.*

Proof. Consider $F \subseteq K$ and $g \in K$ such that $g \in \overline{F}$. We need to find countable $F_0 \subseteq F$ such that $g \in \overline{F_0}$. We first consider the case when g is equal to the constant function $\bar{0}$ and when all members of F are nonnegative functions, and we show only that there is a countable $F_0 \subseteq F$ such that for every $\varepsilon > 0$ and $x \in X$ there is an $f \in F_0$ such that $f(x) < \varepsilon$. For $H \subseteq F$, set

$$X_H = \{x \in X : (\forall f \in H)\ f(x) > \varepsilon)\}.$$

We need to find a countable $H \subseteq F$ such that $X_H = \emptyset$. Otherwise, since X is a separable metric space, we can find a countable $G \subseteq F$ such that $\overline{X_H} = \overline{X_G}$ for all $H \supseteq G$. Let $P = \overline{X_G}$. Pick a countable dense subset S of X_G and choose a sequence (f_n) of elements of F that pointwise converges to $\bar{0}$ on S. Going to a subsequence, we may actually assume that (f_n) converges pointwise everywhere on X to some Baire Class 1 function h. Let $H = G \cup \{f_n : n \in \mathbb{N}\}$. Then $\overline{X_H} = P$, and therefore, $h(x) \geq \varepsilon$ for all $x \in X_H$. On the other hand, $h(x) = 0$ for all $x \in S$. Since S and X_H are two dense subsets of the closed set P, we see that h has no point of continuity in P, a contradiction.

It remains to see that the general case can be reduced to the case just considered. To see this, assuming further, as we may, that g is in fact a continuous function, for every positive integer n, we define the mapping

$$\Psi : \mathbb{R}^X \to \mathbb{R}^{(X^n)}$$

as follows:

$$\Psi(f)(x_1, \ldots, x_n) = |f(x_1) - g(x_1)| + \ldots + |f(x_n) - g(x_n)|.$$

It is easily seen that Ψ is a continuous map from the first Baire class on X into the first Baire class on X^n. Thus, by the special case of the proof, we can find a countable subset F_n of F such that for every $\varepsilon > 0$ and $x_1, \ldots, x_n$ in X there is an $f \in F_n$ such that

$$|f(x_1) - g(x_)| < \varepsilon, \ldots, |f(x_n) - g(x_n)| < \varepsilon.$$

Let $F_\infty = \bigcup_{n=1}^\infty F_n$. Then F_∞ is a countable subset of F and $g \in \overline{F_\infty}$, as required.   □

Corollary 7.51 *Every separable compact set K of Borel functions defined on some Polish space X is countably tight and sequentially compact.*

Proof. Changing the topology of X, we may assume that there is a countable dense subset D of K consisting of continuous functions on X. Note that under this additional assumption, the compact set K contains only Baire Class 1 functions on X, and so the result follows from Theorems 7.49 and 7.50. For if there is $g \in K$ not of first Baire class, then we can find two countable sets $P, Q \subseteq X$ dense inside the same perfect subset of X such that $\sup(g \upharpoonright P) < \inf(g \upharpoonright Q)$. Finding a sequence (f_n) of elements of D that pointwise converges on $P \cup Q$, we get a sequence of continuous functions with no converging subsequence,[6] contradicting Corollary 5.57.   □

Corollary 7.52 *Every coideal on $\mathbb{N}$ represented by a compact set of Borel functions defined on some Polish space P is coanalytic and selective.*

Proof. This follows from Theorem 7.48 and Corollary 7.51.   □

7.4 SOME APPLICATIONS OF ULTRA-RAMSEY THEORY

A coideal $\mathcal{H}$ on $\mathbb{N}$ is *bisequential* if for every ultrafilter $\mathcal{U} \subseteq \mathcal{H}$, there is a sequence (A_n) of elements of $\mathcal{U}$ such that

$$(\forall A \notin \mathcal{H})\,(\exists\, n)\ A \cap A_n = \emptyset. \tag{7.45}$$

Having in mind the representation $\mathcal{H} = \mathcal{H}_K(x, (x_n))$, this is a strong way of saying that every subsequence of (x_n) that accumulates to x has a further subsequence that converges to x. It follows that $\mathcal{H}$ is comeager in 2^A for every set $A \in \mathcal{H}$. The following result is a basis for all our applications of local Ramsey theory in this section.

Theorem 7.53 *Every coanalytic (or analytic) selective coideal $\mathcal{H}$ on $\mathbb{N}$ is bisequential.*

---

[6]Since its pointwise limit would be at the same time a Baire Class 1 function and a function without a point of continuity on $\overline{P} = \overline{Q}$.

Proof. Choose an ultrafilter $\mathcal{U}$ on $\mathbb{N}$ consisting only of sets from $\mathcal{H}$. Applying Theorem 7.42, we get a $\mathcal{U}$-tree T with root $\emptyset$ such that $[T] \subseteq \mathcal{H}$ or $[T] \cap \mathcal{H} = \emptyset$. For $s \in T$, let

$$A_s = \{n \in \mathbb{N} : s \cup \{n\} \in T\}. \qquad (7.46)$$

Then $A_s \in \mathcal{U}$ for all $s \in T$. So, in particular, $\bigcap_{s \in F} A_s \in \mathcal{H}$ for every finite $F \subseteq T$. Since $\mathcal{H}$ is selective, there is a $B \in \mathcal{H}$ such that

$$B/s \subseteq A_s \text{ for all } s \in T, \ s \subseteq B. \qquad (7.47)$$

If follows that $B \in [T] \cap \mathcal{H}$. This shows that we must have the alternative $[T] \subseteq \mathcal{H}$. Note that any set $C \subseteq \mathbb{N}$ that has a nonempty intersection with all A_s $(s \in T)$ contains an infinite branch of T. It follows that any such set $C \subseteq \mathbb{N}$ contains an element of $\mathcal{H}$. This finishes the proof. $\qquad \square$

Corollary 7.54 *Every coanalytic (or analytic) coideal on $\mathbb{N}$ represented by a countably tight compact space is bisequential.*

Proof. This follows from Theorems 7.48 and 7.53. $\qquad \square$

Corollary 7.55 *Every coideal on $\mathbb{N}$ represented by a compact set of Borel functions defined on some Polish P is bisequential.*

Proof. This follows from Corollary 7.52 and Theorem 7.53. $\qquad \square$

Corollary 7.56 (Bourgain-Fremlin-Talagrand)   *Every compact set K of Baire Class 1 functions defined on some Polish space is Fréchet.*[7]

Proof. Consider a subset F of K that accumulates to some $x \in K$. By Theorem 7.50, there is a sequence (x_n) of elements of F that accumulates to x. By Corollary 7.55, the coideal $\mathcal{H}(x, (x_n))$ is bisequential, so in particular there is a subsequence (x_{n_k}) of (x_n) converging to x. $\qquad \square$

Corollary 7.57 (Odell-Rosenthal) *A separable Banach space X contains a subspace isomorphic to ℓ_1 iff there is an $x^{**} \in X^{**}$ so that there is no sequence $(x_n) \subseteq X$ with the properly that $y^*(x_n) \to x^{**}(y^*)$ for all $y \in X^*$.*

Proof. A simple interpretation of Rosenthal's ℓ_1-theorem states that if a separable Banach space X contains no isomorphic copy of ℓ_1, then the double dual ball $B_{X^{**}}$ consists of Baire Class 1 functions defined on the dual ball B_{X^*} in its weak*-topology. Since B_X is pointwise dense in $B_{X^{**}}$, we get the conclusion of the theorem from the Bourgain-Fremlin-Talagrand theorem. $\square$

Our next example shows that Theorem 7.26 can naturally be deduced from Corollary 7.56. This may not be so surprising, since both of these two results are based on a single basic fact (Corollary 7.25) from the local Ramsey theory.

---

[7]A topological space X is said to have the *Fréchet property* if the closure $\overline{Y}$ of a subset Y of X is obtained by taking the limits of all converging sequences (y_n) of elements of Y.

Example 7.4.1   Let $\mathcal{A}$ be an infinite family of infinite almost disjoint subsets of $\mathbb{N}$ and let

$$\mathcal{A}^* = \mathcal{A} \cup \mathrm{FIN}. \tag{7.48}$$

Identifying sets with the corresponding characteristic functions, we consider $\mathcal{A}^*$ as a subset of the Cantor cube $2^{\mathbb{N}}$ from which it inherits its separable metric topology. For $n \in \mathbb{N}$, let $\pi_n : \mathcal{A}^* \to \{0,1\}$ be the restriction of the projection map of $2^{\mathbb{N}}$, i.e., $\pi_n(X) = 1$ iff $n \in X$. For $A \in \mathcal{A}$, let $\delta_A : \mathcal{A}^* \to \{0,1\}$ be the characteristic function of the singleton $\{A\}$:

$$\delta_A(x) = 1 \quad \text{iff} \quad x = A.$$

Let $\bar{0} : \mathcal{A}^* \to \{0,1\}$ be the constant 0 function. Note that $\bar{0}$ and π_n are continuous functions on $\mathcal{A}^*$ and that $\delta_A (A \in \mathcal{A})$ are Baire Class 1 functions on $\mathcal{A}^*$. Note also that

$$K_{\mathcal{A}} = \{\bar{0}\} \cup \{\pi_n : n \in \mathbb{N}\} \cup \{\delta_A : A \in \mathcal{A}\} \tag{7.49}$$

is a compact subset of the Tychonov cube $\{0,1\}^{\mathcal{A}^*}$. Note that the sequence (π_n) accumulates to $\bar{0}$ and that its subsequence $(\pi_n)_{n \in B}$ converges to $\bar{0}$ iff $B \cap A$ is finite for all $A \in \mathcal{A}$. Thus if $\mathcal{A}$ is taken to be a maximal almost disjoint family of infinite subsets of $\mathbb{N}$, the conclusion of the Bourgain-Fremlin-Talagrand theorem fails for $K_{\mathcal{A}} \subseteq \mathcal{B}_1(\mathcal{A}^*)$. Note that if $\mathcal{A}$ is analytic, so is $\mathcal{A}^*$. In this case $\mathcal{A}^*$ would be a continuous image of the irrationals $\mathbb{R} \setminus \mathbb{Q}$ and, therefore, $\mathcal{B}_1(\mathcal{A}^*)$ equipped with the topology of pointwise convergence would be homeomorphic to a subset of $\mathcal{B}_1(\mathbb{R} \setminus \mathbb{Q})$. Hence, the compactum $K_{\mathcal{A}}$ would be a subject of the Bourgain-Fremlin-Talagrand theorem, and so, in particular there would be a subsequence (π_{n_k}) of (π_n) converging to $\bar{0}$. So as pointed out above, in this case $\mathcal{A}$ cannot be a maximal almost disjoint family of infinite subsets of $\mathbb{N}$. This shows that Bourgain-Fremlin-Talagrand theorem can be considered an extension of the theorem of Mathias presented above (see Theorem 7.26) and saying that there are no maximal almost disjoint analytic families of infinite subsets of $\mathbb{N}$.

Recall that a coideal $\mathcal{H}$ on $\mathbb{N}$ as a subset of $2^{\mathbb{N}}$ is comeager whenever it has the Baire property in $2^{\mathbb{N}}$. We have also seen that the comeagerness of $\mathcal{H}$ is equivalent to the existence of a strictly increasing sequence $(n_k) \subseteq \mathbb{N}$ such that $\bigcup_{k \in M} [n_k, n_{k+1}) \in \mathcal{H}$ for all infinite $M \subseteq \mathbb{N}$. In the proof of Theorem 7.47 we used such a property to prove that $\mathcal{H}$ is selective. To move from selectiveness of $\mathcal{H}$ to bisequentiality we needed $\mathcal{H}$ to be $\mathcal{H}$-Ramsey. The following result connects these two requirements although the "Ramsey" is to be interpreted in the classical sense, i.e. relative to the coideal of infinite subsets of $\mathbb{N}$.

Theorem 7.58 *Suppose Γ is a collection of subsets of $\mathbb{N}^{[\infty]}$ closed under taking continuous preimages. If every set from Γ is Ramsey then every coideal belonging to Γ is comeager.*

Proof. Given a coideal $\mathcal{H} \in \Gamma$, set

$$\mathcal{X} = \left\{ M = (m_k) \in \mathbb{N}^{[\infty]} : \bigcup_{k=0}^{\infty} [m_{2k}, m_{2k+1}) \in \mathcal{H} \right\}. \qquad (7.50)$$

Clearly, $\mathcal{X}$ is a continuous preimage of $\mathcal{H}$, so by our assumption $\mathcal{X} \in \Gamma$. It follows that $\mathcal{X}$ is Ramsey, so we can find an infinite $M \subseteq \mathbb{N}$ such that

$$M^{[\infty]} \subseteq \mathcal{X} \text{ or } M^{[\infty]} \cap \mathcal{X} = \emptyset. \qquad (7.51)$$

Since $\mathcal{H}$ is a coideal, it is easily seen that the second alternative is impossible, so we are left with the first. Choose a strictly increasing sequence $(n_i) \subseteq M$ that takes every second member of M. Then from $M^{[\infty]} \subseteq \mathcal{X}$ we conclude that $\bigcup_{i \in B} [n_i, n_{i+1}) \in \mathcal{H}$ for all infinite $B \subseteq \mathbb{N}$. So $\mathcal{H}$ is comeager in $2^{\mathbb{N}}$. $\square$

Corollary 7.59 *For every coanalytic (or analytic) coideal $\mathcal{H}$ in $\mathbb{N}$, there is an infinite increasing sequence $(m_k) \subseteq \mathbb{N}$ such that $\bigcup_{k \in B} [m_k, m_{k+1}) \in \mathcal{H}$ for all infinite $B \subseteq \mathbb{N}$.*

Proof. Let Γ be the minimal field of subsets of $2^{\mathbb{N}}$ that contains all open sets and that is closed under the Souslin operation. So in particular Γ is closed under continuous images, contains all analytic and coanalytic subsets of $2^{\mathbb{N}}$, and every set from Γ is Ramsey. By Theorem 7.58 every coideal $\mathcal{H}$ belonging to Γ is comeager, and by its proof it satisfies the conclusion of the Corollary. $\square$

We finish this section with an application to abstract Ramsey theory itself. It depends also on the following well-known result.

Theorem 7.60 (Balcar-Pelant-Simon) *There is a family $\mathcal{T} \subseteq \mathbb{N}^{[\infty]}$ with the following properties*

(1) $\left(\forall A \in \mathbb{N}^{[\infty]} \right) \left(\exists B \in \mathcal{T} \right) \; B \subseteq^* A,$

(2) $(\forall A, B \in \mathcal{T}) \; [A \subseteq^* B \vee B \subseteq^* A \vee |A \cap B| < \aleph_0].$

A family $\mathcal{T}$ with these properties is usually in the literature called *base-matrix*. It is really a family of infinite subsets of $\mathbb{N}$ forming a tree in the ordering of reverse almost-inclusion and being dense in $\mathbb{N}^{[\infty]}$ relative to the same ordering.

Theorem 7.61 *Let $\mathcal{H}$ be a coanalytic selective coideal on $\mathbb{N}$. Then the corresponding Ramsey space $(\mathbb{N}^{[\infty]}, \mathcal{H}, \subseteq, \subseteq, r)$ is topological.*

Proof. By Theorem 7.60, we can fix a base-matrix $\mathcal{T} \subseteq \mathbb{N}^{[\infty]}$ and assume it is closed under finite changes of its elements. Let $\mathcal{T}_0 = \mathcal{T} \cap \mathcal{H}$ and consider the refinement of the metrizable topology of $\mathbb{N}^{[\infty]}$ with the basic open sets of the form

$$[s, M] \quad (s \in \mathbb{N}^{[<\infty]}, M \in \mathcal{T}_0). \qquad (7.52)$$

That this collection forms a basis for a topology on $\mathbb{N}^{[\infty]}$ follows easily from properties (1) and (2) of $\mathcal{T}$. Thus we have the usual notions of nowhere dense, meager, and property of Baire relative to this new $\mathcal{T}_0$-topology.

We need to check that the family of $\mathcal{T}_0$-meager sets coincides with the family of $\mathcal{H}$-meager sets and similarly, that the families of $\mathcal{T}_0$-Baire and $\mathcal{H}$-Baire subsets of $\mathbb{N}^{[\infty]}$ coincide. In both cases, it follows from the fact that

$$[s, M] \cap \mathcal{T}_0 \neq \emptyset \tag{7.53}$$

for all $s \in \mathbb{N}^{[<\infty]}$ and $M \in \mathcal{H}$, so let us show that this is indeed true. Since $\mathcal{T}_0$ is closed under finite changes, it suffices to show that for every $M \in \mathcal{H}$ there is an $N \in \mathcal{T}_0$ such that $N \subseteq^* M$. To see this, choose an ultrafilter $\mathcal{U} \subseteq \mathcal{H}$ such that $M \in \mathcal{U}$. By Theorem 7.53, there is a sequence $(A_n) \subseteq \mathcal{U}$ such that

$$(\forall A \notin \mathcal{H}) \, (\exists \, n) \, A \cap A_n = \emptyset. \tag{7.54}$$

Choose an infinite set $B \subseteq \mathbb{N}$ such that $B \setminus A_n$ is finite for all n. Then B has finite intersection with every infinite subset of $\mathbb{N}$ that does not belong to $\mathcal{H}$, or in other words, $B^{[\infty]} \subseteq \mathcal{H}$. By (1) we can find a $C \in \mathcal{T}$ such that $C \subseteq^* B$. Then $C \in \mathcal{H}$ and therefore $C \in \mathcal{T}_0$, as required. This finishes the proof. □

7.5 LOCAL RAMSEY THEORY AND ANALYTIC TOPOLOGIES ON $\mathbb{N}$

Sequential convergence is a subject matter that shows up in several areas of mathematics, and one usually studies it for utilitarian reasons. In this section we give one such study as it is really very closely tied with the concepts and results of the local Ramsey theory developed so far. More precisely, the purpose of this section is to stress the usefulness of the Ramsey-theorecic view on sequential convergence in countable topological spaces (X, τ), especially when the topology τ viewed as a subset of the Cantor space 2^X is assumed to be *analytic*. In the previous section, we have already encountered an effective use of local Ramsey theory in proving a result of this sort (see, for example, Corollary 7.56).

We start by recalling the notions that we are already familiar with. We shall say that a point $x \in X$ is a *Fréchet point* (or, X is *Fréchet at x*) if for every $F \subseteq X$ with $x \in \overline{F}$ there is a sequence $x_n \in F$ converging to x. Analogously, we define the notion of a *bisequential point* by requiring that if an ultrafilter converges to x, then it contains a sequence of sets converging to x. We say that a space X is Fréchet (bisequential) if every point of X is a Fréchet (bisequential) point in X. Taking analogies with some notions from local Ramsey theory, we arrive at the following two definitions.

Definition 7.62 *A point x in a topological space X is a q^+-point if for every F with $x \in \overline{F}$ and every partition $F = \bigcup_n F_n$ of A into finite sets, there is a subset H of F such that $x \in \overline{H}$ and $|H \cap F_n| \leq 1$ for all n.*

Definition 7.63 *A point x in a topological space X is a p^+-point if for any decreasing sequence P_n of subsets of $X \setminus x$ such that $x \in \overline{P_n}$ for all n, there is $P \subseteq X \setminus x$ such that $x \in \overline{P}$ and $P \subseteq^* P_n$ (i.e., $P \setminus P_n$ is finite) for all n.*

We note that a point that is at the same time a Fréchet point and a p^+-point is what is called a *strongly Fréchet point* in the literature. This hints that the following analog of an important notion of local Ramsey Theory might be of some topological interest as well.

Definition 7.64 *We say that a point x in a topological space X is a selective point if it is both a p^+- and q^+-point.*

As we have seen, points in countable topological spaces correspond to ideals and coideals on the countable index set, and the notion of a selective point is meant to correspond to the selectivity of the corresponding coideal. So, we have the following reformulation of Theorem 7.53.

Theorem 7.65 *A point in a countable analytic space is selective if and only if it is bisequential.*

It is also worth pointing out the following reformulation of Corollary 7.54.

Theorem 7.66 *Every countable analytic space with a countably tight compactification is bisequential.*

It thus appears that in the realm of countable analytic spaces selectivity is a rather strong requirement in comparison with the Fréchet property. For example, from Theorem 7.65, we learn that selectivity is a productive property, while the Fréchet property is not. A typical countable analytic Fréchet space whose square is not Fréchet is the *sequential fan*, $S(\omega)$, the space defined over $\mathbb{N} \times \mathbb{N} \cup \{\infty\}$, where all points in $\mathbb{N} \times \mathbb{N}$ are isolated and the neighborhood filter of ∞ is generated by the sets of the form
$$U_f = \{(n,m) \in \mathbb{N} \times \mathbb{N} : m \geq f(n)\} \cup \{\infty\}$$
for $f \in \mathbb{N}^{\mathbb{N}}$. To see that $S(\omega)^2$ is not Fréchet, note that the set
$$\{((m,n),(0,m)) : m, n \in \mathbb{N}\}$$
accumulates to (∞, ∞), but it contains no sequence converging to (∞, ∞). We prove below a general fact from which it follows that the point ∞ must fail to be a p^+-point or a q^+-point in $S(\omega)$. Clearly, every point of a Fréchet space is a q^+-point,[8] so ∞ in $S(\omega)$ must fail to be a p^+-point. There is of course a more direct way to see that ∞ is not a p^+-point of $S(\omega)$, by simply noticing that
$$P_n = \{(x,y) : x > n)\}, \ (n \in \mathbb{N})$$
is a sequence of sets that accumulate to ∞ although no set that is almost included in all the P_n accumulates to ∞. This example suggests that a deeper reason for this phenomenon lies in the fact that $S(\omega)$ fails to have either of the following two standard diagonal-sequence properties.

---

[8]In fact, more is true: every point in a countable *sequential space* is a q^+-point. Recall that a space X is sequential if sequentially closed subsets of X are in fact closed.

Definition 7.67 *A point x in X has the* diagonal-sequence property *if for any double-indexed sequence $\{x_{nk}\}$ in X such that*

$$x_{nk} \to_k x \text{ for all } n$$

then for each n, we can choose $k(n)$ such that

$$x_{nk(n)} \to_n x.$$

If we require that some infinite subsequence of $\{x_{nk(n)}\}$ converges to x rather than the sequence itself, we say that x has the weak diagonal sequence property *in X.*

Lemma 7.68 *The following are equivalent for every Fréchet point x of any countable topological space X,*

(1) *x is not a p^+-point in X,*

(2) *x fails to have the weak diagonal sequence property in X,*

(3) *X contains a closed copy of $S(\omega)$ with x as its point at infinity.*

Proof. The only nontrivial implication is from (1) to (3). Suppose (P_n) is a decreasing sequence of subsets of $X \setminus \{x\}$ that accumulate to x, but no set that is almost included in every P_n accumulates to x. Since x is a Fréchet point, for each n we can find a sequence $\{x_{nk} : k \in \mathbb{N}\} \subseteq P_n$ converging to x. We may further assume that every $y \in X \setminus \{x\}$ has a neighborhood that has an empty intersection with all but finitely many of these sequences. Then $\{x_{nk} : (n, k) \in \mathbb{N} \times \mathbb{N}\} \cup \{x\}$ is a closed copy of $S(\omega)$ in X. $\square$

Corollary 7.69 *A countable analytic Fréchet space is bisequential if and only if it has the weak diagonal sequence property.*

Proof. This follows from Theorem 7.65 and Lemma 7.68. $\square$

Corollary 7.70 *A countable analytic Fréchet space is bisequential if and only if it contains no closed copy of $S(\omega)$.*

Proof. This also follows from Theorem 7.65 and Lemma 7.68. $\square$

Corollary 7.71 *A countable analytic space is bisequential if and only if its square is Fréchet.*

Proof. This follows from the fact that the square of $S(\omega)$ is not Fréchet. $\square$

We now give another result about preservation of topological properties when taking products in the class of countable analytic spaces.

Theorem 7.72 *Suppose X and Y are two countable analytic spaces with the weak diagonal sequence property. Then their product $X \times Y$ also has the weak diagonal sequence property.*

Proof. Consider a double-indexed sequence (x_{mn}) in $X \times Y$ such that

$$x_{mn} \to_n x \text{ for all } m.$$

We need to find an increasing sequence (m_k) and a sequence (n_k) such that $x_{m_k n_k} \to_k x$. Identifying (m, n) with $x_{m,n}$ and ∞ with x, we may assume we have two analytic topologies σ and τ on $\mathbb{N} \times \mathbb{N} \cup \{\infty\}$ with the weak diagonal sequence property, with ∞ as the only nonisolated point and that for each m the set $C_m = \{m\} \times \mathbb{N}$ converges to ∞ in both topologies. We need to find an infinite set $M \subseteq \mathbb{N} \times \mathbb{N}$ such that $M \to_\sigma \infty$ and $M \to_\tau \infty$ (or in other words, M converges to ∞ in both topologies) and such that $M \cap C_m$ is finite for all m.

Let $\mathcal{H}$ be the collection of all subsets A of $\mathbb{N}^2$ such that $A \cap C_m$ is finite for all m and such that A contains an infinite sequence that converges to ∞ relative to the topology σ as well as an infinite sequence that converges to ∞ relative to τ. We claim that $\mathcal{H}$ is a selective coideal on $\mathbb{N}^2$. The fact that $\mathcal{H}$ has property (q) is clear, so let us check that $\mathcal{H}$ has property (p). So let (P_n) be a given decreasing sequence of elements of $\mathcal{H}$. For each n, pick an infinite subset $M_n \subseteq P_n$ such that $M_n \to_\sigma \infty$. Since σ has the weak diagonal sequence property, we can find an infinite set $M \subseteq \bigcup_{n=0}^\infty M_n$ such that $M \cap M_n$ is finite for all n and still converging to ∞. This gives us an infinite subset M of P_0 such that $M \to_\sigma \infty$ and such that $M \subseteq^* P_n$ for all n. Similarly, we find an infinite subset N of P_0 such that $N \to_\tau \infty$ and such that $N \subseteq^* P_n$ for all n. Let $P = M \cup N$. Then $P \in \mathcal{H}$ and $N \subseteq^* P_n$ for all n, as required.

Applying Corollary 7.23 to the selective coideal $\mathcal{H}$ and the analytic family

$$\mathcal{A} = \{M \subseteq \mathbb{N} \times \mathbb{N} : \infty \notin \overline{M}^\sigma\},$$

we get a set $B \in \mathcal{H}$ such that $B^{[\infty]} \subseteq \mathcal{A}$, or else $B^{[\infty]} \cap \mathcal{A} = \emptyset$. Note that the first alternative is eliminated by the fact that the set B itself does not belong to $\mathcal{A}$, as it contains a sequence that converges to ∞ relative to the topology σ. So we must have the second alternative, $B^{[\infty]} \cap \mathcal{A} = \emptyset$, which in particular means that $B \to_\sigma \infty$, as all of its infinite subsets accumulate to ∞ relative to σ. By the definition of $\mathcal{H}$, the set B must contain an infinite subset A such that $A \to_\tau \infty$. So we have arrived at an infinite set that converges to ∞ relative to both topologies and that has the finite intersection with C_m for all m. This finishes the proof. □

We now give an application of local Ramsey theory to get an interesting metrizability criterion for a class of topological groups.

Theorem 7.73 *A countable analytic group is metrizable if and only if it is Fréchet.*

Proof. To prove the nontrivial implication from right to left, let G be a given countable Fréchet analytic topological group. By a well-known topological group metrization theorem of Birkhof-Kakutani (see Theorem 9.33), it suffices to show that the identity e of G has a countable neighborhood base.

Let us first show that the identity e of the group G is a selective point in G, or in other words, that the coideal

$$\mathcal{H} = \left\{ A \subseteq G : e \in \overline{A \setminus \{e\}} \right\} \tag{7.55}$$

is selective. This reduces to showing that $\mathcal{H}$ satisfies conditions (p) and (q) of Lemma 7.4. The property (q) follows easily, since if $A \in \mathcal{H}$ is decomposed into a sequence (F_n) of finite subsets by the Fréchet property of G, we can find a $B \subseteq A$ forming a sequence converging to e, so by going to a sufficiently thin subset of B, we can get a member of $\mathcal{H}$ that takes at most one point from each of the finite sets F_n. So, let us check that $\mathcal{H}$ satisfies (p). Let (A_n) be a given decreasing sequence of elements of $\mathcal{H}$. If there is a sequence (x_k) converging to e such that for all n there are infinitely many k such that $x_k \in A_n$, we can find the desired set $B \in \mathcal{H}$ such that $B \setminus A_n$ is finite for all n. So assume such a converging sequence cannot be found. Then, going perhaps to a subsequence of (A_n) for each n we can find a converging sequence $x_k^n \longrightarrow_k e$ consisting entirely of elements of $A_n \setminus (A_{n+1} \cup \{e\})$. Consider the set

$$X = \left\{ x_n^0 \cdot x_k^n : k, n \in \mathbb{N} \right\}. \tag{7.56}$$

Clearly, X accumulates to e, since its closure contains the sequence $(x_n^0)_{n=0}^{\infty}$, which converges to e. By our assumption about the group, the set X contains a converging sequence

$$x_{n_i}^0 \cdot x_{k_i}^{n_i} \longrightarrow_i e. \tag{7.57}$$

Note that the infinite sequence $(n_i)_{i=0}^{\infty}$ of integers cannot take a constant value infinitely many times, so by going to a subsequence, we may assume that $n_i < n_j$ whenever $i < j$. It follows that

$$\left(x_{n_i}^0 \right)^{-1} \longrightarrow_i e. \tag{7.58}$$

Taking the products of terms of the two converging sequences, we conclude that

$$x_{k_i}^{n_i} \longrightarrow_i e. \tag{7.59}$$

Let $B = \left\{ x_{k_i}^{n_i} : i \in \mathbb{N} \right\}$. Then $B \in \mathcal{H}$ and $B \setminus A_n$ is finite for all n. This finishes the proof that $\mathcal{H}$ is a selective coideal.

By homogeneity of topological groups, we conclude that every point of G is a selective point in G, so applying Theorem 7.65, we conclude that G is a bisequential space. So in particular, the coideal $\mathcal{H}$ itself is bisequential. Choose an ultrafilter $\mathcal{U}$ on G containing the neighborhood filter of e such that no nowhere dense subset of G belongs to $\mathcal{U}$. So, in particular $\mathcal{U}$ is nonprincipal and is included in the coideal $\mathcal{H}$. Since $\mathcal{H}$ is bisequential, there is a decreasing sequence $(A_n) \subseteq \mathcal{U}$ such that

$$(\forall A \notin \mathcal{H}) (\exists n\ ) A \cap A_n = \emptyset. \tag{7.60}$$

For each n, we let V_n be the interior of the closure of A_n. Note that since $\mathcal{U}$ contains no nowhere dense set each V_n is open and nonempty. Since G is regular, the sequence (V_n) still converges to e, i.e., every neighborhood of e contains all but finitely many V_n. If follows that.

$$\{V_n \cdot V_n^{-1} : n \in \mathbb{N}\} \tag{7.61}$$

is a countable neighborhood base of e in G. This finishes the proof. $\quad\square$

Corollary 7.74 *A separable topological group* H *is metrizable if and only if it satisfies the following two conditions:*

(a) H *has the Fréchet property.*

(b) H *induces an analytic topology on one of its countable dense subgroups.*

7.6 ULTRA-HALES-JEWETT SPACES

Let $L = \bigcup_{n=0}^{\infty} L_n$ be a given *alphabet* written as an increasing union of finite alphabets L_n and let $v \notin L$ be a given *variable*. Let W_L be the semigroup of *words* over L and let W_{Lv} be the semigroup of *variable-words* over L, i.e., words over $L \cup \{v\}$ in which v occurs at least once. As in the proof of the infinite Hales-Jewett theorem (see Section 2.5), we choose nonprincipal ultrafilters $\mathcal{W}$ and $\mathcal{V}$ on W_L and W_{Lv}, respectively, such that

(1) $\mathcal{W}^\frown \mathcal{W} = \mathcal{W}$,

(2) $\mathcal{V}^\frown \mathcal{V} = \mathcal{V}$,

(3) $\mathcal{V}^\frown \mathcal{W} = \mathcal{W}^\frown \mathcal{V} = \mathcal{V}$,

(4) $\mathcal{V}[\lambda] = \mathcal{W}$ for all $\lambda \in L$,

(5) $\{w \in W_L : |w| \geq n\} \in \mathcal{W}$ for all $n \in \mathbb{N}$.

(Recall that $\mathcal{V}[\lambda]$ is the image of $\mathcal{V}$ under the substitution map $x \to x[\lambda]$, where $x[\lambda]$ is obtained from x by replacing every occurrence of v by the letter λ.) The set $W_L^{<\infty}$ of finite sequences of words is considered a tree under end-extension. To avoid the confusion between words and finite sequences of words, we avoid using the concatenation symbol $\frown$ when dealing with members of $W_L^{<\infty}$. Thus an immediate successor of a node t of $W_L^{<\infty}$ is uniquely determined by a word $w \in W_L$ and is denoted by

$$(t, w) = t \cup \{\langle |t|, w \rangle\}. \tag{7.62}$$

A $\mathcal{W}$-*tree* is now defined as in Section 7.2 above, i.e., as a downward closed subset T that has a *stem*, $\mathrm{st}(T)$, the maximal node comparable to all other nodes of T, and such that for all $t \in T$ equal or end-extending $\mathrm{st}(T)$,

$$\{w \in W_L : (t, w) \in T\} \in \mathcal{W}. \tag{7.63}$$

For a $\mathcal{W}$-tree T, by $[T]$ we denote the set of all $(w_n) \in W_L^{\infty}$ such that

$$(w_0, \ldots, w_{n-1}) \in T \text{ for all } n. \tag{7.64}$$

The collection of sets of the form $[T]$ for T a $\mathcal{W}$-tree forms a basis for a zero-dimensional topology on the set W_L^∞ of all infinite sequences of words which refines the metrizable product topology. We refer to this topology on W_L^∞ as the $\mathcal{W}$-*topology*. As in Section 7.2 we define the ordering

$$T' \leq^0 T \text{ iff } T' \subseteq T \text{ and st} (T') = \text{st} (T) \qquad (7.65)$$

and its variations $\leq^n$ for all $n \geq 0$. We use $\leq^0$ to define when a subset $\mathcal{X}$ of W_L^∞ is $\mathcal{W}$-*Ramsey*, i.e., whenever for every $\mathcal{W}$-tree T there is a $T' \leq^0 T$ such that

$$[T'] \subseteq \mathcal{X} \text{ or } [T'] \cap \mathcal{X} = \emptyset. \qquad (7.66)$$

If for every $\mathcal{W}$-tree T we can find a $\mathcal{W}$ tree $T' \leq^0 T$ such that $[T'] \cap \mathcal{X} = \emptyset$, we call $\mathcal{X}$ a $\mathcal{W}$-*Ramsey null set*. Then referring to Theorem 7.42 (or its proof), we get the following result.

Theorem 7.75 *Every subset of W_L^∞ that has the property of Baire relative to the $\mathcal{W}$-topology is $\mathcal{W}$-Ramsey and vice versa. Every subset of W_L^∞ that is meager relative to the $\mathcal{W}$-topology is $\mathcal{W}$-Ramsey null and vice versa.*

Corollary 7.76 *For every finite $\mathcal{W}$-Baire-measurable coloring of W_L^∞, there is a $\mathcal{W}$-tree T with stem $\emptyset$ such that $[T]$ is monochromatic.*

Corollary 7.77 *For every finite metrically Souslin-measurable coloring of W_L^∞, there is a $\mathcal{W}$-tree T with stem $\emptyset$ such that $[T]$ is monochromatic.*

We now need a lemma to relate these results to the standard applications of the Ramsey space

$$(W_L^{[\infty]}, W_{Lv}^{[\infty]}, \leq, \leq^0, r) \qquad (7.67)$$

of rapidly increasing sequences of words and variable-words developed above in Section 4.4. So let us recall some definitions from Section 4.4. A sequence (x_n) from $W_L^{\leq\infty}$ or $W_{Lv}^{\leq\infty}$ is said to be *rapidly increasing* if

$$|x_n| > \sum_{i=0}^{n-1} |x_i| \qquad (7.68)$$

for all n in the domain of the sequence. By $W_L^{[<\infty]}, W_{Lv}^{[<\infty]}, W_L^{[\infty]},$ and $W_{Lv}^{[\infty]}$ we denote the corresponding families of rapidly increasing finite or infinite sequences. For an $X = (x_n) \in W_{Lv}^{[\infty]}$, set

$$[X]_L = \{x_{n_0}[\lambda_0]^\frown \dots ^\frown x_{n_k}[\lambda_k] \in W_L : n_0 < \dots < n_k, \quad \lambda_i \in L_{n_i} (i \leq k)\}$$

$$[X]_{Lv} = \{x_{n_0}[\lambda_0]^\frown \dots ^\frown x_{n_k}[\lambda_k] \in W_{Lv} : n_0 < \dots < n_k,$$

$$\lambda_i \in L_{n_i} \cup \{v\}(i \leq k)\}.$$

From the assumption that the sequence $X = (x_n)$ is rapidly increasing, for every $w \in [X]_L$ and $x \in [X]_L$, the choices of sets of integers $\{n_0 < \ldots < n_k\}$ and $\{m_0 < \ldots < m_\ell\}$ such that the equalities

$$w = x_{n_0} [\lambda_0] \,^\frown \ldots \,^\frown x_{n_k} [\lambda_k],$$
$$x = x_{m_0} [\mu_0] \,^\frown \ldots \,^\frown x_{m_\ell} [\mu_\ell]$$

hold for some $\lambda_i \in L_{n_i}$ $(i \leq k)$ and $\mu_i \in L_{n_i} \cup \{v\}$ $(i \leq \ell)$ are unique. So we denote these sets by $\mathrm{supp}_X (w)$ and $\mathrm{supp}_X (x)$, respectively. Recall that for two finite sets F and G of integers, we let $F < G$ denote the fact that $m < n$ whenever $m \in F$ and $n \in G$. For $X = (x_n) \in W_L^{[\leq \infty]}$ and $Y = (y_n) \in W_{Lv}^{[\leq \infty]}$, we say that X is a block subsequence of Y and write $X \leq Y$ if $X \subseteq [Y]_L$ and $\mathrm{supp}_Y (x_m) < \mathrm{supp}_Y (x_n)$ whenever $m < n$.

The reader will notice that in Section 4.4, for some technical reasons, we have defined some other variations on the block-subsequence order. Since in this section we do not need these variations, we use only one symbol $\leq$ for this relation. For example, we consider $X \leq Y$ for X and Y sequences of variable-words, as well as when X is a sequence of nonvariable words and Y is a sequence of variable-words.

Lemma 7.78 *For every W-tree T with stem $\emptyset$ there is a $Y \in W_{Lv}^{[\infty]}$ such that*

$$[Y]_L^{[\infty]} = \left\{ X \in W_L^{[\infty]} : X \leq Y \right\} \subseteq [T]. \tag{7.69}$$

Proof. For $t \in T$, let

$$W_t = \{w \in W_L : (t, w) \in T\}. \tag{7.70}$$

Then $W_t \in W$ for all $t \in T$. Starting with the choice $P_W^0 = W_\emptyset$, we proceed as in the proof of the infinite Hales-Jewett theorem given above in Section 2.5. We build decreasing sequences (P_W^n) and (P_V^n) of elements of W and V, respectively, and a rapidly increasing sequence $Y = (y_n)$ of variable-words such that for all n,

$(a)_n \quad y_n \in P_V^n,$
$(b)_n \quad (\forall \lambda \in L_n)(\forall x \in P_V^n) \quad x [\lambda] \in P_W^n,$
$(c)_n \quad (Vx) (\forall \lambda \in L_n \cup \{v\}) \ y_n [\lambda] \,^\frown x \in P_V^n,$
$(d)_n \quad (W_t) \ y_n \,^\frown t \in P_V^n.$

As $P_W^0 = W_\emptyset$ has been chosen, let

$$P_V^0 = \{x \in W_{Lv}; \forall \lambda \in L_0 \quad x [\lambda] \in P_W^0\}. \tag{7.71}$$

By properties (1) and (2) of W and V, we conclude that $P_V^0 \in V$ and that $\dot{V}$-almost all choices of $y_0 \in P_V^0$ satisfy $(a)_0 - (d)_0$. The inductive step from n to $n + 1$ is done as follows. Let P_W^{n+1} be the intersection of P_W^n, the set

$$Q_W^n = \{w \in W_L : y_n \,^\frown w \in P_V^n\}, \tag{7.72}$$

and all sets of the form

$$W_t \quad (t \in T, \ t \le (y_i)_{i \le n}). \tag{7.73}$$

Note that the space $[(y_i)_{i \le n}]_L$ that the finite sequence $(y_i)_{i \le n}$ generates inside W_L is finite, as we allow only substitutions with letters that belong to one of the finite alphabets L_i ($i \le n$). Since $W_t \in \mathcal{W}$ for all $t \in T$, we conclude that $P_W^{n+1} \in \mathcal{W}$. By $(c)_n$ the set

$$Q_V^n = \{x \in P_V^n \ (\forall \lambda \in L_n \cup \{v\}) \ y_n [\lambda] \ ^\frown x \in P_V^n\} \tag{7.74}$$

belongs to $\mathcal{V}$. So using (1) and (2) again, we conclude that

$$P_V^{n+1} = \{x \in Q_V^n : (\forall \lambda \in L_{n+1}) \ x [\lambda] \in P_W^{n+1}\} \tag{7.75}$$

belongs to $\mathcal{V}$. By properties $(1), (2),$ and (3) of our ultrafilters, $\mathcal{V}$-almost all choices of $y_{n+1} \in P_V^{n+1}$ satisfy $(c)_{n+1}, (d)_{n+1}$, as well as

$$|y_{n+1}| > \sum_{i=0}^{n} |y_i|. \tag{7.76}$$

This describes the recursive construction. Then as in the proof of the infinite Hales-Jewett theorem, we verify the following claim.

Claim 7.78.1 *For nonnegative integers k and $n_0 < \ldots < n_k$ and a choice of letters $\lambda_i \in L_{n_i}$ ($i \le k$), we have that $y_{n_0} [\lambda_0] \ ^\frown \ldots \ ^\frown y_{n_k} [\lambda_k] \in P_W^{n_0}$.*

Using Claim 7.78.1 and the way the recursive construction has been done, we conclude that every finite rapidly increasing sequence $s = (w_i)_{i<j} \le [Y]_L$ is a member of T. This is clear for $s = \emptyset$, so let us consider the case $s = (t, w_{j-1})$ and $t = (w_i)_{i<j-1} \in T$. Let

$$w_{j-1} = y_{n_0} [\lambda_0] \ ^\frown \ldots \ ^\frown y_{n_k} [\lambda_k] \tag{7.77}$$

for some (actually unique) $n_0 < \ldots < n_k$ and $\lambda_\ell \in L_{n_\ell}$ ($\ell \le k$). By Claim 7.78.1, $w_{i-1} \in P_W^{n_0}$. If $n_0 = 0$, then $t = \emptyset, j = 1$, and $P_W^0 = W_\emptyset$ so $s = \langle w_0 \rangle$ is a member of tree T. If $n_0 > 0$, then from the fact that

$$\text{supp}_Y (w_{j-2}) < \text{supp}_Y (w_{j-1}) = \{n_0, \ldots, n_k\}, \tag{7.78}$$

we conclude that $t \le (y_\ell)_{\ell \le n_0 - 1}$. By the way that we have done the recursive step from $n_0 - 1$ to n_0, we know that $P_W^{n_0} \subseteq W_t$. It follows that $w_{i-1} \in W_t$ and therefore $s = (t, w_{i-1}) \in T$. This finishes the proof. $\qquad \square$

Combining Corollary 7.76 and Lemma 7.78, we obtain the following result.

Corollary 7.79 *For every finite W-Baire-measurable coloring of the space $W_L^{[\infty]}$ of all infinite rapidly increasing sequences of words over the alphabet L, there is an infinite rapidly increasing sequence $Y = (y_n)$ of variable-words such that $\{X \in W_L^{[\infty]} : X \le Y\}$ is monochromatic.*

Recall that in Section 4.4 we built the Hales-Jewett space

$$\left(W_L^{[\infty]}, W_{Lv}^{[\infty]}, \le, \le^0, r \right)$$

and used it to show among other things the following fact, which we also are deduce from Corollary 7.79.

Corollary 7.80 *Every subset of $W_L^{[\infty]}$ that is Souslin-measurable relative to the usual metrizable topology of $W_L^{[\infty]}$ is $W_{Lv}^{[\infty]}$-Ramsey.*

Proof. Let $[s, Y]$ be a given basic set of the space $(W_L^{[\infty]}, W_{Lv}^{[\infty]}, \leq, \leq^0, r)$ and let $\mathcal{X}$ be a given metrically Souslin-measurable subset of $W_L^{[\infty]}$. Thus $Y = (y_k) \in W_{Lv}^{[\infty]}$, $s = (w_k)_{k<n} \in W_L^{[\infty]}$ and

$$[s, Y] = \{X \in W_L^{[\infty]} : X \restriction n = s \text{ and } X \leq Y\}. \qquad (7.79)$$

We need to find $Z \leq Y$ in $W_{Lv}^{[\infty]}$ such that $Z \restriction m = Y \restriction m$ for $m = \min \{\ell : s \leq (y_k)_{k<\ell}\}$ and such that $[s, Z]$ is either included in or disjoint from $\mathcal{X}$. Choose ultrafilters $\mathcal{W}$ and $\mathcal{V}$ concentrating on $[Y]_L$ and $[Y]_{Lv}$, respectively, satisfying conditions (1),(2),(3), and (4) above and the following variation of (5):

$$\{w \in [Y]_L : \operatorname{supp}_Y(w) \cap \{0, \ldots, n\} = \emptyset\} \in \mathcal{W} \text{ for all } n. \qquad (7.80)$$

Then

$$T = \{t \in [Y]_L^{[<\infty]} : t \text{ either is extended by or it end-extends } s\} \qquad (7.81)$$

is a $\mathcal{W}$-tree with stem s. By Corollary 7.79 there is a $\mathcal{W}$-tree $T' \subseteq T$ with stem s such that $[T'] \subseteq \mathcal{X}$ or $[T'] \cap \mathcal{X} = \emptyset$. By (the proof of) Lemma 7.78, we can find a $Z = (z_k) \in W_{Lv}^{[\infty]}$ such that $Z \leq Y$ and $Z \restriction m = Y \restriction m$ and such that

$$\{X \in W_L^{[\infty]} : X \leq Z \& X \restriction n = s\} \subseteq [T']. \qquad (7.82)$$

It follows that $[s, Z] \subseteq \mathcal{X}$ or $[s, Z] \cap \mathcal{X} = \emptyset$, as required. $\qquad \square$

Corollary 7.81 *For every finite Souslin-measurable coloring of the set $W_L^{[\infty]}$ of all infinite rapidly increasing sequences of words over the alphabet L, there is an infinite rapidly increasing sequence $Z = (z_n)$ of variable-words such that $\{X \in W_L^{[\infty]} : X \leq Y\}$ is monochromatic.*

Let us now turn to the semigroup W_{Lv} and the idempotent $\mathcal{V}$ and consider the collection of all $\mathcal{V}$-subtrees of $W_{Lv}^{<\infty}$ and the corresponding $\mathcal{V}$-topology on W_{Lv}^{∞} generated by sets $[T]$ of infinite branches through $\mathcal{V}$-trees T. We also consider the corresponding notions of $\mathcal{V}$-*Ramsey* and $\mathcal{V}$-*Ramsey null* for subsets of W_{Lv}^{∞} and as before prove the following facts.

Theorem 7.82 *Every subset of the set W_{Lv}^{∞} of infinite sequences of variable-words that has the property of Baire relative to the $\mathcal{V}$-topology is $\mathcal{V}$-Ramsey and vice versa. Every subset of W_{Lv}^{∞} that is meager relative to the topology is $\mathcal{V}$-Ramsey null and vice versa.*

The proof of the following lemma is analogous to that of Lemma 7.78.

Lemma 7.83 *For every $\mathcal{V}$-tree T with stem $\emptyset$, there is an infinite rapidly increasing sequence $Z = (z_n)$ of variable-words such that every infinite rapidly increasing block-subsequence of Z is a branch through T.*

Corollary 7.84 *For every finite $\mathcal{V}$-Baire-measurable coloring of the space $W_{Lv}^{[\infty]}$ of all infinite rapidly increasing sequences of variable-words over the alphabet L, there is a $Z \in W_{Lv}^{[\infty]}$ such that $\{Y \in W_{Lv}^{[\infty]} : Y \leq Z\}$ is monochromatic.*

Corollary 7.85 *For every finite metrically Souslin-measurable coloring of the space $W_{Lv}^{[\infty]}$ of all infinite rapidly increasing sequences of variable-words over L, there is a $Z \in W_{Lv}^{[\infty]}$ such that the set of all rapidly increasing block-subsequences of Z is monochromatic.*

Remark 7.86 In connection with Corollaries 7.84 and 7.85, we remind the reader that the relation $\leq$ here allows substitutions for some (although not all) occurrences of v, so these corollaries are not just earlier results (like Corollaries 7.80 and 7.81) with v added to the alphabet.

We finish this section by remarking that while the notions of $\mathcal{W}$-Ramsey and $\mathcal{V}$-Ramsey refer to subsets of W_L^∞ and W_{Lv}^∞, the corresponding notions for the Hales-Jewett space refer to seemingly smaller sets $W_L^{[\infty]}$ and $W_{Lv}^{[\infty]}$. The difference however is not essential since, as it is easily seen, the set $W_L^\infty \setminus W_L^{[\infty]}$ is $\mathcal{W}$-Ramsey null and the set $W_{Lv}^\infty \setminus W_{LV}^{[\infty]}$ is $\mathcal{V}$-Ramsey null.

7.7 ULTRA-RAMSEY SPACES OF BLOCK SEQUENCES OF LOCATED WORDS

We start again with an alphabet $L = \bigcup_{n=0}^\infty L_n$ written as an increasing union of finite alphabets L_n and a variable $v \notin L$. Recall that a *located word* is a function from a finite nonempty subset of $\mathbb{N}$ into L and that a *located variable-word* is such a function with range $L \cup \{v\}$ with the value v achieved at least once. Let

$\quad$ FIN$_L$= the collection of located words over L,
$\quad$ FIN$_{Lv}$= the collection of located variable-words over L.

Let FIN$_L^*$ and FIN$_{Lv}^*$ be the collections of all cofinite ultrafilters on FIN$_L$ and FIN$_{Lv}$ respectively, i.e., ultrafilters that contain all sets of the form $\{x : \text{dom}(x) > n\}$ $(n \in \mathbb{N})$. Then the partial operation of taking the union of two functions, provided their domains are disjoint, extends to the full semigroup operation on FIN$_L^*$ and FIN$_{LV}^*$ which we denote $\cup$. In Section 2.5 we show how to produce idempotents $\mathcal{W} \in$ FIN$_L^*$ and $\mathcal{V} \in$ FIN$_{Lv}^*$ such that

(1) $\mathcal{V} \cup \mathcal{W} = \mathcal{W} \cup \mathcal{V} = \mathcal{V}$,

(2) $\mathcal{V}[\lambda] = \mathcal{W}$ for all $\lambda \in L$,

where $\mathcal{U} \mapsto \mathcal{U}[\lambda]$ refers to the extension of the substitution map $x \mapsto x[\lambda]$ from FIN_{Lv} into FIN_L. A *block sequence* is a finite or infinite sequence $X = (x_n)$ of members of FIN_L or FIN_{Lv} such that

$$\mathrm{dom}\,(x_m) < \mathrm{dom}\,(x_n) \text{ whenever } m < n. \qquad (7.83)$$

Let $\mathrm{FIN}_L^{[<\infty]}$ and $\mathrm{FIN}_L^{[\infty]}$ be the collection of all finite and infinite block sequence of members of FIN_L, respectively. Similarly, $\mathrm{FIN}_{Lv}^{[<\infty]}$ and $\mathrm{FIN}_{Lv}^{[\infty]}$ are the collections of finite and infinite block sequences of located variable-words. Recall that $\mathrm{FIN}_L^{[<\infty]}$ and $\mathrm{FIN}_{Lv}^{[<\infty]}$ can be considered as trees under the ordering $\sqsubseteq$ of end-extension. For t in $\mathrm{FIN}_L^{[<\infty]}$ and w in FIN_L we use the notation (t, w) to denote the sequence of length $|t| + 1$ that extends t and has w as its last term. Thus,

$$t \sqsubseteq (t, x)$$

is an immediate successor of t in the tree $\mathrm{FIN}_L^{[<\infty]}$ determined by w. We use similar notation for the immediate successors of nodes of the tree $\mathrm{FIN}_{Lv}^{[<\infty]}$.

A $\mathcal{W}$-tree ($\mathcal{V}$-tree) is a downward closed subtree T of the tree $\mathrm{FIN}_L^{[<\infty]}$ (of the tree $\mathrm{FIN}_{Lv}^{[<\infty]}$) that has a $\sqsubseteq$-maximal node, call it, $\mathrm{st}(T)$, that is comparable to all other nodes and such that every $t \in T$ with property $t \sqsupseteq \mathrm{st}(T)$ satisfies the following condition:

$$\{w \in \mathrm{FIN}_L : (t, w) \in T\} \in \mathcal{W} \text{ (respectively, } \{x \in \mathrm{FIN}_{Lv} : (t, x) \in T\} \in \mathcal{V}).$$

Taking the sets $[T]$ of infinite branches through $\mathcal{W}$-trees ($\mathcal{V}$-trees) T, we obtain a $\mathcal{W}$-topology on $\mathrm{FIN}_L^{[\infty]}$ and a $\mathcal{V}$-topology on $\mathrm{FIN}_{Lv}^{[\infty]}$. Similarly, one defines the notion of $\mathcal{W}$-Ramsey and $\mathcal{V}$-Ramsey for subsets of $\mathrm{FIN}_L^{[\infty]}$ and $\mathrm{FIN}_{Lv}^{[\infty]}$, respectively, and the corresponding respective σ-ideal of $\mathcal{W}$-Ramsey null and $\mathcal{V}$-Ramsey null sets. Then, as in the previous sections, we have the following facts.

Theorem 7.87 *Every $\mathcal{W}$-Baire subset of $\mathrm{FIN}_L^{[\infty]}$ is $\mathcal{W}$-Ramsey and vice versa. Every $\mathcal{W}$-meager subset of $\mathrm{FIN}_L^{[\infty]}$ is $\mathcal{W}$-Ramsey null and vice versa.*

Theorem 7.88 *Every $\mathcal{V}$-Baire subset of $\mathrm{FIN}_{Lv}^{[\infty]}$ is $\mathcal{V}$-Ramsey and vice versa. Every $\mathcal{V}$-meager subset of $\mathrm{FIN}_{Lv}^{[\infty]}$ is $\mathcal{V}$-Ramsey null and vice versa.*

As indicated above, these results can be used to reprove some of the corollaries of the basic fact that

$$\left(\mathrm{FIN}_L^{[\infty]}, \mathrm{FIN}_{Lv}^{[\infty]}, \leq, \leq^0 r\right)$$

is a Ramsey space (see Section 4.5). For this we need first to recall the definitions of the relations $\leq$ and $\leq^0$ for a *block-subsequence*. Recall that a sequence $X = (x_n) \in \mathrm{FIN}_{Lv}^{[\infty]}$ generates two partial subsemigroups, one inside FIN_L and the other inside FIN_{Lv}

$$[X]_L = \{x_{n_0}[\lambda_0] \cup \ldots \cup x_{n_k}[\lambda_k] \in \mathrm{FIN}_L : n_0 < \ldots < n_k, \lambda_i \in L_{n_i}(i \leq k)\}$$

$$[X]_{Lv} = \{x_{n_0}[\lambda_0] \cup \ldots \cup x_{n_k}[\lambda_k] \in \text{FIN}_{Lv} : n_0 < \ldots < n_k,$$

$$\lambda_i \in L_{n_i} \cup \{v\} \, (i \le k)\}.$$

We say that $X = (x_n) \in \text{FIN}_{Lv}^{[\infty]}$ is a block subsequence of $Y = (y_n) \in \text{FIN}_{Lv}^{[\infty]}$ and write $X \le Y$ if $x_n \in [Y]_{Lv}$ for all n. Similarly for $Q = (q_n) \in \text{FIN}_L^{[\infty]}$ and $Y \in \text{FIN}_{Lv}^{[\infty]}$, we say that Q is a block-subsequence of Y and write $Q \le^0 Y$ if $q_n \in [Y]_L$ for all n.

Arguing as in the previous section one shows by using properties (1) and (2) of the idempotent ultrafilters $\mathcal{W}$ and $\mathcal{V}$ that for every $\mathcal{W}$-tree T with stem $\emptyset$, there exists a $Z \in \text{FIN}_{Lv}^{[\infty]}$ such that

$$\{Q \in \text{FIN}_L^{[\infty]} : Q \le^0 Z\} \subseteq [T] \tag{7.84}$$

and that similarly for any $\mathcal{V}$-tree T with stem $\emptyset$ there exists a $Z \in \text{FIN}_{Lv}^{[\infty]}$ such that

$$\{X \in \text{FIN}_{Lv}^{[\infty]} : X \le Z\} \subseteq [T].$$

In fact the proof of this given above shows that we can have a simultaneous version of these two reduction theorems.

Lemma 7.89 *Suppose S is a $\mathcal{W}$-tree with stem $\emptyset$ and T is $\mathcal{V}$-tree with stem $\emptyset$. Then there is a $Z \in \text{FIN}_{Lv}^{[\infty]}$ such that $\{Q \in \text{FIN}_L^{[\infty]} : Q \le^0 Z\} \subseteq [S]$ and $\{X \in \text{FIN}_{Lv}^{[\infty]} : X \le Z\} \subseteq [T].$*

Corollary 7.90 *Suppose we are given a finite coloring of $\text{FIN}_L^{[\infty]} \cup \text{FIN}_{Lv}^{[\infty]}$ whose restrictions on each of these two sets are Baire-measurable relative to the $\mathcal{W}$-topology and $\mathcal{V}$-topology, respectively. Then there is a $Z \in \text{FIN}_{Lv}^{[\infty]}$ such that the sets*

$$\{Q \in \text{FIN}_L^{[\infty]} : Q \le^0 Z\} \text{ and } \{X \in \text{FIN}_{Lv}^{[\infty]} : X \le Z\}$$

are monochromatic.

Corollary 7.91 (Bergelson-Blass-Hindman) *For every metrically Souslin-measurable coloring of $\text{FIN}_L^{[\infty]}$ there is a $Z \in \text{FIN}_{LV}^{[\infty]}$ such that the sets*

$$\{Q \in \text{FIN}_L^{[\infty]} : Q \le^0 Z\} \text{ and } \{X \in \text{FIN}_{Lv}^{[\infty]} : X \le Z\}$$

are monochromatic.

The case $L = \emptyset$ of these results is of independent interest. In fact it was this case that initiated the use of idempotent ultrafilters in this part of Ramsey theory. Note that in this case we have only one partial semigroup $\text{FIN}_{\emptyset v}$ that could be better viewed as the collection FIN of all nonempty finite subsets of $\mathbb{N}$. The ultrafilter $\mathcal{V}$ is simply Glazer's union-idempotent ultrafilter, i.e., an ultrafilter $\mathcal{V}$ on FIN such that $\mathcal{V} \cup \mathcal{V} = \mathcal{V}$, where $\cup$ is the semigroup operation on the space of all cofinite ultrafilters on FIN that extends the partial union operation on FIN. Using the $\mathcal{V}$-subtrees of $\text{FIN}^{[<\infty]}$, we can define two $\mathcal{V}$-topologies, one on the set $\text{FIN}^{[<\infty]}$ of finite block sequences, and the other on the set $\text{FIN}^{[\infty]}$ of infinite block sequences. We call these two topologies *Glazer topologies*. We have the following consequence of Corollary 7.91.

Corollary 7.92 *For every finite coloring of* $\mathrm{FIN}^{[\infty]}$ *that is Baire-measurable relative to the Glazer topology of* $\mathrm{FIN}^{[\infty]}$, *there is a* $Y \in \mathrm{FIN}^{[\infty]}$ *such that* $\{X \in \mathrm{FIN}^{[\infty]} : X \leq Y\}$ *is monochromatic.*

Corollary 7.93 (Milliken) *For every finite metrically Souslin-measurable coloring of* $\mathrm{FIN}^{[\infty]}$, *there is a* $Y \in \mathrm{FIN}^{[\infty]}$ *such that* $\{X \in \mathrm{FIN}^{[\infty]} : X \leq Y\}$ *is monochromatic.*

7.8 ULTRA-RAMSEY SPACE OF INFINITE BLOCK SEQUENCES OF VECTORS

Recall that $\mathrm{FIN} = \mathrm{FIN}_1$ is also the initial term of yet another sequence

$$\mathrm{FIN}_k \ (k = 1, 2, 3, ...)$$

of partial semigroups, the sequence of partial semigroups considered in Section 2.3. Since the key ingredients of the combinatorial analysis of FIN_k are certain idempotent ultrafilters on FIN_k it is natural to consider the corresponding ultra-Ramsey spaces as well. But let us first recall the basic definitions. Keeping in mind the development that led to the discovery of the space FIN_k, it is more natural to think of FIN_k as the set of all maps $p : \mathbb{N} \to \{0, 1, \ldots, k\}$ with finite support that are achieving the maximal value k rather than the set of all maps from finite subsets of $\mathbb{N}$ into $\{1, 2, \ldots, k\}$ that take the value k. This leaves us with a dilemma on how to denote the partial semigroup operation of FIN_k. We choose to denote the partial operation by $+$ rather than $\cup$ since, as originally intended, FIN_k is meant to model a net of the positive part of the unit sphere of the Banach space c_0. Recall also the operation on the partial semigroup FIN_k,

$$T : \mathrm{FIN}_k \to \mathrm{FIN}_{k-1},$$

that distinguishes it even more from the partial subsemigroups considered above. This operation

$$T(p)(n) = \max\{0, p(n) - 1\}$$

is the discrete analog of the scalar multiplication on c_0. Recall that in the course of proving Lemma 2.24, we have constructed a sequence $(\mathcal{U}_k)_{k \geq 1}$ of cofinite ultrafilters such that

(1) $\mathrm{FIN}_k \in \mathcal{U}_k$,

(2) $\mathcal{U}_k + \mathcal{U}_\ell = \mathcal{U}_\ell + \mathcal{U}_k = \mathcal{U}_\ell$ whenever $k \leq \ell$,

(3) $T^{(\ell - k)}(\mathcal{U}_\ell) = \mathcal{U}_k$ whenever $k \leq \ell$.

It is of course possible to fix a positive integer k and work with $\mathcal{U}_k$-subtrees of the set $\mathrm{FIN}_k^{[<\infty]}$ of all finite *block sequences* (p_n) of elements of FIN_k, i.e., sequences with the property that

$$\mathrm{supp}(p_m) < \mathrm{supp}(p_n) \text{ whenever } m < n. \tag{7.85}$$

We prefer, however, to work with the set

$$\text{FIN}_* = \bigcup_{k=1}^{\infty} \text{FIN}_k \tag{7.86}$$

and a family of $\vec{\mathcal{U}}$-subtrees of the set $\text{FIN}_*^{[<\infty]}$ of all finite block sequences of members of FIN_*, where

$$\vec{\mathcal{U}} = (\mathcal{U}_k)_{k \geq 1}. \tag{7.87}$$

More precisely, for the purpose of this section, for $\vec{\mathcal{U}} = (\mathcal{U}_k)_{k \geq 1}$, a $\vec{\mathcal{U}}$-tree is a downward closed subset $\mathcal{U}$ of $\text{FIN}_*^{[<\infty]}$ such that for every $t \in \mathcal{U}$ end-extending the stem,

$$\{p \in \text{FIN}_* : (t, p) \in U\} \in \mathcal{U}_{|t|+1}. \tag{7.88}$$

The sets $[U]$ of infinite branches through various $\vec{\mathcal{U}}$-trees determine a zero-dimensional topology on $\text{FIN}_*^{[\infty]}$ that extends the usual metrizable product topology on this set. As before, we have the notions of $\vec{\mathcal{U}}$-Ramsey and $\vec{\mathcal{U}}$-Ramsey null subsets of $\text{FIN}_*^{[\infty]}$ and the following result, which identifies them with the standard topological notions relative to the $\vec{\mathcal{U}}$-topology on $\text{FIN}_*^{[\infty]}$.

Theorem 7.94 *Every subset of $\text{FIN}_*^{[\infty]}$ that has the property of Baire relative to the $\vec{\mathcal{U}}$-topology is $\vec{\mathcal{U}}$-Ramsey, and vice versa. Every meager subset of $\text{FIN}_*^{[\infty]}$ relative to the $\vec{\mathcal{U}}$-topology is $\vec{\mathcal{U}}$-Ramsey null, and vice versa.*

As before, one needs to relate the sets $[U]$ of infinite branches through $\vec{\mathcal{U}}$-trees to the basic-sets of the ordinary Ramsey space on $\text{FIN}_*^{[\infty]}$. To state this relationship, we need to recall some definitions. For $P = (p_n) \in \text{FIN}_*^{[\infty]}$ and $k \geq 1$, let

$$[P]_k = \text{FIN}_k \cap \{T^{(j_0)}(p_{n_0}) + \ldots + T^{(j_\ell)}(p_{n_\ell}) : n_0 < \ldots < n_\ell, j_0, \ldots, j_\ell \geq 0\}.$$

For $P = (p_n)$ and $Q = (q_n)$ members of $\text{FIN}_*^{[\infty]}$ we say that P is a block-subsequence of Q and write $P \leq Q$ if $p_n \in [Q]_{n+1}$ for all n. The argument given in the proof of Lemma 7.78 with a slight adjustment will give us the following.

Lemma 7.95 *Suppose U is a $\vec{\mathcal{U}}$-tree with stem $\emptyset$. Then there is a $Q \in \text{FIN}_*^{[\infty]}$ such that every $P \in \text{FIN}_*^{[\infty]}$, where $P \leq Q$ is an infinite branch through U.*

Combining Theorem 7.94 and Lemma 7.95, we obtain the following conclusions about finite colorings of the spaces $\text{FIN}_k^{[\infty]}$ and $\text{FIN}_*^{[\infty]}$.

Theorem 7.96 *For every finite coloring of $\text{FIN}_*^{[\infty]}$ that is Baire-measurable relative to the $\vec{\mathcal{U}}$-topology, there is a $Q \in \text{FIN}_*^{[\infty]}$ such that $\{P \in \text{FIN}_*^{[\infty]} : P \leq Q\}$ is monochromatic.*

Corollary 7.97 *For every finite metrically Souslin-measurable coloring of* $\text{FIN}_*^{[\infty]}$, *there is a* $Q \in \text{FIN}_*^{[\infty]}$ *such that* $\{P \in \text{FIN}_*^{[\infty]} : P \leq Q\}$ *is monochromatic.*

Corollary 7.98 *For every positive integer* k *and every finite metrically Souslin-measurable coloring of* $\text{FIN}_k^{[\infty]}$, *there is* $Q \in \text{FIN}_k^{[\infty]}$ *such that* $\{P \in \text{FIN}_k^{[\infty]} : P \leq Q\}$ *is monochromatic.*

Proof. Let c be a given Souslin-measurable coloring of $\text{FIN}_k^{[\infty]}$. Define a coloring c^* on $\text{FIN}_*^{[\infty]}$ by $c^*(Q) = c(\pi(Q))$, where $P = \pi(Q)$ is determined as follows:

$$p_n = T^{(n)}(q_{k+n}). \tag{7.89}$$

Clearly c^* is Souslin-measurable, so Corollary 7.97 applies, giving us a $Q \in \text{FIN}_*^{[\infty]}$ such that c^* is monochromatic on

$$\{P \in \text{FIN}_*^{[\infty]} : P \leq Q\}. \tag{7.90}$$

Let $Q' = \pi(Q) \in \text{FIN}_k^{[\infty]}$. Then every $P' \leq Q'$ in $\text{FIN}_k^{[\infty]}$ is of the form $\pi(P)$ for some $P \leq Q$. From the way c^* is defined from c, one concludes that c is monochromatic on

$$\{P' \in \text{FIN}_k^{[\infty]} : P' \leq Q'\}. \tag{7.91}$$

This finishes the proof.   □

Corollary 7.99 (Milliken) *For every finite metrically Souslin-measurable coloring of* $\text{FIN}^{[\infty]}$, *there is a* $Y \in \text{FIN}^{[\infty]}$ *such that* $\{X \in \text{FIN}^{[\infty]} : X \leq Y\}$ *is monochromatic.*

NOTES TO CHAPTER SEVEN

Selective coideals were introduced by Mathias ([72],[74]) starting thus the whole area of local Ramsey theory. The semiselective coideals were introduced more recently by Farah [31], while proving a version of the Local Ellentuck Theorem via a different line of reasoning from the one that we chose to reproduce here. Farah's work was motivated by a question of the author asking for a description of the Ramsey theoretic properties of the coideal on $\mathbb{N}$ living in the perfect-set forcing extension but generated by the ground model infinite subsets of $\mathbb{N}$. Interestingly, this metamathematical question was actually motivated by the parametrized Galvin-Prikry theorem (Theorem 5.49). A step toward the abstract version of the local Ramsey theory was made recently by Mijares [75]. The $\mathcal{U}$-topology on the set $\mathbb{N}^{[<\infty]}$ of all finite subsets of $\mathbb{N}$ was first considered by S. Sirota [100] although its current form is due to Louveau [67], who used it to derive in a topological manner the results of Mathias and Silver. In particular Louveau [68] and [69] was the first to consider the $\mathcal{U}$-topology on the set $\mathbb{N}^{[\infty]}$, although he did not formulate the notion of a $\mathcal{U}$-tree to describe this topology but instead worked with the set of all diagonalizations of sequences $(A_n)_{n=0}^{\infty} \subseteq \mathcal{U}$. This formulation did not allow him to anticipate the notion of $\vec{\mathcal{U}}$-topology on $\mathbb{N}^{[\infty]}$ for an arbitrary sequence $\mathcal{U}_s$ ($s \in \mathbb{N}^{[<\infty]}$) of nonprincipal ultrafilters on $\mathbb{N}$. The full notion of a $\vec{\mathcal{U}}$-tree appears in the paper of Blass [10], which, however, does not consider the corresponding $\vec{\mathcal{U}}$-topology on $\mathbb{N}^{[\infty]}$ and therefore does not anticipate the ultra-Ellentuck theorem, or in other words, the fact that the triple $(\mathbb{N}^{[\infty]}, \vec{\mathcal{U}}, \leq)$ forms a topological Ramsey space. As already mentioned in Chapter 1, Galvin's lemma appears originally in his announcement [33]. It is a nontrivial extension of an earlier generalization of Ramsey's theorem due to Nash-Williams [82]. The result of Bourgain, Fremlin, and Talagrand appears in their paper [11] and the result of Odell and Rosenthal appears in their paper [83]. Regarding the Bourgain-Fremlin-Talagrand theorem, it should be mentioned that the fact that points in separable compact sets of Baire Class 1 functions lead to bisequential coideals was first pointed out by R. Pol [86] (see also [57]), and this fact can also be deduced using reasoning from the paper of Debs [18]. Theorem 7.60 is due to Balcar-Pelant-Simon [5]. Corollaries 7.69, 7.70, 7.71 and Theorem 7.73 appear in the paper of Todorcevic and Uzcategui [109]. Theorem 7.72 also appears in [109], but under the extra assumption that the product $X \times Y$ is Fréchet. That this assumption is in fact not needed was also observed in the paper of Dodos and Kanellopoulos [24]. Corollary 7.91 of Bergelson, Blass, and Hindman was proved in their paper [8] using the result from [10] in place of the topological argument given above.

Chapter Eight

Infinite Products of Finite Sets

8.1 SEMICONTINUOUS COLORINGS OF INFINITE PRODUCTS OF FINITE SETS

The Ramsey theory of infinite products of finite sets has several aspects. The first aspect is in describing a field $\mathcal{M}$ of subsets[1] of $\mathbb{N}^\infty$ with the following property: For every sequence (m_i) of positive integers, there is a sequence (n_i) of positive integers such that for every $\mathcal{M}$-measurable coloring

$$c : \prod_{i=0}^{\infty} H_i \rightarrow \{0, 1\} \tag{8.1}$$

such that $H_i \subseteq \mathbb{N}$ and $|H_i| = n_i$ for all i, there exist $J_i \subseteq H_i$ with $|J_i| = m_i$ for all i such that c is constant on the subproduct $\prod_{i=0}^{\infty} J_i$. We use the symbol

$$\begin{pmatrix} n_0 \\ n_1 \\ n_2 \\ \vdots \end{pmatrix} \rightarrow_{\mathcal{M}} \begin{pmatrix} m_0 \\ m_1 \\ m_2 \\ \vdots \end{pmatrix} \tag{8.2}$$

as a shorthand for this statement. Another aspect of the theory is in finding out how fast the sequence (n_i) has to grow in terms of the sequence (m_i). It turns out that the problem already shows its full weight for the constant sequence $m_i = 2$ for all i.

We shall start with a fact which shows that the field $\mathcal{M}$, if it exists, will not have much to do with the familiar fields of Baire and Lebesgue measurable subsets of $\mathbb{N}^\infty$. The measure we take on $\mathbb{N}^\infty$ is the product measure relative to the standard atomic counting probability measure of $\mathbb{N}$. There is perhaps some abuse of terminology in using the word "Lebesgue" for this measure, but this is justified by using a standard fact (see Theorem 9.41) from real analysis which says that there is a Borel isomorphism between $\mathbb{N}^\infty$ and the unit interval transferring the product measure of $\mathbb{N}^\infty$ to the Lebesgue measure of $[0, 1]$.

---

[1] The collection of all infinite sequences $(n_i) \subseteq \mathbb{N}$. Thus we may identify $\mathbb{N}^\infty$ with the power $\mathbb{N}^\mathbb{N}$ whenever we are interested in considering the natural product topology or the product measure on this set.

Lemma 8.1 *There is a coloring $c : \mathbb{N}^\infty \to \{0,1\}$ that is measurable in the sense of Baire as well as Lebesgue that is not constant on any set of the form $\prod_{i=0}^\infty J_i$ such that $|J_i| \geq 2$ for all but finitely many i.*

Proof. Let E_0 be the equivalence relation on $\mathbb{N}^\infty$ defined as follows:

$$(x_i)E_0(y_i) \quad \text{iff} \quad x_i = y_i \text{ for all but finitely many } i.$$

For each $e \in \mathbb{N}^\infty/E_0$ we fix a representative $(x_i^e) \in e$. The coloring c is defined according to the following two cases.
Case 1. $x_i = x_{i+1}$ for infinitely many i. Let $c((x_i)) = 0$.
Case 2. $x_i \neq x_{i+1}$ for all but finitely many i. In this case put

$$c((x_i)) = |\{i : x_i \neq x_i^e\}| (\mathrm{mod}\ 2), \tag{8.3}$$

where $e = [(x_i)]_{E_0}$.
Note that the set of $(x_i) \in \mathbb{N}^\infty$ which fall into Case 1, is at the same time of full Lebesgue measure and of full Baire category. So, the coloring c is Lebesgue- as well as Baire-measurable.

Consider a product $\prod_{i=0}^\infty J_i \subseteq \mathbb{N}^\infty$ such that $|J_i| \geq 2$ for all but finitely many i. Then we can find $(x_i) \in \prod_{i=0}^\infty J_i$ such that $x_i \neq x_{i+1}$ for all but finitely many i. Let e be the E_0-equivalence class of (x_i). Since $x_i = x_i^e$ and $|J_i| \geq 2$ for almost all i, we can find $(y_i) \in \prod_{i=0}^\infty J_i$ and i_0 such that $y_i = x_i$ for all $i \neq i_0$ and $y_{i_0} \neq x_{i_0} = x_{i_0}^e$. It follows that the cardinalities of the sets

$$\{i : x_i \neq x_i^e\} \text{ and } \{i : y_i \neq x_i^e\} \tag{8.4}$$

are of different parity and therefore $c((x_i)) \neq c((y_i))$. This finishes the proof.
$\square$

The second lemma gives us some idea about the rate of growth of any sequence (n_i) such that $(n_i) \to (2)$ in the realm of continuous colorings.

Lemma 8.2 *Suppose $n_i \leq 2^{2^{i-1}}$ for all but finitely many i. Then there is a continuous coloring $c : \prod_{i=0}^\infty n_i \to \{0,1\}$ that is not constant on any subproduct $\prod_{i=0}^\infty J_i$ such that $|J_i| = 2$ for all i.*

Proof. Pick ℓ such that $n_i \leq 2^{2^{i-1}}$ for all $i > \ell$. For $k > \ell$, the probability[2] that a given coloring

$$c : \prod_{i=0}^k n_i \to \{0,1\} \tag{8.5}$$

is constant on a given subproduct $\prod_{i=0}^k J_i$ such that $|J_i| = 2$ for all $i \leq k$ is equal to $2/2^{2^{k+1}}$. Hence, the probability that there is such a subproduct on which c is constant is equal to

---

[2]With respect to the uniform distribution on the set of all such colorings.

$$\frac{2}{2^{2k+1}} \prod_{i=0}^{k} \binom{n_i}{2}.$$ (8.6)

This quantity is dominated by

$$M \frac{2}{2^{2k+1}} \prod_{i=0}^{k} \binom{2^{2^{i-1}}}{2} \le M \frac{2}{2^{2k+1}} \cdot \frac{1}{2^k} 2^{\sum_{i=0}^{k} 2^i} = \frac{M}{2^{k+1}},$$ (8.7)

where $M = \prod_{i=0}^{\ell} \binom{n_i}{2}$. So for any k for which $M/2^{k+1} < 1$, we can find a coloring $c_k : \prod_{i=0}^{k} n_i \to \{0,1\}$ that is not constant on any subproduct $\prod_{i=0}^{k} J_i$ such that $|J_i| = 2$ for all $i \le k$. Defining $c : \prod_{i=0}^{\infty} n_i \to 2$ by letting $c(x) = c_k(x \restriction k)$, we get a continuous coloring which is not monochromatic on any subproducts of 2-element sets.   □

This lemma shows that the assumption about the rate of growth in the following result is in some sense optimal.

Theorem 8.3 *If $n_i = 2^{2^{2i+1}}$ for all i, then*

$$\begin{pmatrix} n_0 \\ n_1 \\ n_2 \\ \vdots \end{pmatrix} \to \begin{pmatrix} 2 \\ 2 \\ 2 \\ \vdots \end{pmatrix}$$ (8.8)

holds in the realm of continuous colorings.

This result follows from a more general fact that treats the sequences (m_i) simultaneously and which gives us a solid basis for solving the general problem.

Definition 8.4 *Let $S : \mathbb{N}^{<\infty} \to \mathbb{N}$ be defined as follows:*

$$S(m_0) = 2m_0 - 1,$$
$$S(m_0, \ldots, m_{i+1}) = 2(m_{i+1} - 1) \left[\prod_{i=0}^{i} \binom{S(m_0, \ldots, m_k)}{m_k} \right] + 1.$$ (8.9)

This particular choice of S is made in order to have the following immediate property.

Lemma 8.5 *Let $(m_i) \in \mathbb{N}^{\infty}$ and let $n_i = S(m_0, \ldots, m_i)$ for all i. Then for every k and every coloring*

$$c : \prod_{i=0}^{k} n_i \to \{0,1\}$$ (8.10)

there exist $J_i \subseteq n_i, |J_i| = m_i (i \le k)$ such that c is constant on $\prod_{i=0}^{k} J_i$.

Notation. $S(2_k) = S(m_0, \ldots, m_k)$, where $m_i = 2$ for all $i \leq k$.

Lemma 8.6 $S(2_n) \leq 2^{2^{2n+1}}$ *for all* n.

Proof. Note that the first two values $S(2_0) = 3$ and $S(2_1) = 7$ satisfy the inequality. For $n \geq 2$, we have

$$S(2_n) = 2(2-1) \prod_{i=0}^{n-1} \binom{S(2_i)}{2} + 1$$

$$\leq 2^2 \frac{1}{2^n} 2^{\sum_{i<n} 2^{2i+2}} \leq 2^{2^{2n+1}}. \tag{8.11}$$

□

Note that Theorem 8.3 follows from Lemmas 8.5 and 8.6 via König's infinity lemma, which ensures that every continuous coloring is really a coloring of a finite product. It turns out that the same sequence works in the realm of semicontinuous colorings, as the following application of Lemma 8.5 shows.

Lemma 8.7 *Let* $(m_i) \in \mathbb{N}^\infty$ *and let* $n_i = S(m_0, \ldots, m_i)$ *for all* i. *Then for every closed set* $F \subseteq \prod_{i=0}^{\infty} n_i$ *there is a subproduct* $\prod_{i=0}^{\infty} J_i \subseteq \prod_{i=0}^{\infty} n_i$ *such that* $|J_i| = m_i$ *for all* i *that is either included in or is disjoint from* F.

Proof. For $k \geq 0$ define $c_k : \prod_{i=0}^{k} n_i \rightarrow \{0,1\}$ by letting $c_k(t) = 0$ iff t has no extensions in F. By Lemma 8.5 for each k, there exist $J_i^k \subseteq n_i (i \leq k)$ such that $|J_i^k| = m_i (i \leq k)$ and such that c_k is constant on $\prod_{i=0}^{k} J_i^k$. If for some k the constant value of c_k is equal to 0, we have that the conclusion of Lemma 8.7 holds. So we are left with the case that for every k the constant value of c_k is equal to 1. Pick a nonprincipal ultrafilter $\mathcal{U}$ on $\mathbb{N}$. Then for every i, there is a $J_i \subseteq n_i$ such that

$$A_i = \{k \geq i : J_i^k = J_i\} \in \mathcal{U}. \tag{8.12}$$

Then $|J_i| = m_i$ for all i and $\prod_{i=0}^{\infty} J_i \subseteq F$, as required. □

Definition 8.8 *The* pth *iterate* $S^{(p)} : \mathbb{N}^{<\infty} \rightarrow \mathbb{N}$ *of* S *is defined recursively as follows:*

$$S^{(0)}(m_0, \ldots, m_i) = S(m_0, \ldots, m_i)$$
$$S^{(p+1)}(m_0, \ldots, m_i) = S(S^{(p)}(m_0), S^{(p)}(m_0, m_1), \ldots, S^{(p)}(m_0, \ldots, m_i)). \tag{8.13}$$

The following fact is an immediate consequence of the definition.

Lemma 8.9 *If* $p' \geq p$ *and* $m_j' \geq m_j$ *for all* $j \leq i$ *then* $S^{(p')}(m_0', \ldots, m_i') \geq S^{(p)}(m_0, \ldots, m_i)$.

Definition 8.10 *For* $M \subseteq \mathbb{N}$, *define* $S_M : \mathbb{N}^{<\infty} \rightarrow \mathbb{N}$ *recursively as follows:*

$$S_M(m_0) = S(m_0),$$
$$S_M(m_0, \ldots, m_i) = S^{(p)}(m_0, \ldots, m_i), \tag{8.14}$$

where

$$p = |\bigcup_{k \in M \cap i} \prod_{j=0}^{k} S_M(m_0, \ldots, m_j)|. \tag{8.15}$$

It is worth observing that the following monotonicity property follows immediately from the definitions.

Lemma 8.11 *If $M \subseteq M' \subseteq \mathbb{N}$ and $m'_i \geq \leftarrow m_i$ for all i, then*

$$S_{M'}(m'_0, \ldots, m'_i) \geq S_M(m_0, \ldots, m_i).$$

Note that $S_\emptyset = S$ and that for a given infinite sequence (m_i) of positive integers the sequence

$$(S_M(m_0, \ldots, m_i))_{i=0}^{\infty} \tag{8.16}$$

is composed of steps of the form

$$(S^{(p)}(m_0, \ldots, m_i))_{i=\bar{k}+1}^{\bar{\ell}}, \tag{8.17}$$

where $\bar{k} < \bar{\ell}$ are two consecutive members of M and where

$$p = |\bigcup_{k \in M \cap (\bar{k}+1)} \prod_{j=0}^{k} S_M(m_0, \ldots, m_j)|. \tag{8.18}$$

Notation. For $R = S^{(p)}$ for some $p \in \mathbb{N}$ or $R = S_M$ for some $M \subseteq \mathbb{N}$ and for an infinite sequence $(m_i) \in \mathbb{N}^\infty$, we denote by $R((m_i))$ the infinite sequence of positive integers whose ith term is equal to $R(m_0, \ldots, m_i)$.

Definition 8.12 *For an infinite sequence (n_i) of positive integers, an (n_i)-product or a product of type (n_i) is any product of the form $\prod_{i=0}^{\infty} H_i$, where $|H_i| = n_i$ for all i.*

The following consequence of Lemma 8.7 that uses this terminology is worth noting.

Lemma 8.13 *Let $(m_i) \in \mathbb{N}^\infty, M \subseteq \mathbb{N}, k \in M$, and $N = M \setminus \{k\}$. Then for every $S_M((m_i))$-product $\prod_{i=0}^{\infty} H_i$ and every closed subset $\mathcal{X} \subseteq \prod_{i=0}^{\infty} H_i$, there is an $S_N((m_i))$-subproduct $\prod_{i=0}^{\infty} J_i$ of $\prod_{i=0}^{\infty} H_i$ such that*

$$(\forall x, y \in \prod_{i=0}^{\infty} J_i) \, [x \upharpoonright k = y \upharpoonright k \Rightarrow (x \in \mathcal{X} \Leftrightarrow y \in \mathcal{X})]. \tag{8.19}$$

Proof. Note that $S_N((m_i)) \upharpoonright k = S_M((m_i)) \upharpoonright k$ and that for two consecutive members $\bar{k} < \bar{\ell}$ of M above k, if

$$S_N((m_i)) \upharpoonright (\bar{k}, \bar{\ell}] = (S^{(p)}(m_0, \ldots, m_j))_{j=\bar{k}+1}^{\bar{\ell}} \tag{8.20}$$

for some p, then

$$S_M((m_i)) \upharpoonright (\bar{k}, \bar{\ell}] = (S^{(p+q)}(m_0, \ldots, m_j))_{j=\bar{k}+1}^{\bar{\ell}}, \tag{8.21}$$

where

$$q = \left| \prod_{i=0}^k S_M(m_0, \ldots, m_i) \right| = \left| \prod_{i=0}^k S_N(m_0, \ldots, m_i) \right|. \tag{8.22}$$

It follows that for $i > k$,

$$S_M(m_0, m_1, \ldots, m_i) \geq S^{(q)}(S_N(m_0), S_N(m_0, m_1), \ldots, S_N(m_0, m_1, \ldots, m_i))$$

$$\geq S^{(q)}(S_N(m_0, \ldots, m_{k+1}), \ldots, S_N(m_0, \ldots, m_i)).$$

So applying Lemma 8.7 successively q times and thereby treating one $s \in \prod_{i=0}^k S_N(m_0, \ldots, m_i)$ at a time, we can choose a decreasing sequence

$$H_i = H_i^0 \supseteq H_i^1 \supseteq \ldots \supseteq H_i^q \quad (i > k) \tag{8.23}$$

such that for each $s \in \prod_{i=0}^k S_N(m_0, \ldots, m_i)$ there is a $p \leq q$ such that the product $\prod_{i=k+1}^\infty H_i^p$ is either included or is disjoint from $\mathcal{X}_s$, where

$$\mathcal{X}_s = \left\{ x \in \prod_{i=k+1}^\infty H_i : s^\frown x \in \mathcal{X} \right\}. \tag{8.24}$$

It is clear that then the corresponding product $\prod_{i=0}^k H_i \times \prod_{i=k+1}^\infty H_i^q$ satisfies the conclusion of Lemma 8.13. $\qquad \square$

8.2 POLARIZED RAMSEY PROPERTY

Throughout this section (m_i) is a fixed infinite nondecreasing sequence of positive integers. An S_M-sequence (or product) refers to the $S_M((m_i))$-sequence (or product) defined in the previous section, i.e., a sequence (H_i) of finite sets such that

$$|H_i| = S_M(m_0, \ldots, m_i) \text{ for all } i. \tag{8.25}$$

We reserve the notation $(J_i), (H_i), (K_i)$, etc., for infinite sequences of non-empty finite subsets of $\mathbb{N}$. We let $(J_i) \leq_n (H_i)$ mean

$$J_i = H_i \text{ for } i < n, \text{ and } J_i \subseteq H_i \text{ for } i \geq n. \tag{8.26}$$

The corresponding infinite products are denoted by $\prod_i J_i, \prod_i H_i, \prod_i K_i$, etc., and we shall sometimes write $\prod_i J_i \leq_n \prod_i H_i$ instead of $(J_i) \leq_n (H_i)$. The letters $A, B, C, D, \ldots$ are reserved for infinite subsets of $\mathbb{N}$.

We let $A \subset_\infty B$ denote the fact that $A \subseteq B$ and that $B \backslash A$ is infinite. A pair (A, B) of infinite subsets of $\mathbb{N}$ is called *regular* if $A \subset_\infty B$. For two regular pairs (A, B) and (C, D), we write $(C, D) \leq (A, B)$ whenever

$$A \subset_\infty C \subset_\infty D \subset_\infty B. \tag{8.27}$$

For an integer $n \geq 0$ and regular pairs (A, B) and (C, D), let $(A, B) \leq_n (C, D)$ denote the fact that $(A, B) \leq (C, D)$ and that

$$A \cap n = C \cap n \text{ and } B \cap n = D \cap n. \tag{8.28}$$

A subset $\mathcal{X} \subseteq \prod_{i=0}^\infty H_i$ *depends only on coordinates* $< n$ if

$$(\forall x, y \in \textstyle\prod_{i=0}^\infty H_i) \, [x \upharpoonright n = y \upharpoonright n \Rightarrow (x \in \mathcal{X} \Leftrightarrow y \in \mathcal{X})]. \tag{8.29}$$

Definition 8.14 *A set $\mathcal{X} \subseteq \mathbb{N}^\infty$ is (m_i)-Ramsey if for every regular pair (A, B), every $n \in B \backslash A$, and every S_B-product $\prod_i H_i$, there are a regular pair $(C, D) \leq_n (A, B)$ and an S_D-product $\prod_i J_i \leq_n \prod_i H_i$ such that $\mathcal{X} \cap \prod_i J_i$ depends only on coordinates $< n$.*

Note the following reformulation of Lemma 8.13, which relates to this notion.

Lemma 8.15 *Every closed subset of $\mathbb{N}^\infty$ is (m_i)-Ramsey.*

Definition 8.16 *A set $\mathcal{X} \subseteq \mathbb{N}^\infty$ is (m_i)-Ramsey null if for every regular pair (A, B), every $n \in B \backslash A$, and every S_B-product $\prod_i H_i$, there are $(C, D) \leq_n (A, B)$ and an S_D-product $\prod_i J_i \leq_n \prod_i H_i$ such that $\mathcal{X} \cap \prod_i J_i = \emptyset$.*

To show that this notion defines a σ-additive ideal on $\mathbb{N}^\infty$ we need the following concept.

Definition 8.17 *A fusion sequence is an infinite sequence (A_k, B_k, n_k, H^k) of quadruples such that for all k,*

(1) *$n_k < n_{k+1}$ and $n_k \in B_k \backslash A_k$,*

(2) *(A_k, B_k) is a regular pair and $H^k = (H_i^k)$ is a S_{B_k}-sequence of finite sets,*

(3) $(A_{k+1}, B_{k+1}) \leq_{n_k} (A_k, B_k)$,

(4) $H^{k+1} \leq_{n_k} H^k$,

(5) $(B_{k+1} \backslash A_{k+1}) \cap (n_k, n_{k+1}) \neq \emptyset$.

We define the limit of a fusion sequence (A_k, B_k, n_k, H^k) to be the triple $(A_\infty, B_\infty, H^\infty = (H_i^\infty))$, where

$$A_\infty = \bigcup_{k=0}^\infty A_k, \ B_\infty = \bigcap_{k=0}^\infty B_k,$$

$$H_i^\infty = \lim_{k \to \infty} H_i^k. \tag{8.30}$$

Note that (A_∞, B_∞) is a regular pair of infinite subsets of $\mathbb{N}$ and that $H^\infty = (H_i^\infty)$ is an S_{B_∞}-sequence. Note moreover that for all k,

$$(A_\infty, B_\infty) \leq_{n_k} (A_k, B_k), \text{ and } H^\infty \leq_{n_k} H^k. \tag{8.31}$$

Lemma 8.18 *The ideal of (m_i)-Ramsey null sets is σ-additive.*

Proof. Let $(\mathcal{X}_k)$ be a given increasing sequence of (m_i)-Ramsey null subsets of $\mathbb{N}^\infty$ and let (A, B), $n \in B \setminus A$ and $\prod_i H_i$ be given inputs as in Definition 8.16. Using the hypothesis that the $\mathcal{X}_k$ are (m_i)-Ramsey null and starting from the initial values $A_0 = A, B_0 = B, n_0 = n$, and $H^0 = (H_i)$, we build a fusion sequence (A_k, B_k, n_k, H^k) such that for all k,

$$\mathcal{X}_k \cap \left(\prod_{i=0}^\infty H_i^{k+1} \right) = \emptyset. \tag{8.32}$$

Moreover, we ensure that n_{k+1} is chosen in $B_{k+1} \setminus A_{k+1}$ above n_k such that $(B_{k+1} \setminus A_{k+1}) \cap [n_k, n_{k+1}) \neq \emptyset$.

Let (A_∞, B_∞) and $H^\infty = (H_i^\infty)$ be the limits of the fusion sequence. Then $(H_i^\infty) \leq_n (H_i), (A_\infty, B_\infty) \leq_n (A, B)$ and

$$\left(\bigcup_{k=0}^\infty \mathcal{X}_k \right) \cap \left(\prod_{i=0}^\infty H_i^\infty \right) = \emptyset \tag{8.33}$$

as required. $\square$

Lemma 8.19 *The (m_i)-Ramsey subsets of $\mathbb{N}^\infty$ form a σ-field.*

Proof. Let $(\mathcal{X}_k)$ be a given sequence of (m_i)-Ramsey sets, and let $(A, B), n \in \mathbb{N}$ and $\prod_i H_i$ be the inputs of Definition 8.14 for testing whether the union

$$\mathcal{X} = \bigcup_{k=0}^\infty \mathcal{X}_k \tag{8.34}$$

is (m_i)-Ramsey. Let n_0 be the minimal integer of $B\backslash A$ above n. Starting from $A_0 = A$, $B_0 = B$, n_0, and $(H_i^0) = (H_i)$, we build a fusion sequence (A_k, B_k, n_k, H^k) as follows. Starting from the regular pair (A_k, B_k) of infinite subsets of $\mathbb{N}$, the integer $n_k \in B_k \backslash A_k$, and the S_{B_k}-product $\prod_i H_i^k$, we apply the hypothesis that the set $\mathcal{X}_k$ is (m_i)-Ramsey to get a regular pair $(A_{k+1}, B_{k+1}) \leq_{n_k} (A_k, B_k)$ and the S_{B_k}-product $\prod_i H_i^{k+1} \leq_{n_k} \prod_i H_i^k$ such that

$$\mathcal{X}_k \cap \left(\prod_i H_i^{k+1}\right) \text{ depends only on coordinates } < n_k. \qquad (8.35)$$

Choose now an $n_{k+1} \in B_{k+1}\backslash A_{k+1}$ above n_k such that

$$(B_{k+1}\backslash A_{k+1}) \cap [n_k, n_{k+1}) \neq \emptyset. \qquad (8.36)$$

This describes the recursive construction. Let (A_∞, B_∞) and $H^\infty = (H_i^\infty)$ be the limits of the fusion sequence. Then, we have that

$$(A_\infty, B_\infty) \leq_n (A, B), \quad (H_i^\infty) \leq_n (H_i), \text{ and} \qquad (8.37)$$

$$\mathcal{X} \cap \left(\prod_{i=0}^\infty H_i^\infty\right) \text{ is an open subset of the } S_{B_\infty}\text{-product } \prod_i H_i^\infty. \qquad (8.38)$$

Applying Lemma 8.15, we get $(C, D) \leq_n (A, B)$ and an S_D-product $\prod_i J_i \leq_n \prod_i H_i^\infty$ such that the intersection

$$\mathcal{X} \cap \left(\prod_{i=0}^\infty J_i\right) \text{ depends only on coordinates } < n, \qquad (8.39)$$

as required. $\qquad\qquad\qquad\qquad\qquad\qquad\qquad\qquad\qquad\qquad\qquad\qquad\qquad\qquad\square$

Corollary 8.20 *All Borel subsets of $\mathbb{N}^\infty$ are (m_i)-Ramsey.*

Notation. For a given product $\prod_i H_i$ of finite sets and a given finite sequence $s \in \mathbb{N}^{[<\infty]}$, set

$$(\textstyle\prod_{i=0}^\infty H_i)[s] = \{x \in \mathbb{N}^\infty : x \restriction |s| = s \text{ and } (\forall i > |s|) \, x(i) \in H_i\}. \qquad (8.40)$$

We are now ready to state and prove the main result of this section.

Theorem 8.21 *The field of (m_i)-Ramsey sets is closed under the Souslin operation.*

Proof. Let $\mathcal{X}_a$ $(a \in \mathbb{N}^{[<\infty]})$ be a given Souslin scheme indexed by finite subsets of $\mathbb{N}$. We assume that $\mathcal{X}_b$ is a subset of $\mathcal{X}_a$ whenever b end-extends a. Recall that for $M \in \mathbb{N}^{[\infty]}$, the restriction $r_n(M)$ is the finite subset of M obtained by taking the first n elements of M according to its increasing enumeration. So we can write the result of the Souslin operation as

$$\mathcal{X} = \bigcup_{M \in \mathbb{N}^{[\infty]}} \bigcap_{n \in \mathbb{N}} \mathcal{X}_{r_n(M)}. \tag{8.41}$$

Similarly, we consider the relativized versions

$$\mathcal{X}_a^* = \bigcup_{M \in [a, \mathbb{N}]} \bigcap_{n \geq |a|} \mathcal{X}_{r_n(M)}, \tag{8.42}$$

where $[a, \mathbb{N}]$ denotes the family of all infinite subsets of $\mathbb{N}$ that end-extend a. Let (A, B) be a given regular pair of infinite subsets of $\mathbb{N}$, let $n \in \mathbb{N}$ and let $\prod_i H_i$ be a given S_B-product. We need to find $(C, D) \leq_n (A, B)$ and an S_D-product $\prod_{i=0}^{\infty} J_i \leq_n \prod_{i=0}^{\infty} H_i$ such that the intersection

$$\mathcal{X} \cap (\prod_{i=0}^{\infty} J_i) \tag{8.43}$$

depends only on coordinates $< n$. Starting from the initial values

$$A_0 = A, \ B_0 = B, \ n_0 = n, \ \text{and } H^0 = H = (H_i),$$

we build a fusion sequence $(A_k, B_k, n_k, H^k = (H_i^k))$ as follows. Suppose we have determined A_k, B_k, n_k and the product $\prod_i H_i^k$. Let

$$p = |\bigcup_{j \in B_k \cap n_k} \prod_{i < j} H_i^k|. \tag{8.44}$$

Then S_{A_k} eventually dominates $S^{(p)}$, so we can choose an integer $\bar{n}_k$ such that $\bar{n}_k \geq n_k$ and such that

$$S_{A_k}(m_0, \ldots, m_i) \geq S^{(p)}(m_0, \ldots, m_i) \text{ for all } i \geq \bar{n}_k. \tag{8.45}$$

Let (s_ℓ, a_ℓ) $(\ell \leq m)$ be an enumeration of

$$(\prod_{i=0}^{\bar{n}_k} H_i^k) \times \mathcal{P}(\{0, 1, \ldots, n_k\}). \tag{8.46}$$

Suppose there exist $(C, D) \leq (A_k, B_k)$ and an S_D-product $\prod_i J_i \subseteq \prod_i H_i^k$ such that

$$(\prod_{i=0}^{\infty} J_i)[s_0] \cap \mathcal{X}_{a_0}^* = \emptyset. \tag{8.47}$$

Since $A_k \subseteq C \subseteq D$ we have that

$$|J_i| = S_D(m_0, \ldots, m_i) \geq S^{(p)}(m_0, \ldots, m_i) \text{ for all } i \geq \bar{n}_k. \tag{8.48}$$

So we can find a sequence $J^0 = (J_i^0)$ such that $J_i^0 \subseteq J_i$ for $i \geq \bar{n}_k$, such that $J_i^0 \subseteq H_i^k$ for $n_k \leq i < \bar{n}_k$, such that $J_i^0 = H_i^k$ for $i < n_k$, and such that the corresponding product $\prod_i J_i^0$ is an S_{D_0}-product for some $(C_0, D_0) \leq_{n_k}$ (A_k, B_k). If such $(C, D) \leq (A_k, B_k)$ and $\prod_i J_i \subseteq \prod_i H_i^k$ do not exist we let $(J_i^0) = (H_i^k)$ and $(C_0, D_0) = (A_k, B_k)$. Now starting from (C_0, D_0) and $\prod_i J_i^0$ we treat the next pair (s_1, a_1). So, suppose there exist $(C, D) \leq (C_0, D_0)$ and an S_D-product $\prod_i J_i \subseteq \prod_i J_i^0$ such that

$$(\textstyle\prod_{i=0}^{\infty} J_i)[s_1] \cap \mathcal{X}_{a_1}^* = \emptyset. \tag{8.49}$$

Then again $A_k \subseteq C_0 \subseteq C \subseteq D$, so

$$|J_i| = S_D(m_0, \ldots, m_i) \geq S^{(p)}(m_0, \ldots, m_i) \text{ for } i \geq \bar{n}_k. \tag{8.50}$$

Then we can find $J^1 = (J_i^1)$ such that $J_i^1 \subseteq J_i$ for $i \geq \bar{n}_k$, such that $J_i^1 \subseteq J_i^0$ for $i \leq \bar{n}_k$, such that $J_i^1 = J_i^0$ for $i < n_k$, and such that the corresponding product is an S_{D_1}-product for some regular pair $(C_1, D_1) \leq_{n_k}$ (C_0, D_0). If such $(C, D) \leq (C_0, D_0)$ and $\prod_i J_i \subseteq \prod_i J_i^0$ cannot be found, we let $(C_1, D_1) = (C_0, D_0)$ and $(J_i^1) = (J_i^0)$, and so on. Proceeding this way and treating each pair (s_ℓ, a_ℓ) for $\ell \leq m$, we arrive at $(C_m, D_m) \leq_{n_k} (A_k, B_k)$ and $\prod_i J_i^m \leq_{n_k} \prod_i H_i^k$, and we define

$$(A^{k+1}, B^{k+1}) = (C_m, D_m), (H_i^{k+1}) = (J_i^m). \tag{8.51}$$

Then we achieve the following property of the regular pair (A^{k+1}, B^{k+1}) and the product $\prod_i H_i^{k+1}$:

$(1)_k$ For every $s \in \prod_{i \leq \bar{n}_k} H_i^{k+1}$ and $a \subseteq \{0, 1, \ldots, n_k\}$, either

$$(\textstyle\prod_{i=0}^{\infty} H_i^{k+1})[s] \cap \mathcal{X}_a^* = \emptyset, \tag{8.52}$$

or it is impossible to find $(C, D) \leq (A_{k+1}, B_{k+1})$ and S_D-product $\prod_{i=0}^{\infty} J_i \subseteq \prod_{i=0}^{\infty} H_i^{k+1}$ such that

$$(\textstyle\prod_{i=0}^{\infty} J_i)[s] \cap \mathcal{X}_a^* = \emptyset. \tag{8.53}$$

Choose $n_{k+1} > \bar{n}_k$ such that $B_{k+1} \backslash A_{k+1}$ has some point in the interval $[n_k, n_{k+1}]$. This describes our fusion sequence $(A_k, B_k, n_k, (H_i^k))$. Let (A_∞, B_∞) and (H_i^∞) be its limits. For $a \in \mathbb{N}^{[<\infty]}$, let

$$T_a = \{s \in \bigcup_{k=0}^{\infty} \textstyle\prod_{i=0}^{\bar{n}_k} H_i^\infty : (\textstyle\prod_{i=0}^{\infty} H_i^\infty)[s] \cap \mathcal{X}_a^* = \emptyset\}, \tag{8.54}$$

$$\Psi(\mathcal{X}_a^*) = \bigcup\{(\textstyle\prod_{i=0}^{\infty} H_i^\infty)[s] : s \in T_a\}, \tag{8.55}$$

and let

$$\Phi(\mathcal{X}_a^*) = (\mathcal{X}_a \cap \textstyle\prod_i H_i^\infty) \backslash \Psi(\mathcal{X}_a^*). \tag{8.56}$$

Then $\Phi(\mathcal{X}_a^*)$ is an (m_i)-Ramsey set, and so is the difference

$$\mathcal{M}_a = \Phi(\mathcal{X}_a^*)\setminus \bigcup_{\ell > \max(a)} \Phi(\mathcal{X}_{a\cup\{\ell\}}^*). \tag{8.57}$$

Note that $\mathcal{M}_a \subseteq \Phi(\mathcal{X}_a^*)\setminus\mathcal{X}_a^*$. Starting with the values $C_0 = A_\infty$, $D_0 = B_\infty$, $p_0 = n$ and $(J_i^0) = (H_i^\infty)$, we now build the second fusion sequence $(C_k, D_k, p_k, (J_i^k))$ such that (p_k) is a subsequence of (n_k) and such that

(2)$_k$ For all $a \subseteq \{0, 1, \ldots, p_k\}$, the intersection $\mathcal{M}_a \cap \prod_i J_i^{k+1}$ depends only on coordinates $< p_k$.

Clearly there is no problem in finding the fusion sequence since all sets $\mathcal{M}_a$ are (m_i)-Ramsey, and we treat only finitely many of them at a given stage. We claim that for all k, in statement $(2)_k$, we actually have that $\mathcal{M}_a \cap \prod_i J_i^{k+1} = \emptyset$. To see this, suppose that for some k we can find $a \subseteq \{0, \ldots, p_k\}$ such that $\mathcal{M}_a \cap \prod_i J_i^{k+1}$ depends only on coordinates $< p_k$ but it is nonempty. Let j be such that $p_k = n_j$ and pick $s \in \prod_{i \leq n_j} J_i^{k+1}$ such that

$$(\textstyle\prod_{i=0}^\infty J_i^{k+1})[s] \subseteq \mathcal{M}_a. \tag{8.58}$$

Since $\mathcal{M}_a \cap \mathcal{X}_a^* = \emptyset$ this means that in $(1)_j$ the second alternative fails as we can put $(C, D) = (C_{k+1}, D_{k+1})$ and $\prod_{i=0}^\infty J_i = \prod_{i=0}^\infty J_i^{k+1}$. So we have the first alternative of $(1)_j$, i.e., that

$$(\textstyle\prod_{i=0}^\infty H_i^{k+1})[s] \cap \mathcal{X}_a^* = \emptyset. \tag{8.59}$$

It follows that $(\prod_{i=0}^\infty H_i^\infty)[s] \subseteq \Psi(\mathcal{X}_a^*)$ and therefore

$$\mathcal{M}_a \cap \Psi(\mathcal{X}_a^*) \neq \emptyset, \tag{8.60}$$

a contradiction.

Let (C_∞, D_∞) and (J_i^∞) be the limits of the fusion sequence

$$(C_k, D_k, p_k, (J_i^k)).$$

Then $\prod_{i=0}^\infty J_i^\infty$ is an S_{D_∞}-product such that

$$(C_\infty, D_\infty) \leq_n (A, B), \tag{8.61}$$

$$\textstyle\prod_{i=0}^\infty J_i^\infty \leq_n \prod_{i=0}^\infty H_i, \tag{8.62}$$

$$(\textstyle\prod_{i=0}^\infty J_i^\infty) \cap \mathcal{M}_a = \emptyset \text{ for all } a \in \mathbb{N}^{[<\infty]}. \tag{8.63}$$

Then, as in the Marczewski proof of the preservation of the Souslin property (see Section 4.1 above), one checks that

$$\mathcal{X}_\emptyset^* \cap (\textstyle\prod_{i=0}^\infty J_i^\infty) = \Phi(\mathcal{X}_\emptyset^*) \cap (\textstyle\prod_{i=0}^\infty J_i^\infty). \tag{8.64}$$

Since $\Phi(\mathcal{X}_\emptyset^*)$ is a (m_i)-Ramsey set, we can find $(C, D) \leq_n (C_\infty, D_\infty)$ and an S_D-product $\prod_{i=0}^\infty J_i \leq_n \prod_{i=0}^\infty J_i^\infty$ such that the intersection

$$\Phi(\mathcal{X}_\emptyset^*) \cap (\textstyle\prod_{i=0}^\infty J_i) \tag{8.65}$$

depends only on coordinates $< n$. It follows that $\mathcal{X} = \mathcal{X}_\emptyset^*$ depends only on coordinates $< n$. This finishes the proof. $\qquad\square$

Corollary 8.22 *For every infinite set $M \subseteq \mathbb{N}$ and every Souslin-measurable coloring*

$$c : \textstyle\prod_{i=0}^\infty S_M(m_0, ..., m_i) \to \{0, 1\} \tag{8.66}$$

there exist $H_i \subseteq S_M(m_0, \ldots, m_i)$ such that $|H_i| = m_i$ for all i and such that c is constant on $\prod_{i=0}^\infty H_i$.

Proof: Using Theorem 8.21 we can find an infinite set $N \subseteq M$ and an S_N-product $\prod_i H_i \subseteq \prod_i S_M(m_0, ..., m_i)$ such that c is continuous on $\prod_i H_i$. Since

$$S_N(m_0, \ldots, m_i) \geq S(m_0, \ldots, m_i) \tag{8.67}$$

for all i, we finish using Lemma 8.7 from the previous section. $\qquad\square$

8.3 POLARIZED PARTITION CALCULUS

Recall the *Ackermann hierarchy* of fast-growing functions from $\mathbb{N}$ into $\mathbb{N}$ defined as follows:

$$A_0(0) = 1, A_0(1) = 2, A_0(x) = 2 + x\,(x > 1)$$
$$A_{n+1}(x) = A_n^{(x)}(1), \tag{8.68}$$

where $A_n^{(x)}$ denotes the xth iterate of A_n. (The 0th iterate of any function is the identity function). Thus,

$$A_1(x) = 2x\ (x > 0)$$
$$A_2(x) = 2^x$$
$$A_3(x) = 2^{2^{\cdot^{\cdot^{\cdot^2}}}}\ x\ \text{times.} \tag{8.69}$$

A useful formula that relates A_n and A_{n+1} is the following:

$$A_{n+1}(x+1) = A_n(A_{n+1}(x)). \tag{8.70}$$

Functions that are eventually dominated by an iterate of A_2 are called *elementary*. Thus the *tower function* A_3 is the first member of the Ackermann hierarchy that is not elementary. We shall use the first few members of the Ackermann hierarchy to give some upper bounds on sequences of the form

$$(S(m_0, \ldots, m_i)), (S^{(p)}(m_0, \ldots, m_i)), S_M(m_0, \ldots, m_i)) \tag{8.71}$$

for various choices of infinite nondecreasing sequences (m_i) of positive integers. Recall, for example, the following upper bound that has been established above in Lemma 8.6.

Lemma 8.23 *If* $m_i = 2$ *for all* i *then* $S(m_0, \ldots, m_i) \le A_2^{(2)}(2i+1)$.

Let us continue this and give some upper bounds on the other two sequences associated with particular choice of the sequence (m_i).

Lemma 8.24 *If* $m_i = 2$ *for all* i, *then for all* i,

$$S^{(p)}(m_0, \ldots, m_i) \le A_2^{(p+2)}(2i+1). \tag{8.72}$$

Proof. As in section 8.1, we use the notation

$$S^{(p)}(2_i) = S^{(p)}(m_0, \ldots, m_i) \tag{8.73}$$

where $m_0 = m_1 = \ldots = m_i = 2$. The case $p = 0$ of the above inequality is given in the previous Lemma. To see the inductive step note that for $p > 0$,

$$S^{(p)}(2_i) = 2(S^{(p-1)}(2_i) - 1) \prod_{j=0}^{i-1} \left(\frac{S^{(p)}(2_j)}{S^{(p-1)}(2_j)} \right) + 1$$

$$\le 2A_2^{(p+1)}(2i+1) \prod_{j=0}^{i-1} \frac{3^{S^{(p-1)}(2_j)}}{[S^{(p-1)}(2_j)]^{S^{(p-1)}(2_j)}} \prod_{j=0}^{i-1} [S^{(p)}(2_j)]^{S^{(p-1)}(2_j)}$$

$$\le 2A_2^{(p+1)}(2i+1)[S^{(p)}(2_{i-1})]^{\sum_{j=0}^{i-1} S^{(p-1)}(2_j)} \tag{8.74}$$

$$\le 2A_2^{(p+1)}(2i+1)[A_2^{(p+2)}(2i-1)]^{A_2^{(p+1)}(2_i)}$$

$$\le A_2^{(p+2)}(2i+1).$$

$\square$

Recall the definition of the sequence $(S_M(m_0, \ldots, m_i))$ for a given subset $M \subseteq \mathbb{N}$ given above in Section 8.1. It is a sequence composed of steps of the form $(S^{(p)}(m_0, ..., m_i))$ in intervals determined by consecutive members of M, which also determine the number p. The fastest of them is the sequence $(S_M(m_0, \ldots, m_i))$ for $M = \mathbb{N}$.

Lemma 8.25 *Suppose* $m_i = 2$ *for all* i. *Then for all* i

$$S_{\mathbb{N}}(m_0, \dots, m_i) \le A_4(i+3). \tag{8.75}$$

Proof. Using the notation

$$S_{\mathbb{N}}(m_0, \dots, m_i) = S_{\mathbb{N}}(2_i), \tag{8.76}$$

we shall actually show that

$$(i+1)[S_{\mathbb{N}}(2_i)]^{i+1} + 2i + 6 \le A_4(i+3) \tag{8.77}$$

holds for all i. Since $S_{\mathbb{N}}(2_0) = 3$ and since $A_4(3) = 2^{16}$, the inequality holds for $i = 0$. The inductive step at $i > 0$ follows from the following sequence of inequalities in which we use the definition of $S_{\mathbb{N}}$ and the estimates of the previous Lemma:

$$
\begin{aligned}
(i+1)[S_{\mathbb{N}}(2_i)]^{i+1} + 2i + 6 \ &\le (i+1)[A_2^{(i[S_{\mathbb{N}}(2_{i-1})]^i + 2)}(2i+1)]^{i+1} + 2i + 6 \\
&\le A_2^{(n[S_{\mathbb{N}}(2_{i-1})]^i + 3)}(2i+1) \\
&\le A_3(i[S_{\mathbb{N}}(2_{i-1})]^i + 2i + 4) \\
&\le A_3(A_4(i+2)) = A_4(i+3).
\end{aligned}
\tag{8.78}
$$

□

Similar arguments give us estimates for other choices of the infinite sequence (m_i). For example, we have proved the following facts along the same lines:

Theorem 8.26 *The following holds for all integers* $p, i \ge 0$:

(1) $S^{(p)}(A_2(0), \dots, A_2(i)) \le A_2^{(p+3)}(i)$,

(2) $S_{\mathbb{N}}(A_2(0), \dots, A_2(i)) \le A_4(i+3)$,

(3) $S^{(p)}(A_3(0), \dots, A_3(i)) \le A_3(p+i+2)$,

(4) $S_{\mathbb{N}}(A_3(0), \dots, A_3(i)) \le A_4(i+3)$.

It follows that we have the same upper bound,

$$S_{\mathbb{N}}(m_0, \dots, m_i) \le A_4(i+3), \tag{8.79}$$

for any choice of nondecreasing infinite sequence (m_i) of positive integers starting from the constant sequence $m_i = 2$ all the way up to the rapidly increasing sequence $m_i = A_3(i)$. For levels $n \ge 3$, we have the following behavior proved again along the same lines.

Theorem 8.27 *The following holds for all integers $n \geq 3$ and $p, i \geq 0$:*

(1) $S^{(p)}(A_n(0), \ldots, A_n(i)) \leq A_n(p + i + 2)$,

(2) $S_{\mathbb{N}}(A_n(0), \ldots, A_n(i)) \leq A_{n+1}(i + 3)$.

Combining this with the main result of Section 8.2, we obtain the following fact.

Theorem 8.28 *For every integer $n \geq 3$ the partition relation*

$$
\begin{pmatrix} A_{n+1}(3) \\ A_{n+1}(4) \\ A_{n+1}(5) \\ \vdots \end{pmatrix} \rightarrow \begin{pmatrix} A_n(0) \\ A_n(1) \\ A_n(2) \\ \vdots \end{pmatrix} \tag{8.80}
$$

holds in the realm of Souslin-measurable colorings.

Corollary 8.29 *The partition relation*

$$
\begin{pmatrix} A_4(3) \\ A_4(4) \\ A_4(5) \\ \vdots \end{pmatrix} \rightarrow \begin{pmatrix} 2 \\ 2 \\ 2 \\ \vdots \end{pmatrix} \tag{8.81}
$$

holds in the realm of Souslin-measurable colorings. □

In the partition calculus just exposed, we have ignored the colorings in more than two colors. It is clear that the same set of results with very similar upper bounds can be obtained for any other number of colors by simply making the appropriate change in the basic function

$$
S = S_2 : \mathbb{N}^{<\infty} \rightarrow \mathbb{N}. \tag{8.82}
$$

It turns out, however, that there is also a direct way to deduce results about $(\ell + 1)$-colorings from results about ℓ-colorings in this kind of partition calculus.

Theorem 8.30 *Suppose that for some integer $\ell \geq 1$ and two nondecreasing infinite sequences (m_i) and (n_i), the partition relation*

$$
\begin{pmatrix} n_0 \\ n_1 \\ n_2 \\ \vdots \end{pmatrix} \rightarrow \begin{pmatrix} m_0 \\ m_1 \\ m_2 \\ \vdots \end{pmatrix}_\ell \tag{8.83}
$$

holds in the realm of Souslin-measurable colorings. Then also the partition relation

$$
\begin{pmatrix} n_1 \\ n_3 \\ n_5 \\ \vdots \end{pmatrix} \rightarrow \begin{pmatrix} m_0 \\ m_1 \\ m_2 \\ \vdots \end{pmatrix}_{\ell+1}
\tag{8.84}
$$

holds in the realm of Souslin-measurable partitions.

Proof. Let

$$
c : \prod_{i=0}^{\infty} n_{2i+1} \rightarrow \{0, 1, ..., \ell\}
\tag{8.85}
$$

be a given Souslin-measurable coloring. Define $d : \prod_{i=0}^{\infty} n_i \rightarrow \mathbb{Z}$ by

$$
d((x_i)) = c((x_{2i+1})) - c((x_{2i})).
\tag{8.86}
$$

Define now $\bar{c} : \prod_{i=0}^{\infty} n_i \rightarrow \{0, ..., \ell - 1\}$ by letting

$$
\begin{aligned}
\bar{c}((x_i)) \quad &= 0 && \text{if} \quad d((x_i)) \in \{0, -2\} \\
&= 1 && \text{if} \quad d((x_i)) \in \{-1, 1, 2\} \\
&= 2 && \text{if} \quad d((x_i)) \in \{-3, 3\} \\
&\vdots \\
&= \ell - 1 && \text{if} \quad d((x_i)) \in \{-\ell, \ell\}.
\end{aligned}
\tag{8.87}
$$

By the hypothesis of the theorem, there exists an infinite subproduct $\prod_{i=0}^{\infty} H_i$ of $\prod_{i=0}^{\infty} n_i$ such that $|H_i| = m_i$ for all i and such that $\bar{c}$ is constant on $\prod_{i=0}^{\infty} H_i$.

Claim 8.30.1 *The original coloring c is constant either on $\prod_{i=0}^{\infty} H_{2i+1}$ or on $\prod_{i=0}^{\infty} H_{2i}$.*

Proof. Suppose c takes two different values p and q on $\prod_{i=0}^{\infty} H_{2i}$ and two different values r and s on $\prod_{i=0}^{\infty} H_{2i+1}$. Since $\bar{c}$ is constant on $\prod_{i=0}^{\infty} H_i$, two of the four differences

$$
r - p, r - q, s - p, s - q
\tag{8.88}
$$

must be equal. Since $r - p \neq r - q, r - p \neq s - p, r - q \neq s - q$, and $s - p \neq s - q$, we have the following two possibilities.
Case 1: $r - p = s - q$. Then $r - q \neq s - p$, since otherwise we get that $r = s$. It follows that d takes at least three different values on $\prod_{i=0}^{\infty} H_i$. Since $\bar{c}$ is constant on this product, it follows that the constant value of $\bar{c}$ on this product is 1 and that d takes only the values $\{-1, 1, 2\}$ on this product. If $r - p = s - q = 1$, then $(r - q) + (s - p) = 2$, but there are no distinct $x, y \in \{-1, 1, 2\}$ satisfying $x + y = 2$. If $r - p = s - q = -1$ then

$(r - q) + (s - p) = -2$, but there are no distinct $x, y \in \{-1, 1, 2\}$ satisfying $x + y = -2$. Finally, if $r - p = s - q = 2$, then $(r - q) + (s - p) = 4$, but again there are no distinct $x, y \in \{-1, 1, 2\}$ satisfying $x + y = 4$.

Case 2: $r - q = s - p$. Note that this is just Case 1 with p and q interchanged, so by symmetry, we are done with this case. □

This finishes the proof of the theorem. □

NOTES TO CHAPTER EIGHT

The Ramsey theory of products exposed in this chapter was developed in the papers of DiPrisco-Llopis-Todorcevic [21] and DiPrisco-Todorcevic [23] although its final form appears in the paper of Todorcevic [108]. In particular, the paper [108] gives a primitive recursive upper bound on the polarized Ramsey theorem for the constant sequence $m_i = 2$. The upper bound given in [21] and [23] is of the order of the Ackermann function. The proof of Theorem 8.30 was adopted from the paper of DiPrisco-Henle [20], where the same implication was proved in the unrestricted case.

Chapter Nine

Parametrized Ramsey Theory

9.1 HIGHER DIMENSIONAL RAMSEY THEOREMS PARAMETRIZED BY INFINITE PRODUCTS OF FINITE SETS

The purpose of this and the next few sections is to show that the Ellentuck space $(\mathbb{N}^{[\infty]}, \subseteq, r)$ can be parametrized by the products of finite sets. The parametrized theory is built in steps starting from the following basic pigeon hole principle proved above in Section 3.3.

Lemma 9.1 *There is an* $R : \mathbb{N}_+^{<\infty} \to \mathbb{N}_+$ *such that for every infinite sequence* (m_i) *of positive integers and for every coloring*

$$c : \bigcup_{k \in \mathbb{N}} \prod_{i<k} R(m_0, \ldots, m_i) \to \{0, 1\} \tag{9.1}$$

there exist $H_i \subseteq R(m_0, \ldots, m_i)$, *with* $|H_i| = m_i$ *for all* i *and an infinite set* $A \subseteq \mathbb{N}$ *such that* c *is monochromatic on*

$$\bigcup_{k \in A} \prod_{i<k} H_i. \tag{9.2}$$

From now on, we fix $R : \mathbb{N}_+^{<\infty} \to \mathbb{N}_+$ satisfying Lemma 9.1 and, modifying R if necessary, we can assume that the quantity $R(m_0, \ldots, m_i)$ dominates all of its arguments and that it is monotonically increasing with respect to each of them. Moreover, we assume that R dominates the mapping S of Definition 8.4.

Definition 9.2 *For an integer* $p \geq 0$, *the pth iterate* $R^{(p)} : \mathbb{N}_+^{<\infty} \to \mathbb{N}_+$ *of* R *is defined as follows:*

$$\begin{aligned}
R^{(0)}(m_0, \ldots, m_i) &= R(m_0, \ldots, m_i), \\
R^{(p+1)}(m_0, \ldots, m_i) &= R(R^{(p)}(m_0), R^{(p)}(m_0, m_1), \ldots, R^{(p)}(m_0, \ldots, m_i)).
\end{aligned} \tag{9.3}$$

Then we have the following monotonicity property which will be frequently and implicitly used in what follows.

Lemma 9.3 $R^{(p')}(m_0', \ldots, m_{i'}') \geq R^{(p)}(m_0, \ldots, m_i)$ *whenever* $i' \geq i$, $p' \geq p$ *and* $m_j' \geq m_j$ *for all* $j \leq i$.

Definition 9.4 *For $M \subseteq \mathbb{N}$ define $R_M : \mathbb{N}^{<\infty} \to \mathbb{N}$ recursively as follows:*

$$R_M(m_0) = R(m_0)$$
$$R_M(m_0, ..., m_i) = R^{(p)}(m_0, ..., m_i), \tag{9.4}$$

where

$$p = |\bigcup_{k \in M \cap i} \prod_{j=0}^{k} R_M(m_0, ..., m_j)|. \tag{9.5}$$

Then we have the following monotonicity property which follows immediately from the definition and Lemma 9.3.

Lemma 9.5 $R_{M'}(m'_j, ..., m'_{i'}) \geq R_M(m_0, ..., m_i)$ *whenever* $M' \supseteq M$, $i' \geq i$, *and* $m'_j \geq m_j$ *for all* $j \leq i$.

As in the case of the function S in Section 8.1, we adopt the notation $R^{(p)}((m_i))$ and $R_M((m_i))$ for the infinite sequences of positive integers whose ith terms are $R^{(p)}(m_0, ..., m_i)$ and $R_M(m_0, ..., m_i)$, respectively. Similarly, for an infinite sequence (n_i) of positive integers, a product $\prod_{i=0}^{\infty} H_i$ is called an (n_i)-*product* or *product of type* (n_i) if $|H_i| = n_i$ for all i. Recall also the notation

$$T(\vec{H}) = \bigcup_{k \in \mathbb{N}} \prod_{i<k} H_i \tag{9.6}$$

for a given infinite sequence $\vec{H} = (H_i)$ of finite subsets of $\mathbb{N}$. We consider $T(\vec{H})$ a tree ordered by end-extension. For $s \in T(\vec{H})$, we let

$$T(\vec{H})[s] = \{t \in T(\vec{H}) : t \text{ is compatible with } s\}. \tag{9.7}$$

Lemma 9.6 *Let* $M \subseteq \mathbb{N}, k \in M$, *and* $N = M \backslash \{k\}$. *Then for every infinite* $B \subseteq \mathbb{N}$, *every* $R_M((m_i))$-*sequence* $\vec{H} = (H_i)$ *of finite subsets of* $\mathbb{N}$, *and every coloring*

$$c : T(\vec{H}) \to \{0, 1\} \tag{9.8}$$

there exist an $R_N((m_i))$-*sequence* $\vec{J} = (J_i) \leq_k \vec{H}$ *and an infinite set* $A \subseteq B$ *such that* c *is constant on*

$$T(\vec{J})[s](A) = \{t \in T(\vec{J})[s] : |t| \in A\} \tag{9.9}$$

for all $s \in \prod_{i<k} H_i$.

Proof. Let $p = |\prod_{i<k} H_i|$ and let $\{s_\ell : \ell < p\}$ be an enumeration of $\prod_{i<k} H_i$. Let

$$q = |\bigcup_{j \in M \cap k} \prod_{i<j} H_i|. \tag{9.10}$$

Then on the interval $[k, \min(M \backslash (k+1)))$, the sequence $R_N((m_i))$ is equal to $R^{(q)}((m_i))$, while the sequence $R_M((m_i))$ is equal to $R^{(p+q)}((m_i))$. For any other interval $[\bar{k}, \bar{\ell})$ determined by two successive members of M (equivalently, members of N), if the sequence $R_N((m_i))$ agrees on $[\bar{k}, \bar{\ell})$ with some sequence of the form $R^{(p+\bar{q})}((m_i))$, then $R_M((m_i))$ agrees on $[\bar{k}, \bar{\ell})$ with the same sequence $R^{(p+\bar{q})}((m_i))$. It follows that

$$|H_i| = R_M((m_0, \ldots, m_i)) \geq R^{(p)}(R_N(m_0), \ldots, R_N(m_0, \ldots, m_i)) \quad (9.11)$$

for all $i \geq k$. We shall apply the basic property of R and successively shrink $(H_i)_{i \geq k}$ to

$$J_i^0 \supseteq J_i^1 \supseteq \ldots \supseteq J_i^{p-1} \quad (i \geq k) \quad (9.12)$$

such that $|J_i^\ell| = R^{(p-\ell-1)}(R_N(m_0), \ldots, R_N(m_0, \ldots, m_i))$ for $i \geq k$ and such that for some sequence $B \supseteq A_0 \supseteq \ldots \supseteq A_{p-1}$ of infinite sets, c is constant on the set

$$T((\bar{J}^\ell))[s_i](A_\ell) \quad (9.13)$$

for all $i \leq \ell < p$, where $\bar{J}^\ell = (H_i)_{i<k} \frown (J_i^\ell)_{i \geq k}$. Setting $B = A_{-1}$, it suffices to see how one obtains $(J_i^{\ell+1})_{i \geq k}$ and $A_{\ell+1}$ starting from $(J_i^\ell)_{i \geq k}$ and A_ℓ. Consider the following coloring:

$$c_\ell : T(\bar{J}^\ell)[s_{\ell+1}] \to \{0, 1\}, \quad (9.14)$$

defined by letting $c_\ell(t) = c(t \restriction n)$ where n is the maximal element of A_ℓ that is $\leq |t|$. By Lemma 9.1 there exist $J_i^{\ell+1} \subseteq J_i^\ell (i \geq k)$ and an infinite set $C \subseteq \mathbb{N}$ such that

$$|J_i^{\ell+1}| = R^{(p-\ell-2)}(R_N(m_0), \ldots, R_N(m_0, \ldots, m_i)) \quad (9.15)$$

for all $i \geq k$ and such that c_ℓ is constant on

$$T(\bar{J}^{\ell+1})[s_{\ell+1}](C), \quad (9.16)$$

where $\bar{J}^{\ell+1} = (H_i)_{i<k} \frown (J_i^{\ell+1})_{i \geq k}$. For $n \in C$ let $m(n) = \max\{m \in A_\ell : m \leq n\}$. Shrinking C, we may assume that $m(n) < m(n')$ whenever $n < n'$ belong to C. Let

$$A_{\ell+1} = \{m(n) : n \in C\}. \quad (9.17)$$

Tracing back through the definitions, we see that the original map c is constant on

$$T(\bar{J}^{\ell+1})[s_{\ell+1}](A_{\ell+1}). \quad (9.18)$$

This completes the inductive step. After p steps we arrive at $(J_i^{p-1})_{i \geq k}$ and an infinite set $A_{p-1} \subseteq B$ such that if we let

$$A = A_{p-1} \text{ and } \bar{J} = (H_i)_{i<k} \frown (J_i^{p-1})_{i \geq k}, \quad (9.19)$$

then $\bar{J}$ is an $R_N((m_i))$-sequence and c is constant on $T(\bar{J})[s](A)$ for all $s \in \prod_{i<k} H_i$. This finishes the proof. $\qquad \square$

The following is an application of the previous lemma that is relevant to the parametrized theory of infinite products of finite sets.

Lemma 9.7 *Suppose B and M are infinite subsets of $\mathbb{N}$ and that $\mathcal{O}_\ell(\ell \in B)$ is a given family of open subsets of $\mathbb{N}^\infty$. Then for every integer n and every $R_M((m_i))$-sequence $\vec{H} = (H_i)$ of finite subsets of $\mathbb{N}$ there exist infinite $N \subseteq_n M$ and an $R_N((m_i))$-sequence $\vec{J} \leq_n \vec{H}$ such that the mapping*

$$\ell \mapsto \mathcal{O}_\ell \cap \prod_{i=0}^\infty J_i \tag{9.20}$$

is constant on an infinite subset A of B.

Proof. Recall that the mapping R of Lemma 9.1 is chosen to dominate the mapping S of Definition 8.4 on which the polarized theory of the previous chapter is based. So using the basic property of the mapping S given in Lemma 8.7 and taking a preliminary fusion sequence in the sense of Definition 8.17 to shrink $\vec{H}$, we may assume that

$$\mathcal{O}_\ell \cap \prod_{i=0}^\infty H_i \tag{9.21}$$

is a relatively clopen subset of $\prod_{i=0}^\infty H_i$ for all $\ell \in B$. So for each $\ell \in B$ we can choose an integer n_ℓ such that the set $\mathcal{O}_\ell \cap \prod_{i=0}^\infty H_i$ depends only on coordinates $< n_\ell$. We may further assume that $n < n_k < n_\ell$ holds for all $k, \ell \in B$ with $k < \ell$. Define

$$c : T(\vec{H}) \to \{0, 1\} \tag{9.22}$$

as follows. If for $t \in T(\vec{H})$ there are no $k \in B$ such that $n_k \leq |t|$, we put $c(t) = 0$. If for a given $t \in T(\vec{H})$ we can find a $k \in B$ such that $n_k \leq |t|$, we let $k(t)$ be the maximal such k and put

$$c(t) = 1 \text{ iff } (\textstyle\prod_{i=0}^\infty H_i)[t \upharpoonright n_{k(t)}] \subseteq \mathcal{O}_{k(t)}. \tag{9.23}$$

Find $\bar{n} \in M$ above n and let $N = M \backslash \{\bar{n}\}$. Applying Lemma 9.6, we find an infinite subset C of $\mathbb{N}$ and an $R_N((m_i))$-sequence $\vec{J} = (J_i) \leq_{\bar{n}} \vec{H}$ such that c is constant on

$$T(\vec{J})[s](C) \tag{9.24}$$

for all $s \in \prod_{i<\bar{n}} H_i$. For each $m \in C$, let $\ell(m)$ be the maximal element ℓ of B such that $n_\ell \leq m$. We may assume that $\ell(m) < \ell(m')$ whenever $m < m'$ belong to C. Let

$$A = \{\ell(m) : m \in C\}. \tag{9.25}$$

Then for each $\ell \in A$, the restriction

$$\mathcal{O}_\ell \cap \left(\prod_{i=0}^\infty J_i\right) \tag{9.26}$$

depends only on coordinates $< n_\ell$. Moreover, the mapping

$$\ell \mapsto \mathcal{O}_\ell \cap \left(\prod_{i=0}^\infty J_i\right) \tag{9.27}$$

is constant on A. This finishes the proof. □

In order to extend the previous lemma, we need to recall the following notion introduced above in Section 1.3.

Definition 9.8 *A family $\mathcal{F}$ of nonempty finite subsets of some infinite set $A \subseteq \mathbb{N}$ is a* barrier *on A if*

(1) *$a \not\sqsubseteq b$ for all $a \neq b$ from $\mathcal{F}$,*

(2) *every infinite $X \subseteq A$ has an initial segment in $\mathcal{F}$.*

We shall also use the following important property of barriers also established above in Section 1.3.

Lemma 9.9 (Nash-Williams) *Suppose $\mathcal{F}$ is a barrier on some infinite set $A \subseteq \mathbb{N}$. Then for every finite coloring*

$$\mathcal{F} = \bigcup_{\ell=0}^{k} \mathcal{F}_\ell, \tag{9.28}$$

there exist infinite $B \subseteq A$ and $\ell \leq k$ such that $\mathcal{F}|B \subseteq \mathcal{F}_\ell$.

We also need to recall from Section 1.3 the notion of rank of a given barrier $\mathcal{F}$ and its following useful property (see Lemma 1.25)

$$\mathrm{rk}(\mathcal{F}) = \sup\{\mathrm{rk}(\mathcal{F}_{\{n\}}) + 1 : n \in A\}, \tag{9.29}$$

where for $n \in A$,

$$\mathcal{F}_{\{n\}} = \{a \subseteq A/n : \{n\} \cup a \in \mathcal{F}\} \tag{9.30}$$

denotes the barrier on $A/n = \{k \in A : k > n\}$ induced by $\mathcal{F}$. A combinatorial explanation of the definition of $\mathrm{rk}(\mathcal{F})$ follows from the fact that

$$T(\mathcal{F}) = \{b \subseteq A : (\exists a \in \mathcal{F})b \sqsubseteq a\} \tag{9.31}$$

considered a tree under the end-extension relation $\sqsubseteq$ is a tree with no infinite branches. So, one way to define the rank of barrier $\mathcal{F}$ is to put $\mathrm{rk}(\mathcal{F}) = \rho(\emptyset)$, where $\rho = \rho_\mathcal{F}$ is the function $\rho : T(\mathcal{F}) \to \mathrm{Ord}$ determined recursively by the following rule:

$$\rho(s) = \sup\{\rho(t) + 1 : t \in T(\mathcal{F}) \text{ and } s \sqsubset t)\}, \tag{9.32}$$

where we use the convention that the supremum of an empty set of ordinals is equal to 0. So the inductive Equation (9.29) now follows rather easily from the fact that

$$\rho(\emptyset) = \sup\{\rho(\{n\}) + 1 : n \in A\} \tag{9.33}$$

and the fact that

$$\rho(\{n\}) = \mathrm{rk}(\mathcal{F}_{\{n\}}) \text{ for all } n \in A. \tag{9.34}$$

We shall also need the following fact from Section 3.1, which relates an arbitrary family of finite subsets of $\mathbb{N}$ to the Nash-Williams notion of barrier.

Lemma 9.10 (Galvin) *For every family $\mathcal{F}$ of nonempty finite subsets of $\mathbb{N}$ and every infinite set $A \subseteq \mathbb{N}$, there is an infinite $B \subseteq B$ such that either*

(1) B *includes no element of* $\mathcal{F}$,

(2) $\mathcal{F}|B = \{a \in \mathcal{F} : a \subseteq B\}$ *contains a barrier on* B.

We are now ready to state and prove the parametrized versions of these results.

Lemma 9.11 *Suppose $\mathcal{O}_b(b \in \mathcal{F})$ is a family of open subsets of $\mathbb{N}^\infty$ indexed by a barrier $\mathcal{F}$ on some infinite set $B \subseteq \mathbb{N}$. Let M be an infinite subset on $\mathbb{N}$ and let n be a given integer. Then for every $R_M((m_i))$-sequence $\vec{H} = (H_i)$ of finite subsets of $\mathbb{N}$, there exist an infinite set $N \subseteq_n M$, an $R_N((m_i))$-sequence $\vec{J} = (J_i) \leq_n \vec{H}_i$, and an infinite $A \subseteq B$ such that the mapping*

$$a \mapsto \mathcal{O}_a \cap \prod_{i=0}^{\infty} J_i \tag{9.35}$$

is constant on $\mathcal{F}|A$ and the constant value is a clopen subset of $\prod_{i=0}^{\infty} J_i$.

Proof. This is proved by induction on the rank of $\mathcal{F}$. The case of $\mathrm{rk}(\mathcal{F}) = 1$ has been taken care of by Lemma 9.7, so let us assume that $\mathrm{rk}(\mathcal{F}) > 1$ and that the conclusion of the lemma is true for barriers of smaller ranks. Assume that $n_0 = \min B > n$. Applying the inductive hypothesis to the mapping

$$a \mapsto \mathcal{O}_{\{n_0\} \cup a} \tag{9.36}$$

defined on the barrier $\mathcal{F}_{\{n_0\}}$ on B/n_0, the integer n_0, and the infinite set M, we obtain an infinite set $M_0 \subseteq_{n_0} M$ and an $R_{M_0}((m_i))$-sequence $\vec{J}^0 = (J_i^0) \leq_{n_0} \vec{H}$ such that the mapping

$$a \mapsto \mathcal{O}_{\{n_0\} \cup a} \cap \prod_{i=0}^{\infty} J_i^0 \tag{9.37}$$

is constant on $\mathcal{F}_{\{n_0\}}|A_0$ for some infinite $A_0 \subseteq B/n_0$. Starting from this, we build a decreasing sequence (A_k) of infinite subsets of B, a decreasing sequence (M_k) of infinite subsets of M, and an infinite sequence $(\vec{J}^k)$ of infinite sequences of finite subsets of $\mathbb{N}$ such that

(1) $A_k \supseteq A_{k+1}$ and $n_k = \min A_k$,

(2) $M_{k+1} \leq_{n_k} M_k$ and $M_{k+1} \cap [n_k, n_{k+1}) \neq \emptyset$,

(3) $\vec{J}^k = (J_i^k)$ is an $R_{M_k}((m_i))$-sequence,

(4) $\vec{J}^{k+1} \leq_{n_k} \vec{J}^k$,

(5) $b \mapsto \mathcal{O}_{\{n_k\} \cup \{b\}} \cap \prod_{i=0}^{\infty} J_i^k$ is constant on $\mathcal{F}_{\{n_k\}}|A_k$.

Having chosen these sequences, let $\vec{J}^\infty = (J_i^\infty)$ be the fusion of $(\vec{J}^k)$, let M_∞ be the intersection of (M_k) and let $A = \{n_k\}_{k=0}^\infty$. It follows that the restriction of the mapping

$$b \mapsto \mathcal{O}_b \cap \prod_{i=0}^\infty J_i^\infty \qquad (9.38)$$

to the barrier $\mathcal{F}\,|A_\infty$ depends only on the minimum of the element b of $\mathcal{F}\,|A_\infty$. In other words, we have arrived at an infinite set $M_\infty \subseteq_n M$, an $R_{M_\infty}((m_i))$-sequence $H^\infty = (H_i^\infty) \leq_n \vec{H}$, an infinite set $A_\infty \subseteq B_i$ and a mapping

$$n \mapsto \mathcal{O}_n^* \cap \prod_{i=0}^\infty J_i^\infty \qquad (9.39)$$

defined on A_∞, where for $n \in A_\infty$, we denote by $\mathcal{O}_n^* \cap \prod_{i=0}^\infty J_i^\infty$ the constant value of the mapping

$$b \mapsto \mathcal{O}_{n \cup \{b\}} \cap \prod_{i=\infty} J_i^\infty \qquad (9.40)$$

when restricted to $\mathcal{F}|A_\infty$. Applying Lemma 9.7 again, we obtain an infinite $M' \subseteq_n M_\infty$, an infinite $A \subseteq A_\infty$, and an $R_{M'}((m_i))$-sequence $J = (J_i) \leq_n J^\infty$ such that

$$n \mapsto \mathcal{O}_n^* \cap \prod_{i=0}^\infty J_i \qquad (9.41)$$

is constant on A. It follows that

$$a \mapsto \mathcal{O}_a \cap \prod_{i=0}^\infty J_i \qquad (9.42)$$

is constant on $\mathcal{F}|A$, as desired. $\square$

9.2 COMBINATORIAL FORCING PARAMETRIZED BY INFINITE PRODUCTS OF FINITE SETS

In this section $\mathcal{O}$ is a fixed open subset of the product $\mathbb{N}^\infty \times \mathbb{N}^{[\infty]}$ relative to the standard complete separable metrizable topology of this product. Moreover, (m_i) will be a fixed infinite nondecreasing sequence of positive integers. It will be convenient to refine the tree notation used in Sections 3.3 and 9.1 as follows. For an infinite sequence $\vec{H} = (H_i)$ of finite subsets of $\mathbb{N}$ and a finite sequence $s \in \mathbb{N}^{<\infty}$ and $k \in \mathbb{N}$, set, ,

$$[s, \vec{H}] = \{x \in \mathbb{N}^\infty : x \restriction |s| = s \text{ and } x(i) \in H_i \text{ for } i \geq |s|\},$$
$$[s, \vec{H}]^k = \{x \in \mathbb{N}^k : x \restriction |s| = s \text{ and } x(i) \in H_i \text{ for } i \in [|s|, k)\}, \qquad (9.43)$$
$$[s, \vec{H}]^{<\infty} = \bigcup_{k \geq |s|} [s, \vec{H}]^k.$$

Let $\mathcal{R}_\infty = \mathcal{R}_\infty((m_i))$ be the collection of all infinite sequences $\vec{H} = (H_i)$ of nonempty finite subsets of $\mathbb{N}$ for which we can find infinite $M \subseteq \mathbb{N}$ such that

$$R_M(m_0, \ldots, m_i) = |H_i| \text{ for all } i. \tag{9.44}$$

For $\vec{H} \in \mathcal{R}_\infty((m_i))$, we let $M(\vec{H})$ denote the infinite subset $M \leq \mathbb{N}$ witnessing this membership.

   Notation. We use $\vec{H}, \vec{J}, \vec{K}, \ldots$ to denote members of $\mathcal{R}_\infty$,
   $x, y, z, \ldots$ to denote members of $\mathbb{N}^\infty$
   $s, t, u, \ldots$ to denote members of $\mathbb{N}^{<\infty}$,
   $A, B, C, \ldots$ to denote members of $\mathbb{N}^{[\infty]}$, and
   $a, b, c, \ldots$ to denote members of $\mathbb{N}^{[<\infty]}$.
   We also adopt the usual notation

$$[a, A] = \{B \in \mathbb{N}^{[\infty]} : a \sqsubseteq B \subseteq A\} \text{ and } [n, A] = [r_n(A), A]$$

for basic open sets of the Ellentuck topology[1] on $\mathbb{N}^{[\infty]}$.

Definition 9.12 *We say that $(\vec{H}, A)$ accepts (s, a) if $[s, \vec{H}] \times [a, A] \subseteq \mathcal{O}$. If there are no $\vec{J} \leq_{|s|} \vec{H}$ and $B \subseteq A$ such that $(\vec{J}, B)$ accepts (s, a), we say that $(\vec{H}, A)$ rejects (s, a). We say that $(\vec{H}, A)$ decides (s, a) if it either accepts or rejects (s, a).*

The following facts follow immediately from the definition.

Lemma 9.13 *(a) If $(\vec{H}, A)$ accepts (s, a), then for every $\vec{J} \leq \vec{H}$ and $B \subseteq A$ the pair $(\vec{J}, B)$ accepts (s, a).*

(b) If $(\vec{H}, A)$ rejects (s, a), then for every $\vec{J} \leq_{|s|} \vec{H}$ and $B \subseteq A$, the pair $(\vec{J}, B)$ rejects (s, a).

(c) If for some $k \geq |s|$, the pair $(\vec{H}, A)$ accepts (t, a) for every $t \in [s, \vec{H}]^k$, then $(\vec{H}, A)$ accepts (s, a).

(d) For every pair $(\vec{H}, A)$ and every pair (s, a), there exist $\vec{J} \leq_{|s|} \vec{H}$ and $B \subseteq A$ such that $(\vec{J}, B)$ decides (s, a).

Lemma 9.14 *For every $\vec{H} \in \mathcal{R}_\infty$, $A \in \mathbb{N}^{[\infty]}$ and $n \in \mathbb{N}$, there exist $\vec{J} \leq_n \vec{H}$ and infinite $B \subseteq A$ such that for every $a \in \mathbb{N}^{[<\infty]}$, every $k \in B$ with $k \geq \max(a)$, and every $s \in \prod_{i<k} J_i$, the pair $(\vec{J}, B)$ decides (s, a).*

Proof. Starting from $\vec{H}$, we build an infinite sequence $(\vec{H}^k)$ of members of $\mathcal{R}_\infty$ and a decreasing infinite sequence $A \supseteq A_0 \supseteq A_1 \supseteq \ldots \supseteq A_k \supseteq \ldots$ of infinite subsets of $\mathbb{N}$ such that if we put $n_k = \min(A_k)$ then for all k,

   (1) $n < n_k < n_{k+1}$,

---

[1] Recall that, for $A \in \mathbb{N}^{[\infty]}$ and $n \in \mathbb{N}$, we denote by $r_n(A)$ the finite set of cardinality n formed by taking the first n elements of A according to its increasing enumeration.

(2) $\vec{H}^{k+1} \leq_{n_k} \vec{H}^k$ and therefore $M(\vec{H}^{k+1}) \subseteq_{n_k} M(\vec{H}^k)$,

(3) $M(\vec{H}^{k+1}) \cap [n_k, n_{k+1}) \neq \emptyset$,

(4) $(\vec{H}^{k+1}, A_{k+1})$ decides (s, a) for $s \in \prod_{i<n_k} H_i^k$ and $a \subseteq \{0, 1, \ldots, n_k\}$.

Let $B = \{n_k : k \in \mathbb{N}\}$ and let $J_i = \lim_{k \to \infty} H_i^k$. Then $\vec{J} = (J_i) \in \mathcal{R}_\infty$, $\vec{J} \leq_n \vec{H}$ and for every a and every $k \in B$ with $k \geq \max(a)$, the pair $(\vec{J}, B)$ decides (s, a) for every $s \in \prod_{i<k} J_i$. $\square$

Lemma 9.15 *Suppose that $\vec{J}$ and B satisfy the conclusion of Lemma 9.14. Then for every $n \in \mathbb{N}$, there exist $\vec{K} \leq_n \vec{J}$ and an infinite set $C \subseteq \mathbb{N}$ such that for all $\ell \in C$, all $s \in \prod_{i<\ell} K_i$, and all $a \subseteq \{0, 1, \ldots, \ell\}$, there exist infinitely many $k \in B/\ell$ such that either*

(1) $(\vec{J}, B)$ *accepts* (t, a) *for all* $t \in [s, \vec{K}]^k$, *or*

(2) $(\vec{J}, B)$ *rejects* (t, a) *for all* $t \in [s, \vec{K}]^k$.

Proof. Starting from $\vec{J}$, we build a fusion sequence $(\vec{J}^k)$ of elements of $\mathcal{R}_\infty$ and a strictly increasing sequence of integers (n_k) above n as follows. Let n_0 be the minimal member of $M(\vec{J})$ above n. Applying Lemma 9.6 to the sequence of mappings

$$c_a : \bigcup_k \prod_{i<k} J_i \to \{0, 1\} \quad (a \subseteq \{0, \ldots, n_0\}) \tag{9.45}$$

defined by

$$c_a(t) = 0 \text{ iff } (\vec{J}, B) \text{ accepts } (t, a), \tag{9.46}$$

we get an infinite sequence $\vec{J}^0 \leq_{n_0} \vec{J}$ such that for all $a \subseteq \{0, \ldots, n_0\}$ and all $s \in \prod_{i<n_0} J_i$ there exist infinitely many $k \in B/n_0$ such that the mapping c_a is constant on $[s, \vec{J}^0]^k$. By the choice of $\vec{J}$ and B, it follows that for all $a \subseteq \{0, \ldots, n_0\}$ and $s \in \prod_{i<n_0} J_i$, there exist infinitely many $k \in B/n_0$ such that either

$$(\vec{J}, B) \text{ accepts } (t, a) \text{ for all } t \in [s, \vec{J}^0]^k, \tag{9.47}$$

or else

$$(\vec{J}, B) \text{ rejects } (t, a) \text{ for all } t \in [s, \vec{J}^0]^k. \tag{9.48}$$

Let n_1 be the second member of $M(\vec{J}^0)$ above n_0. Repeating the same procedure, we find $\vec{J}^1 \leq_{n_1} \vec{J}^0$ such that for all $a \subseteq \{0, \ldots, n_1\}$ and $s \in \prod_{i<n_1} J_i^1$, there exist infinitely many $k \in B/n_1$ such that either

$$(\vec{J}, B) \text{ accepts } (t, a) \text{ for all } t \in [s, \vec{J}^1]^k, \tag{9.49}$$

or else

$$(\vec{J}, B) \text{ rejects } (t, a) \text{ for all } t \in [s, \vec{J}^1]^k. \tag{9.50}$$

It is clear that we can continue this process indefinitely. After $(\vec{J}^k)$ and (n_k) have been constructed, we let $\vec{K}$ be the limit of $(\vec{J}^k)$ and let

$$C = \{n_k : k \in \mathbb{N}\}.$$

Then $\vec{K}$ and C satisfy the conclusion of the lemma. $\qquad\square$

The following is the crucial lemma of this section.

Lemma 9.16 *For every $\vec{H} \in \mathcal{R}, n \in \mathbb{N}, A \in \mathbb{N}^{[\infty]}$ and $a \in \mathbb{N}^{[<\infty]}$, there exist $\vec{J} \leq_n \vec{H}, B \subseteq A$ and $k \geq n$ such that for all $s \in \prod_{i<k} J_i$,*

$$[s, \vec{J}] \times [a, B] \subseteq \mathcal{O} \text{ or } [s, \vec{J}] \times [a, B] \subseteq \mathcal{O}^c. \tag{9.51}$$

Proof. Clearly, we may concentrate on the case $a = \emptyset$. Shrinking $\vec{H}$ and A, we may assume that they satisfy the conclusion of Lemma 9.14. Applying Lemma 9.15 and shrinking $\vec{H}$ and A, we may also assume that for every $\ell \in A$ every $s \in \prod_{i<\ell} H_i$ and every $a \subseteq \{0, \ldots, \ell\}$, there exist infinitely many k such that either

(1) $(\vec{H}, A)$ accepts (t, a) for all $t \in [s, \vec{H}']^k$, or

(2) $(\vec{H}, A)$ rejects (t, a) for all $t \in [s, \vec{H}']^k$.

For $b \in \mathbb{N}^{[<\infty]}$, let

$$\mathcal{O}_b = \bigcup \{[t, \vec{H}] : [t, \vec{H}] \times [b, A] \subseteq \mathcal{O}\}. \tag{9.52}$$

Applying Lemma 9.10 to

$$\mathcal{B} = \{\emptyset \neq b \in \mathbb{N}^{[<\infty]} : \mathcal{O}_b \neq \emptyset\}, \tag{9.53}$$

we may consider the following two cases.

Case 1. There is an infinite set $B \subseteq A$ such that B contains no member of $\mathcal{B}$. We claim that in this case

$$[\emptyset, \vec{H}] \times [\emptyset, B] \cap \mathcal{O} = \emptyset. \tag{9.54}$$

Otherwise, there exist $x \in \prod_i H_i$ and $X \subseteq B$ such that $(x, X) \in \mathcal{O}$. Then for some $k \in \mathbb{N}$ and $\ell \in X$,

$$[x \upharpoonright k] \times [X \cap \{0, ., \ell\}, \mathbb{N}] \subseteq \mathcal{O}. \tag{9.55}$$

Letting $b = X \cap \{0, ., \ell\}$, it follows that $b \in \mathcal{B}$ and $b \subseteq B$, a contradiction.

Case 2. There is an infinite set $B \subseteq A$ such that $\mathcal{B}|B$ contains a barrier $\mathcal{F}$ on B. Applying Lemma 9.11, we can find $C \subseteq B, \vec{J} \leq_n \vec{H}$ and a relatively clopen subset $\mathcal{P} \subseteq \prod_i J_i$ such that

$$\mathcal{O}_b \cap \prod_{i=0}^{\infty} J_i = \mathcal{P} \text{ for all } b \in \mathcal{F}|C. \tag{9.56}$$

Pick $\bar{n} > n$ in A such that $\mathcal{P} \cap \prod_{i=0}^{\infty} J_i$ depends only on coordinates $< \bar{n}$.

Claim 9.16.1 *Suppose $s \in \prod_{i < \bar{n}} J_i$ and $[s, \vec{J}] \cap \mathcal{P} = \emptyset$. Then for every $D \subseteq C$ and every $b \in \mathcal{F}|C$, there exists $E \subseteq D$ such that $([s, \vec{J}] \times [b, E]) \cap \mathcal{O} = \emptyset$.*

Proof. Pick $s \in \prod_{i < \bar{n}} J_i$ avoiding $\mathcal{P}$ and pick $b \in \mathcal{F}|C$. Let

$$\mathcal{B}_b = \{\emptyset \neq c \in (C/b)^{[< \infty]} : \mathcal{O}_{b \cup c} \cap [s, \vec{J}] \neq \emptyset\}. \tag{9.57}$$

Applying Lemma 9.10 to $\mathcal{B}_b$ and D, we have the following two cases.

Case 2.1. There is an infinite set $E \subseteq D$ that contains no members of $\mathcal{B}_b$. Then as in Case 1, we check that

$$([s, \vec{J}] \times [b, E]) \cap \mathcal{O} = \emptyset. \tag{9.58}$$

Case 2.2. There is an infinite set $E \subseteq D/b$ such that $\mathcal{B}_b|E$ contains a barrier $\mathcal{F}_b$ on E. Applying Lemma 9.11, we find $F \subseteq E$, $\vec{K} \leq_n \vec{J}$ and a nonempty relatively clopen subset $\mathcal{Q} \subseteq [s, \vec{K}]$ such that

$$\mathcal{O}_{b \cup c} \cap [s, \vec{K}] = \mathcal{Q} \text{ for all } c \in \mathcal{F}_b|F. \tag{9.59}$$

We claim that $\mathcal{O}_b \cap [s, \vec{K}] \neq \emptyset$, contradicting the choice of $\mathcal{P}$ and s. To see this, choose $\ell \in A'/b$ such that $\mathcal{Q}$ depends only on coordinates $< \ell$ and an $\bar{s} \in \prod_{i < \ell} K_i$ end-extending s such that $[\bar{s}, \vec{K}] \subseteq \mathcal{Q}$. Tracing back through the definitions, one sees that

$$[\bar{s}, \vec{H}] \subseteq \mathcal{O}_{b \cup c} \text{ for all } c \in \mathcal{F}_b|F. \tag{9.60}$$

From the definition of $\mathcal{O}_{b \cup c}$ and the fact that $\mathcal{F}_b|E$ is a barrier on E, we get

$$([\bar{s}, \vec{H}] \times [b, F]) \subseteq \mathcal{O}. \tag{9.61}$$

By the properties of $\vec{H}, A$, and A' we know that $(\vec{H}, A)$ decides $(\bar{s}, b)$. From Lemma 9.13 (b) and the fact that $(\vec{H}, F)$ accepts $(\bar{s}, b)$, we infer that $(\vec{H}, A)$ must also accept $(\bar{s}, b)$. It follows that

$$[\bar{s}, \vec{H}] \times [b, A] \subseteq \mathcal{O} \tag{9.62}$$

and therefore $[\bar{s}, \vec{H}] \subset \mathcal{O}_b$. Since $\bar{s}$ end-extends s, which was chosen to avoid the relatively clopen subset $\mathcal{P}$ of $\prod_{i=0}^{\infty} J_i$ depending on coordinates $< \bar{n} < \ell$, we infer that

$$\mathcal{O}_b \cap \prod_{i=0}^{\infty} J_i \neq \mathcal{P}, \tag{9.63}$$

a contradiction. This shows that Case 2.2 is impossible. This also finishes the proof of the claim. $\square$

Using Claim 9.16.1, we build a decreasing sequence

$$C = C_0 \supseteq C_1 \supseteq \dots \supseteq C_k \supseteq \dots$$

of infinite subsets of C such that if $n_k = \min C_k$, then (n_k) is a strictly increasing sequence of positive integers and such that for every k, every $b \in \mathcal{F}$, $b \subseteq \{n_0, \dots, n_k\}$, and every $s \in \prod_{i < \bar{n}} J_i$,

$$[s, \vec{J}] \cap \mathcal{P} = \emptyset \text{ implies } ([s, \vec{J}] \times [b, C_{k+1}]) \cap \mathcal{O} = \emptyset. \tag{9.64}$$

Let $C_\infty = \{n_k : n \in \mathbb{N}\}$. Then $\vec{J} \leq_n \vec{H}$, $C_\infty \subseteq A$, and $\bar{n} \geq n$ satisfy the condition of Lemma 9.16, or in other words, for every $s \in \prod_{i<\bar{n}} J_i$, either

$$([s, \vec{J}] \times [\emptyset, C_\infty]) \subseteq \mathcal{O} \tag{9.65}$$

or

$$([s, \vec{J}] \times [\emptyset, C_\infty]) \cap \mathcal{O} = \emptyset. \tag{9.66}$$

This finishes the proof.   □

9.3 PARAMETRIZED RAMSEY PROPERTY

Throughout this section, (m_i) is a fixed infinite nondecreasing sequence of positive integers. We adopt the notation and definitions from the previous section, as well as the notation and definitions from the previous chapter. For example, a *regular* pair (N, M) is a pair of infinite subsets of $\mathbb{N}$ such that $N \subseteq M$ and $M \setminus N$ is infinite, a fact which we denote by

$$N \subset_\infty M. \tag{9.67}$$

For regular pairs (Q, P) and (N, M), we write $(Q, P) \leq (N, M)$ whenever

$$N \subset_\infty Q \subset_\infty P \subset_\infty M. \tag{9.68}$$

Let $(Q, P) \leq_n (N, M)$ denote the fact that $(Q, P) \leq (N, M)$ and

$$Q \cap n = N \cap n \text{ and } P \cap n = M \cap n. \tag{9.69}$$

Definition 9.17 *A subset $\mathcal{X} \subseteq \mathbb{N}^\infty \times \mathbb{N}^{[\infty]}$ is* para-Ramsey *if for every regular pair (N, M), every $R_M((m_i))$-sequence $\vec{H}$ of finite subsets of $\mathbb{N}$, and every $a \in \mathbb{N}^{[<\infty]}$, there exist a regular pair $(Q, P) \leq_n (N, M)$, an $R_P((m_i))$-sequence $\vec{J} \leq_n \vec{H}$, an integer $k \geq n$, and an infinite set $B \subseteq A$ such that for all $s \in \prod_{i<k} J_i$*

$$[s, \vec{J}] \times [a, B] \subseteq \mathcal{X} \text{ or } [s, \vec{J}] \times [a, B] \subseteq \mathcal{X}^c. \tag{9.70}$$

Note the following reformulation of Lemma 9.16 from the previous section.

Lemma 9.18 *Every open subset of $\mathbb{N}^\infty \times \mathbb{N}^{[\infty]}$ is para-Ramsey.*

To show that the field of para-Ramsey subsets at $\mathbb{N}^\infty \times \mathbb{N}^{[\infty]}$ is a σ-field, we need to adopt the notion of fusion to this context.

Definition 9.19 *A fusion sequence is an infinite sequence of the form*

$$(N_k, M_k, \vec{H}^k, n_k, A_k) \tag{9.71}$$

such that for all k,

(1) *(N_k, M_k) is a regular pair,*

(2) *$\vec{H}^k$ is a $R_{M_k}((m_i))$-sequence of finite subsets of $\mathbb{N}$,*

(3) $n_k < n_{k+1}$,

(4) $(N_{k+1}, M_{k+1}) \leq_{n_k} (N_k, M_k)$,

(5) $\vec{H}^{k+1} \leq_{n_k} \vec{H}^k$,

(6) $(M_{k+1} \setminus N_{k+1}) \cap [n_k, n_{k+1}) \neq \emptyset$,

(7) $A_k \supseteq A_{k+1}$ and $\min(A_{k+1}) > \min(A_k)$.

The limits of the fusion sequence $(N_k, M_k, \vec{H}^k, n_k, A_k)$ are defined as follows:

$$N_\infty = \bigcup_{k=0}^{\infty} N_k,$$

$$M_\infty = \bigcap_{k=0}^{\infty} M_k,$$

$$H_i^\infty = \lim_{k \to \infty} H_i^k,$$

$$A_\infty = \{\min(A_k) : k \in \mathbb{N}\} \cup a$$

for some finite set $a \subseteq \mathbb{N}$, which may or may not be added to A_∞, depending on the context. Note that (N_∞, M_∞) is a regular pair of infinite subsets of $\mathbb{N}$ and that

$$(N_\infty, M_\infty) \leq_{n_k} (N_k, M_k) \text{ for all } k. \tag{9.72}$$

Note moreover that $\vec{H}^\infty = (H^\infty)$ is an $R_{M_\infty}((m_i))$-sequence and that

$$\vec{H}^\infty \leq_{n_k} \vec{H}^k \text{ for all } k. \tag{9.73}$$

Lemma 9.20 The para-Ramsey subsets of $\mathbb{N}^\infty \times \mathbb{N}^{[\infty]}$ form a σ-field.

Proof. Let $(\mathcal{X}_k)$ be a given sequence of para-Ramsey sets and let (N, M), n, $\vec{H}, [a, A]$ be the given inputs of Definition 9.17 needed for checking that

$$\mathcal{X} = \bigcup_{k=0}^{\infty} \mathcal{X}_k \tag{9.74}$$

is para-Ramsey. Starting from the initial values $N_0 = N, M_0 = M_0, n_0 = n, \vec{H}^0 = \vec{H}$ and $A_0 = A$ we build a fusion sequence $(N_k, M_k, \vec{H}^k, n_k, A_k)$, as follows. Suppose $N_k, M_k, \vec{H}^k, n_k$, and A_k have been determined. Applying the assumption that $\mathcal{X}_k$ is para-Ramsey, successively over all

$$b \subseteq a \cup \{\min(A_0), \ldots, \min(A_k)\}, \tag{9.75}$$

we get an $(N_{k+1}, M_{k+1}) \leq_{n_k} (N_k, M_k)$, an $R_{M_{k+1}}((m_i))$-sequence $\vec{H}^{k+1} \leq_{n_k} \vec{H}^k$, an integer $\bar{n} \geq n_k$, and an infinite set $A_{k+1} \subseteq A_k$ such that for all $b \subseteq a \cup \{\min(A_0), \ldots, \min(A_k)\}$ and all $s \in \prod_{i < \bar{n}} \vec{H}_i^{k+1}$, either

$$[s, \vec{H}^{k+1}] \times [b, A_{k+1}] \subseteq \mathcal{X}_k \text{ or } [s, \vec{H}^{k+1}] \times [b, A_{k+1}] \subseteq \mathcal{X}_k^c. \tag{9.76}$$

We assume that $\min(A_{k+1}) > \min(A_k), \max(a)$. Having constructed the fusion sequence we take its limits $(N_\infty, M_\infty), \vec{H}^\infty$ and

$$A_\infty = a \cup \{\min(A_k) : k \in \mathbb{N}\}.$$

It follows that

$$\mathcal{X} \cap \left(\prod_{i=0}^{\infty} H_i^\infty \times [a, A_\infty] \right) \tag{9.77}$$

is a relatively open subset of $(\prod_i H_i^\infty) \times [a, A_\infty]$, so we can finish using Lemma 9.18. □

Corollary 9.21 *All Borel subsets of $\mathbb{N}^\infty \times \mathbb{N}^{[\infty]}$ are para-Ramsey.*

Corollary 9.22 (Galvin-Prikry) *All Borel subsets of $\mathbb{N}^{[\infty]}$ are Ramsey.*

We are now ready to state and prove the main result of this section.

Theorem 9.23 *The field of para-Ramsey subsets of $\mathbb{N}^\infty \times \mathbb{N}^{[\infty]}$ is closed under the Souslin operation.*

Proof. Let $\mathcal{X}_v$ $(v \in \mathbb{N}^{[<\infty]})$ be a given Souslin scheme of para-Ramsey subsets of $\mathbb{N}^\infty \times \mathbb{N}^{[\infty]}$. We assume that $\mathcal{X}_v \subseteq \mathcal{X}_u$ whenever v end-extends u. Let[2]

$$\mathcal{X} = \bigcup_{V \in \mathbb{N}^{[\infty]}} \bigcap_{n \in \mathbb{N}} \mathcal{X}_{r_n(V)}. \tag{9.78}$$

Similarly as before, we consider the relativized versions

$$\mathcal{X}_v^* = \bigcup_{V \in [v, \mathbb{N}]} \bigcap_{n \geq |v|} \mathcal{X}_{r_n(V)} \tag{9.79}$$

for $v \in \mathbb{N}^{[<\infty]}$.

Let $(N, M), \vec{H}, n$ and $[a, A]$ be given inputs as in Definition 9.17 for checking that the set $\mathcal{X}$ is para-Ramsey. Starting from the initial values

$$N_0 = N, M_0 = M, n_0 = n, \vec{H}^0 = \vec{H}, \text{ and } A_0 = A,$$

we build a fusion sequence

$$(N_k, M_k, \vec{H}^k, n_k, A_k) \tag{9.80}$$

as follows. Suppose that $N_k, M_k, \vec{H}^k, n_k$, and A_k have been determined. Let

$$p = 2^{n_k+1} \left| \bigcup_{j \in M_k \cap n_k} \prod_{i<j} H_i^k \right|. \tag{9.81}$$

Since $R_{N_k}((m_i))$ eventually dominates $R^{(p)}((m_i))$, we find a minimal integer $\bar{n}_k \geq n_k$ such that

$$R_{N_k}(m_0, \ldots, m_i) \geq R^{(p)}(m_0, \ldots, m_i) \text{ for all } i \geq \bar{n}_k. \tag{9.82}$$

---

[2]Recall that $r_n(V)$ is the set formed by taking the first n members of V.

Let (s_ℓ, v_ℓ, b_ℓ) $(\ell \le m)$ be an enumeration of

$$(\textstyle\prod_{i < \bar{n}_k} H_i^k) \times \mathcal{P}(n_k) \times \mathcal{P}(a \cup \{\min(A_0), \dots, \min(A_k)\}). \qquad (9.83)$$

Suppose there exist $(Q, P) \le (N_k, M_k)$, an $S_P((m_i))$-sequence $\vec{J} \le \vec{H}^k$ and infinite $B \subseteq A_k$ such that

$$[s_0, \vec{J}] \times [b_0, B] \cap \mathcal{X}_{v_0}^* = \emptyset. \qquad (9.84)$$

Since $N_k \subseteq Q \subseteq P$ we have that

$$|J_i| = R_P(m_0, \dots, m_i) \ge R_{N_k}(m_0, \dots, m_i) \ge R^{(p)}(m_0, \dots, m_i) \text{ for } i \ge \bar{n}_k. \qquad (9.85)$$

So, we can find $(Q_0, P_0) \le_{n_k} (N_k, M_k)$ and an $R_{P_0}((m_i))$-sequence $\vec{J}^0$ such that

(1) $J_i^0 \subseteq J_i$ for $i \ge \bar{n}_k$

(2) $J_i^0 \subseteq H_i^k$ for $n_k \le i < \bar{n}_k$

(3) $J_i^0 = H_i^k$ for $i < n_k$.

Note that $\vec{J}^0 \le_{n_k} \vec{H}^k$ and $[s_0, \vec{J}^0] \subseteq [s_0, \vec{J}]$, so we have that

$$[s_0 \ \vec{J}^0] \times [b_0, B] \cap \mathcal{X}_{v_0}^* = \emptyset. \qquad (9.86)$$

Let $B_0 = B$.
If such $(Q, P) \le (N_k, M_k)$, $J \le \vec{H}^k$ and $B \subseteq A_n$ cannot be found, set

$$\vec{J}^0 = \vec{H}^k, B_0 = A_k, (Q_0, P_0) = (N_k, M_k). \qquad (9.87)$$

Now starting from $\vec{J}^0$, (Q_0, P_0), and B_0 in place of $\vec{H}^k$, (N_k, P_k), and A_k, we find a regular pair $(Q_1, P_1) \le_{n_k} (Q_0, P_0)$, an $R_{P_1}((m_i))$-sequence $\vec{J}^0$, and an infinite $B_1 \subseteq B$ such that either

$$[s_1, \vec{J}^1] \times [b_1, B_1] \cap \mathcal{X}_{v_1}^* = \emptyset \qquad (9.88)$$

or else it is impossible to find a $(Q, P) \le (Q_0, P_0)$, an $R_p((m_i))$-sequence $J \le \vec{J}^0$, and an infinite set $B \subseteq B_0$ such that

$$[s_1, \vec{J}] \times [b_1, B] \cap \mathcal{X}_{v_1}^* = \emptyset, \qquad (9.89)$$

and so on. Proceeding this way, we successively treat each triple (s_ℓ, v_ℓ, b_ℓ) $(\ell \le m)$ and arrive at $(Q_m, P_m) \le_{n_k} (N_k, M_k)$, an $R_{P_m}((m_i))$-sequence $\vec{J}^m \le_{n_k} \vec{H}^k$ and infinite $B_m \subseteq A$. Let

$$(N_{k+1}, M_{k+1}) = (Q_m, P_m),$$
$$\vec{H}^{k+1} = \vec{J}^m,$$
$$A_{k+1} = B_m / \max\{\min A_k, \max(a)\}.$$

Choose $n_{k+1} > \bar{n}_k$ such that $M_{k+1} \setminus N_{k+1}$ has a point in the interval $[n_k, n_{k+1})$. This ends the description of the inductive step in the fusion construction.

Note for future reference that a triple N_{k+1}, $(M_{k+1}, \vec{H}^{k+1})$, and A_{k+1} constructed in this way has the following property:

$(4)_k$ For every $s \in \prod_{i < n_k} H_i^{k+1}$, every $v \subseteq \{0, 1, \dots, n_k\}$ and every

$$b \subseteq a \cup \{\min(A_0), \dots, \min(A_k)\},$$

either

$$([s, H^{k+1}] \times [b, A_{k+1}]) \cap X_v^* = \emptyset,$$

or else it is impossible to find a regular pair $(Q, P) \le (N_{k+1}, M_{k+1})$, an $R_P((m_i))$-sequence $\vec{J} \le \vec{H}^{k+1}$, and an infinite set $B \subseteq A_{k+1}$ such that

$$([s, \vec{J}] \times [b, B]) \cap X_v^* = \emptyset.$$

Let $(N_\infty, M_\infty), \vec{H}^\infty$ and $A_\infty = a \cup \{\min(A_k) : k \in \mathbb{N}\}$ be the limits of the fusion sequence $(N_k, M_k, n_k, H^k, A_k)$.

For $v \in N^{[<\infty]}$, we form the set

$$\Psi(X_v^*)) = \bigcup \left\{ [s, \vec{H}^\infty] \times [b, A_\infty] : [s, \vec{H}^\infty] \times [b, A_\infty] \cap X_V^* = \emptyset \right\}. \quad (9.90)$$

Then put

$$\Phi(X_v^*) = (X_v \cap \prod_{r=0}^\infty H_i^\infty \times [a, A_\infty]) \setminus \Psi(X_v^*) \quad (9.91)$$

Then $\Phi(X_v^*)$ is a para-Ramsey set for all $v \in N^{[<\infty]}$. Therefore, for each $v \in N^{[<\infty]}$, the difference

$$\mathcal{M}_v = \Phi(X_v^*) \setminus \bigcup_{\ell > \max(v)} \Phi(X_{v \cup \{\ell\}}^*) \quad (9.92)$$

is also para-Ramsey. Note also that

$$\mathcal{M}_v \subseteq \Phi(X_v^*) \setminus X_v^* \text{ for all } v \in N^{[<\infty]}. \quad (9.93)$$

Now we go on to define the second fusion sequence

$$(Q_k, P_k, p_k, \vec{J}^k, B_k), \quad (9.94)$$

starting from the initial values

$$Q_0 = N_\infty, P_0 = M_\infty, p_0 = n, \vec{J}^0 = \vec{H}^\infty, \text{ and } B_0 = A_\infty$$

and making sure that (p_k) is a subsequence of (n_k). Applying the fact that the sets $\mathcal{M}_v$ are para-Ramsey, it is clear that we can choose the second fusion to have the following property.

$(5)_k$ For some $\bar{p}_k \in [p_k, p_{k+1})$ it holds that whenever $s \in \prod_{i < \bar{p}_k} J_i^{k+1}$,

$$b \subseteq a \cup \{\min(B_0), \ldots, \min(B_k)\}$$

and $v \subseteq \{0, 1, \ldots, p_k\}$ are given, then either

$$[s, \vec{J}^{k+1}] \times [b, B_{k+1}] \subseteq \mathcal{M}_v \text{ or } [s, \vec{J}^{k+1}] \times [b, B_{k+1}] \cap \mathcal{M}_v = \emptyset. \quad (9.95)$$

Moreover, besides the assumption that the sequence $(\bar{p}_k)$ is a subsequence of the sequence $(\bar{n}_k)$, we require that $\bar{p}_k = \bar{n}_\ell$ implies $n_\ell \geq p_k$. We claim that in $(5)_k$ we always have the second alternative. Suppose not and fix an integer k, a sequence $s \in \prod_{i < \bar{p}_k} J_i^{k+1}$ and a set

$$b \subseteq \cup \{\min(B_0), \ldots, \min(B_k)\} \quad (9.96)$$

end-extending a such that

$$[s, \vec{J}^{k+1}] \times [b, B_{k+1}] \subseteq \mathcal{M}_v. \quad (9.97)$$

Let ℓ be such that $\bar{p}_k = \bar{n}_\ell$. Then $n_\ell \geq p_k$, so we have that

$$s \in \prod_{i < n_\ell} H_i^{\ell+1}, \quad v \subseteq \{0, \ldots, n_\ell\}, \quad (9.98)$$

$$b \subseteq a \cup \{\min(A_0), \ldots, \min(A_\ell)\}. \quad (9.99)$$

Note that $(Q_{k+1}, P_{k+1}) \leq (N_{\ell+1}, M_{\ell+1}), \vec{J}^{k+1} \leq \vec{H}^{\ell+1}$ and $B_{k+1} \subseteq A_{\ell+1}$ show that the second alternative of $(4)_\ell$ does not hold, so we have the first, i.e.,

$$[s, \vec{H}^{\ell+1}] \times [b, A_{\ell+1}] \cap \mathcal{X}_v^* = \emptyset. \quad (9.100)$$

Then by our definition,

$$[s, \vec{H}^\infty] \times [b, A_\infty] \subseteq \Psi(\mathcal{X}_v^*). \quad (9.101)$$

So $\mathcal{M}_v$ intersects $\Psi(X_v^*)$, contradicting that the fact $\mathcal{M}_v$ is a subset of $\Phi(\mathcal{X}_v^*)$ that is disjoint from $\Psi(\mathcal{X}_v^*)$.

Having performed the fusion procedure, we arrive at a fusion sequence $(Q_k, P_k, p_k, \vec{J}^k, B_k)$ satisfying $(5)_k$ for all k. Let

$$(Q_\infty, P_\infty), \vec{J}^\infty \text{ and } B_\infty = a \cup \{\min(B_k) : k \in \mathbb{N}\}$$

be the limits of this fusion sequence. Then from the fact just established, which says that in $(5)_k$ the second alternative always holds, we conclude that

$$((\prod_{i=0}^\infty J_i^\infty) \times [a, B_\infty]) \cap \mathcal{M}_v = \emptyset \text{ for all } v \in \mathbb{N}^{[\infty]}. \quad (9.102)$$

As in the previous proofs of this nature, one easily checks that (9.102) has the following consequence:

$$\mathcal{X}_\emptyset^* \cap \left(\prod_{i=0}^\infty J_i^\infty \times [a, B_\infty]\right) = \Phi(\mathcal{X}_\emptyset^*) \cap \left(\prod_{i=0}^\infty J_i \times [a, B_\infty]\right). \qquad (9.103)$$

Since $\Phi(\mathcal{X}_\emptyset^*)$ is para-Ramsey, we can find $k > n$, a $(Q, P) \leq_n (Q_\infty, P_\infty)$, an $R_P((m_i))$-sequence $\vec{J} \leq_n \vec{J}^\infty$, and an infinite $B \subseteq B_\infty$ such that for all $s \in \prod_{i<k} J_i$, either

$$[s, \vec{J}] \times [a, B] \subseteq \Phi\left(\mathcal{X}_\emptyset^*\right) \text{ or}$$
$$([s, \vec{J}] \times [a, B]) \cap \Phi\left(\mathcal{X}_\emptyset^*\right) = \emptyset.$$

Since $\mathcal{X} = \mathcal{X}_\emptyset^*$ this, together with Equation ((9.103)), gives us the conclusion of the theorem. $\square$

Corollary 9.24 *Every analytic subset of* $\mathbb{N}^\infty \times \mathbb{N}^{[\infty]}$ *is para-Ramsey.*

Corollary 9.25 (Silver) *Every analytic subset of* $\mathbb{N}^{[\infty]}$ *is Ramsey.*

9.4 INFINITE-DIMENSIONAL RAMSEY THEOREM PARAMETRIZED BY INFINITE PRODUCTS OF FINITE SETS

In this section we state and prove what appears to be the optimal parametrized form of the infinite-dimensional Ramsey theorem.

Theorem 9.26 (Parametrized Infinite-Dimensional Ramsey Theorem) *Let* (m_i) *be an infinite nondecreasing sequence of positive integers and define* $n_i = R_\mathbb{N}(m_0, \ldots, m_i)$. *Then for every Souslin-measurable coloring*

$$c : \left(\prod_{i=0}^\infty n_i\right) \times \mathbb{N}^{[\infty]} \to 2, \qquad (9.104)$$

there exist $H_i \subseteq n_i$, *with* $|H_i| = m_i$ *for all* i, *and an infinite set* $H \subseteq \mathbb{N}$ *such that* c *in constant on the product* $\left(\prod_{i=0}^\infty H_i\right) \times H^{[\infty]}$.

Proof. Applying Theorem 9.23, we find an infinite set $M \subseteq \mathbb{N}$, an $R_M((m_i))$-sequence $J_i \subseteq n_i (i \in \mathbb{N})$, an integer k, and an infinite set $H \subseteq \mathbb{N}$ such that for all $s \in \prod_{i<k} J_i$, the coloring c is constant on $[s, \vec{J}] \times H^{[\infty]}$. Since $R_M((m_i))$ dominates $S_M((m_i))$, which in turn dominates $S((m_i))$, by Lemma 8.5, there exist $H_i \subseteq J_i, |H_i| = m_i$ for all i and $\varepsilon < 2$ such that for all $s \in \prod_{i<k} H_i$, the constant value of c on $[s, \vec{J}] \times H^{[\infty]}$ is equal to ε. It follows that c is constant on the product $\left(\prod_{i=0}^\infty H_i\right) \times H^{[\infty]}$. $\square$

Corollary 9.27 *For every infinite sequence* (m_i) *of positive integers and every integer* ℓ, *there is an infinite sequence* (n_i) *of positive integers such that for every Souslin-measurable coloring*

$$c : \left(\prod_{i=0}^\infty n_i\right) \times \mathbb{N}^{[\infty]} \to \{0, 1, \ldots, \ell - 1\}, \qquad (9.105)$$

there exist $H_i \subseteq n_i$ with $|H_i| = m_i$ for all i, and an infinite set $H \subseteq \mathbb{N}$ such that the product

$$(\textstyle\prod_{i=0}^{\infty} H_i) \times H^{[\infty]} \tag{9.106}$$

is monochromatic.

Proof. By the previous theorem, the conclusion is true for $\ell = 2$ and the multicoloring version follows by successive application of this version.   □

However, working as in the proof of Theorem 8.30 from the previous chapter, we have the following general fact which shows that in this context the number of colors can always be increased by going to an appropriate subsequence of (n_i).

Lemma 9.28 *Let ℓ be an integer ≥ 2. Suppose that (m_i) is an infinite sequence of positive integers and that (n_i) is an infinite increasing sequence of integers with the property that for every Souslin-measurable ℓ-coloring of the product*

$$(\textstyle\prod_{i=0}^{\infty} n_i) \times \mathbb{N}^{[\infty]} \tag{9.107}$$

there exist $H_i \subseteq n_i$ with $|H_i| = m_i$ for all i, and an infinite set $H \subseteq \mathbb{N}$ such that the product $(\prod_{i=0}^{\infty} H_i) \times H^{[\infty]}$ is monochromatic. Then for every Souslin-measurable $(\ell + 1)$-coloring of the product

$$(\textstyle\prod_{i=0}^{\infty} n_{2i+1}) \times \mathbb{N}^{[\infty]} \tag{9.108}$$

there exist $H_i \subseteq n_i$ with $|H_i| = m_i$ for all i, and an infinite set $H \subseteq \mathbb{N}$ such that the product $(\prod_{i=0}^{\infty} H_i) \times H^{[\infty]}$ is monochromatic

Proof. Since the proof is quite analogous to that of Theorem 8.30, we only give a sketch. Let

$$c : (\textstyle\prod_{i=0}^{\infty} n_{2i+1}) \times \mathbb{N}^{[\infty]} \to \{0, 1, ..., \ell\} \tag{9.109}$$

be a given Souslin-measurable coloring. Define

$$d : (\textstyle\prod_{i=0}^{\infty} n_i) \times \mathbb{N}^{[\infty]} \to \mathbb{Z} \tag{9.110}$$

by $d((x_i), X) = c((x_{2i+1}), X) - c((x_{2i}), X)$. Define now

$$\bar{c} : (\textstyle\prod_{i=0}^{\infty} n_i) \times \mathbb{N}^{[\infty]} \to \{0, ..., \ell - 1\} \tag{9.111}$$

by letting

$$
\begin{aligned}
\bar{c}((x_i), X) \quad &= 0 & &\text{if} \quad d((x_i), X) \in \{0, -2\} \\
&= 1 & &\text{if} \quad d((x_i), X) \in \{-1, 1, 2\} \\
&= 2 & &\text{if} \quad d((x_i), X) \in \{-3, 3\} \\
&\ \ \vdots \\
&= \ell - 1 & &\text{if} \quad d((x_i), X) \in \{-\ell, \ell\}.
\end{aligned}
\tag{9.112}
$$

Since this coloring uses no more than ℓ colors, there exist an infinite subproduct $\prod_{i=0}^{\infty} H_i$ of $\prod_{i=0}^{\infty} n_i$ such that $|H_i| = m_i$ for all i and an infinite set $H \subseteq \mathbb{N}$ such that $\bar{c}$ is constant on the product

$$(\textstyle\prod_{i=0}^{\infty} H_i) \times H^{[\infty]}. \tag{9.113}$$

Then working as in the proof of Theorem 8.30 one shows that the original coloring c is constant either on $(\prod_{i=0}^{\infty} H_{2i+1}) \times H^{[\infty]}$ or on $(\prod_{i=0}^{\infty} H_{2i}) \times H^{[\infty]}$. Details are left to the interested reader. $\qquad \square$

Note that in the case of polarized partition calculus of Chapter 8, taking for example, m_i to be the constant sequence $m_i = 2$, we know that the sequence $n_i = A_4(3 + i)$ satisfies the partition relation

$$\begin{pmatrix} n_0 \\ n_1 \\ n_2 \\ \vdots \end{pmatrix} \longrightarrow \begin{pmatrix} 2 \\ 2 \\ 2 \\ \vdots \end{pmatrix} \tag{9.114}$$

in the realm of Souslin-measurable colorings. In the case of parametrized partition properties, nothing of this sort is known. For example, the following question is wide open.

Question 9.29 *Is there a primitive recursive sequence (n_i) such that for every Souslin-measurable coloring*

$$c : (\textstyle\prod_{i=0}^{\infty} n_i) \times \mathbb{N}^{[\infty]} \to 2 \tag{9.115}$$

there exist $H_i \subseteq n_i, |H_i| = 2$ for all i, and infinite $H \subseteq \mathbb{N}$ such that the product

$$(\textstyle\prod_{i=0}^{\infty} H_i) \times H^{[\infty]} \tag{9.116}$$

is monochromatic?

There are indications that the infinite-dimensional Ramsey theorem parametrized by products of finite sets is considerably stronger than the corresponding partition theorem for the products of finite sets alone. Consider, for example, the following simple observation.

Lemma 9.30 *For every analytic subset $\mathcal{A}$ of some product $\prod_{i=0}^{\infty} H_i$ of finite subsets of $\mathbb{N}$ there is a G_δ-subset $\mathcal{G}$ of the product*

$$(\textstyle\prod_{i=0}^{\infty} H_i) \times \mathbb{N}^{[\infty]} \tag{9.117}$$

such that if for some subproduct $\prod_{i=0}^{\infty} J_i \subseteq \prod_{i=0}^{\infty} H_i$ and infinite $H \subseteq \mathbb{N}$ we have that $(\prod_{i=0}^{\infty} J_i) \times H^{[\infty]}$ is either included in or disjoint from $\mathcal{G}$, then the product $\prod_{i=0}^{\infty} J_i$ is either included or is disjoint from $\mathcal{A}$.

Proof. Take a continuous map $f : \mathbb{N}^{[\infty]} \to \prod_{i=0}^{\infty} H_i$ such that $\mathcal{A} = \mathrm{rang}(f)$. Let[3]

$$\mathcal{G} = \Big\{ (x, X) : \ (\exists Y \in \mathbb{N}^{[\infty]}) \, [f(Y) = x \ \& \ (\forall n \in \mathbb{N}) \, (Y(n) \le X(n))] \Big\}$$

Then $\mathcal{G}$ is a G_δ-subset of the product $(\prod_{i=0}^{\infty} H_i) \times \mathbb{N}^{[\infty]}$. Consider a subproduct $\prod_{i=0}^{\infty} J_i \subseteq \prod_{i=0}^{\infty} H_i$ and infinite $H \subseteq \mathbb{N}$ such that $(\prod_{i=0}^{\infty} J_i) \times H^{[\infty]}$ is

---

[3] Here $Y(n)$ and $X(n)$ denote the nth members of Y and X, respectively, according to the increasing enumeration of these subsets of $\mathbb{N}$.

either included or is disjoint from $\mathcal{G}$. First of all, note that if $(\prod_{i=0}^{\infty} J_i) \times H^{[\infty]}$ is included in $\mathcal{G}$ then clearly $\prod_{i=0}^{\infty} J_i \subseteq \mathcal{A}$. Suppose now that

$$((\textstyle\prod_{i=0}^{\infty} J_i) \times H^{[\infty]}) \cap \mathcal{G} = \emptyset. \tag{9.118}$$

We claim that in this case $(\prod_{i=0}^{\infty} J_i) \cap \mathcal{A} = \emptyset$. To see this, consider an $x \in \prod_{i=0}^{\infty} J_i$ and suppose that $x \in \mathcal{A}$. Let $Y \in \mathbb{N}^{[\infty]}$ be such that $f(Y) = x$. Find an $X \in H^{[\infty]}$ such that $Y(n) \leq X(n)$ for all $n \in \mathbb{N}$. Then by definition of $\mathcal{G}$, we conclude that $(x, X) \in \mathcal{G}$, and therefore, (x, X) belongs to the interesction $((\prod_{i=0}^{\infty} J_i) \times H^{[\infty]}) \cap \mathcal{G}$, which is a contradiction. This finishes the proof.   □

It follows that for a given infinite sequence (m_i) of positive integers, the existence of infinite sequence (n_i) of positive integers such that the polarized partition relation

$$\begin{pmatrix} n_0 \\ n_1 \\ n_2 \\ \vdots \end{pmatrix} \longrightarrow \begin{pmatrix} m_0 \\ m_1 \\ m_2 \\ \vdots \end{pmatrix} \tag{9.119}$$

holds in the realm of Souslin-measurable colorings follows from the G_σ-case of the Parametrized Infinite-Dimensional Ramsey Theorem, a fact essentially of the same level of difficulty as the result that open sets belong to the field of para-Ramsey sets (proved above in Section 9.2). This way of proving polarized partition relations, however, gives no bounds on (n_i) in terms of (m_i). This was one of the reasons for asking Question 9.29 above. It is also worth restating that question in the following finitary form.

Question 9.31 *Is there a primitive recursive sequence (n_i) of positive integers such that for every coloring*

$$c : \bigcup_k \prod_{i<k} n_i \to 2,$$

there exist $H_i \subseteq n_i$ such that $|H_i| = 2$ for all i and such that the set

$$\left\{ k : c \upharpoonright \prod_{i<k} H_i \text{ is constant} \right\}$$

is infinite?

NOTES TO CHAPTER NINE

The parametrized Ramsey theory of this chapter was developed in the papers of DiPrisco-Llopis-Todorcevic [22] and DiPrisco-Todorcevic [23] although its final form presented above appears in the paper of Todorcevic [108] containing an essential improvement which allowed us to prove the Parametrized

Infinite-Dimensional Ramsey Theorem (Theorem 9.26) without the use of any additional set-theoretic assumptions. The previous work needed the additional set-theoretic assumption that subsets of $\mathbb{N}^{[\infty]}$ representable as continuous images of coanalytic subsets of $\mathbb{N}^{\mathbb{N}}$ have the Ramsey property. The finite-dimensional Ramsey theorem has been parametrized by products of finite sets in an earlier paper by Henle [46]. It is natural to expect that similar methods will extend the Parametrized Infinite-Dimensional Ramsey Theorem into the context where monochromatic products of finite sets of prescribed cardinalities are replaced by sets of prescribed masses relative to a fixed submeasure on the power-set of $\mathbb{N}$. These extensions will be guided by applications of the Parametrized Infinite-Dimensional Ramsey Theorem which at this point are mostly in the analysis of the set-theoretic forcing (see for example [23] and [115]) but this is likely to change with time.

Appendix

SET THEORETIC NOTATION

The cardinality of a set S is denoted by $|S|$. If X is a subset of X, we shall sometimes use the notation X^c for its complement in S, the set $S \setminus X$.

When $\sigma = (\sigma_n)$ is a (typically finite) sequence of objects, we use the notation $|\sigma|$ to denote the *length* of the sequence which is indeed its cardinality provided we identify σ with the corresponding set of ordered pairs of the form $\langle n, \sigma_n \rangle$.

$\mathbb{N} = \{0, 1, 2, \ldots\}$ is the set of non negative integers.
$\mathbb{N}_+ = \{1, 2, \ldots\}$ is the set of positive integers.
Each integer $n \in \mathbb{N}$ is identified with the set $\{0, 1, \ldots, n-1\}$.
$\mathbb{R}$ is the set of real numbers.
$\mathbb{Q}$ is the set of rational numbers.

The usual notion of the cartesian power M^k for $k \in \mathbb{N}$ or $k = \mathbb{N}$ is frequently restricted to some "symmetric" subset $M^{[k]}$ of M^k, and this restriction depends on the structure present on M and the Ramsey theoretic results one wishes to obtain. For example, when M is simply a set with no structure, then its symmetric power is

$$M^{[k]} = \{S \subseteq M : |S| = k\}.$$

We shall use $M^{[\infty]}$ instead of $M^{[\mathbb{N}]}$, and similarly, sometimes we use M^∞ instead of $M^\mathbb{N}$. Whenever we define the symmetric power $M^{[k]}$, we have the variations

$$M^{[\leq l]} = \bigcup_{k \leq l} M^{[k]}, \quad M^{[<l]} = \bigcup_{k < l} M^{[k]}, \text{ and } M^{[<\infty]} = \bigcup_{k \in \mathbb{N}} M^{[k]}. \tag{9.120}$$

Of course, we use the similar notation $M^{\leq l}$, $M^{<l}$, and $M^{<\infty}$ in the case of full Cartesian powers. We shall consider $M^{<\infty}$ also a tree with the ordering $\sqsubseteq$ of end-extension. In this case we shall typically work with the complete binary tree $2^{<\infty}$ or the tree $\mathbb{N}^{<\infty}$ of finite sequences of integers. When we refer to the set $\mathbb{N}^{[<\infty]}$ as a tree, we identify it with a subtree of $\mathbb{N}^{<\infty}$ by identifying a finite set with its increasing enumeration. The Cantor set $2^\mathbb{N}$ and the Baire space $\mathbb{N}^\mathbb{N}$ have their metrics given by $2^{-\Delta(x,y)}$, where for distinct x and y,

$$\Delta(x, y) = \min\{n : x(n) \neq y(n)\}. \tag{9.121}$$

The function Δ is also giving us a lexicographical ordering

$$x <_{\text{lex}} y \text{ iff } x(\Delta(x,y)) < y(\Delta(x,y)). \tag{9.122}$$

Note that this formula (9.121) makes sense even for finite sequences x and y provided that they are incomparable in the ordering $\sqsubseteq$ of end-extension. It follows that the formula (9.122) makes sense also when x and y are two finite sequences incomparable in the ordering $\sqsubseteq$. If we supplement this definition by the requirement that

$$x <_{\text{lex}} y \quad \text{whenever} \quad x \sqsubset y,$$

we get a lexicographical ordering on $2^{<\infty}$ and $\mathbb{N}^{<\infty}$, still denoted the same way, $<_{\text{lex}}$. This of course is not the only way one can supplement the definition (9.122). For example, we can supplement the formula (9.122), which applies when x and y are incomparable, with the following definition

$$x <^0_{\text{lex}} y \quad \text{whenever} \quad y \sqsubset x$$

when x and y are comparable. As indicated this gives us another version of the lexicographical ordering on $2^{<\infty}$ and $\mathbb{N}^{<\infty}$ denoted by $<^0_{\text{lex}}$. Note that two linear orderings $<_{\text{lex}}$ and $<^0_{\text{lex}}$ coincide on antichains of these two trees.

By Ord we denote the class of ordinals and by $<$ the well-ordering of this class. We shall identify an ordinal β with the set of its predecessors, i.e, $\beta = \{\alpha : \alpha < \beta\}$. A structure of the form (X, R), where R is a binary relation on X,[4] is *well-founded* if there is no infinite sequence (x_n) of pairwise distinct elements of X such that $(x_{n+1}, x_n) \in R$ for all n. Thus is equivalent to saying that every subset Y of X has an R-minimal element, an element y of Y such that $(x, y) \notin R$ for all $x \in Y$, $x \neq y$. Given a well-founded binary structure (X, R), one can define its *rank function*

$$\rho_R : X \to \text{Ord},$$

recursively as follows:

$$\rho_R(y) = \sup(\{\rho_R(x) + 1 : x \in X \setminus \{y\} \text{ and } (x, y) \in R\}),$$

where we use the convention $\sup(\emptyset) = 0$. Thus in particular $\rho_R(y) = 0$, whenever y is a minimal element of X, i.e, an element of X with the property that there is no $x \in X \setminus \{y\}$ such that $(x, y) \in R$. Let

$$\text{rk}(X, R) = \sup\{\rho_R(x) + 1 : x \in X\},$$

the *rank* of the well-founded structure (X, R).

TOPOLOGICAL NOTIONS

Here we follow standard terminology and notation (see, for example, [58]). Recall that a topological space is a pair $(X, \mathcal{T})$ where X is some set and $\mathcal{T}$ a collection of subsets of X containing the set X and closed under taking

---

[4]A binary relation on X is any subset R of X^2.

finite intersections and arbitrary unions. The sets from $\mathcal{T}$ are called *open* and their complements in X are called *closed*. Given a subset A of X, we let $\overline{A}$ denote the *closure* of A in $(X, \mathcal{T})$, the minimal closed superset of X.

Given a topological space $(X, \mathcal{T})$, we say that two disjoint subsets A and B of X can be *separated* if there exist open sets $U \supseteq A$ and $V \supseteq B$ such that $U \cap V = \emptyset$. This leads us to some standard separation axioms that one can put on a given space $(X, \mathcal{T})$. For example, if every pair of distinct points can be separated, the space is T_2, or *Hausdorff*. If every point can be separated from a closed subset of X that does not contain the point, then the space is *regular*. This separation axiom is of course stronger than T_2 when points in X are assumed to be closed. A *basis* for a topological space $(X, \mathcal{T})$ is a collection $\mathcal{B} \subseteq \mathcal{T}$ with the property that $\mathcal{T} = \{\bigcup \mathcal{X} : \mathcal{X} \subseteq \mathcal{B}\}$. A *subbasis* is a collection $\mathcal{B} \subseteq \mathcal{T}$ with the property that $\mathcal{B}^* = \{\bigcap \mathcal{X} : \mathcal{X} \subseteq \mathcal{B} \text{ and } \mathcal{X} \text{ finite}\}$ forms a basis of $(X, \mathcal{T})$. A *neighborhood basis* of a point x in X is a collection $\mathcal{B}_x$ of open subsets of X containing the point x such that every other open set containing x includes a member of $\mathcal{B}_x$. A topological space $(X, \mathcal{T})$ is *separable* if it has a countable dense subset, i.e., a countable set $D \subseteq X$ such that $\overline{D} = X$.

A *metric space* is a pair (X, ρ), where X is a set and ρ is a function mapping ordered pairs of elements of X into nonnegative reals such that

(1) $\rho(x, y) = 0$ if and only if $x = y$,

(2) $\rho(x, y) = \rho(y, x)$,

(3) $\rho(x, z) \leq \rho(x, y) + \rho(y, z)$.

The topology of a metric space is generated by open balls of (X, d), the sets of the form $B_\delta(x) = \{y \in X : \rho(x, y) < \delta\}$, where $x \in X$ and $\delta > 0$. If for a topological space $(X, \mathcal{T})$ we can find a metric ρ on X such that the open ρ-balls form a basis for the topology $\mathcal{T}$, then we call $(X, \mathcal{T})$ a *metrizable space*. The following two well-known results that give sufficient conditions for the existence of such a metric are quite useful (see, for example, [58] and [54]).

Theorem 9.32 (Urysohn) *Every regular Hausdorff space with a countable basis is metrizable.*

Theorem 9.33 (Birkhoff-Kakutani) *Every Hausdorff topological group with a countable neighborhood basis at the identity is metrizable.*

A sequence (x_n) of elements of some metric space (X, ρ) is a *Cauchy sequence* if for every $\varepsilon > 0$ there is m such that $\rho(x_k, x_l) < \varepsilon$ for all $k, l \geq m$. Recall that a metric space (X, ρ) is said to be *complete* if every *Cauchy sequence* has a limit. A topological space X is *Polish* if it is separable and metrizable by a complete metric.

Recall the separation axiom of Tychonov that guarantees that a given space $(X, \mathcal{T})$ can be embedded as a subspace of a *Tychonov cube* $[0, 1]^I$: For every $x \in X$ and every closed subset F of X such that $x \notin F$, there is a continuous function $f : X \to [0, 1]$ such that $f(x) = 0$ and $f(y) = 1$ for all $y \in F$. A space that satisfies this axiom is called a *Tychonov space*. It follows that every Tychonov space X has a compactification γX, a compact space containing X as a dense subspace. In particular, every Tychonov space X has a maximal such compactification called *Čech-Stone compactification* βX charaterized by the fact that every continuous mapping $f : X \to [0, 1]$ extends to a continuous $\beta f : \beta X \to [0, 1]$. Particularly interesting is the Čech-Stone compactification $\beta \mathbb{N}$ of $\mathbb{N}$. One useful representation of $\beta \mathbb{N}$ is obtained by taking the closure of the sequence of projections $\pi_n : 2^{\mathbb{N}} \to \{0, 1\}$ inside the Tychonov cube $\{0, 1\}^{2^{\mathbb{N}}}$,

$$\beta \mathbb{N} = \overline{\{\pi_n : n \in \mathbb{N}\}}.$$

From this representation, one can readily get another representation of $\beta \mathbb{N}$ as the set, denoted again by $\beta \mathbb{N}$, of all ultrafilters on $\mathbb{N}$ with the topology generated by the basic open sets of the form

$$\overline{A} = \{\mathcal{U} \in \beta \mathbb{N} : A \in \mathcal{U}\},$$

where A is an arbitrary subset of $\mathbb{N}$. The fact that the space of all ultrafilters on $\mathbb{N}$ is one representation of the Čech-Stone compactification of $\mathbb{N}$ explains the meaning behind the notation

$$y = \lim_{n \to \mathcal{U}} x_n$$

for a sequence (x_n) of points of some compact Hausdorff space K and $y \in K$. Namely, if we let $f : \mathbb{N} \to K$ denote the map $n \mapsto x_n$, then $y = \beta f(\mathcal{U})$.

There are several "hyperspaces" that one can associate with a given topological space $(X, \mathcal{T})$. The first one is the space $\mathcal{C}(X)$ of all continuous real-valued functions defined on X with the topology of *pointwise convergence*, or in other words the topology $\mathcal{C}(X)$ inherits when considered as a subspace of $\mathbb{R}^X$ equipped with the Tychonov product topology, i.e., the topology generated by the subbasis $U \times \mathbb{R}^{X \setminus \{x\}}$ $(x \in X)$. Another space one can associate with $(X, \mathcal{T})$ is the *exponential space* $\exp(X)$, the collection of all closed subsets of X with topology generated by basic sets of the form

$$\left\{ F \in \exp(X) : F \subseteq \bigcup_{k=0}^{n} U_k \text{ and } F \cap U_k \neq \emptyset \text{ for all } k = 0, \ldots, n \right\},$$

where $U_0, \ldots, U_n$ is a finite sequence of open subsets of X. It is known that $\exp(X)$ is compact when X is compact, but $\exp(X)$ may not be metrizable when X is metrizable (see [58]). For example, the exponential space $\exp(\mathbb{N})$ is not metrizable although $\mathbb{N}$ with its discrete topology is metrizable. In fact it has been known since the 1950's (see, for example, [50]) that even the closed subspace $\mathbb{N}^{[\infty]}$ of $\exp(\mathbb{N})$ contains two disjoint closed subsets that cannot be separated by two disjoint open sets.[5]

---

[5] The reader will have no difficulty verifying that indeed $\mathbb{N}^{[\infty]}$ with the subspace topology is just another way to introduce the Ellentuck space considered in Chapter One.

MEASURE AND CATEGORY

We follow standard books on this subject, such as [84] and [98]. A subset N of a topological space X is *nowhere dense* if every nonempty open subset of X can be refined to a nonempty open subset that is disjoint from N. A subset M of some topological space X is of the *first category*, or *meager*, if it can be covered by a sequence (N_k) of nowhere dense sets. Sets that are not meager in X are called *second category*. These Baire category notions become quite effective when X is assumed to be either a compact Hausdorff space or a separable and completely metrizable space (a *Polish space*). For example, the following characterization of meager subsets of the compact Hausdorff space $F^{\mathbb{N}}$ for every finite nonempty set F is particularly useful.

Lemma 9.34 *For a finite set $F \neq \emptyset$, a subset M of $F^{\mathbb{N}}$ is meager if and only if there is an infinite sequence (σ_n) of finite disjointly supported functions from $\mathbb{N}$ into F such that for every x in M, the set $\{n \in \mathbb{N} : x \supseteq \sigma_n\}$ is finite.*

The following fact is central to the Baire category because of its many uses.

Theorem 9.35 (Baire Category Theorem) *No nonempty open subset of a compact Hausdorff space or a complete metric space is meager. Equivalently, if X is either a compact Hausdorff space or a complete metric space then the intersection of any countable family of dense open subsets of X is dense in X.*

A subset A of a topological space X is *locally meager* if for every $x \in X$, there is an open neighborhood $U \ni x$ such that $A \cap U$ is meager in X. Clearly, every meager subset of X is locally meager. The following well-known fact shows that the converse is true as well.

Theorem 9.36 (Banach Category Theorem) *For every topological space X, locally meager subsets of X are in fact meager in X.* $\square$

A subset A of a topological space X has the *property of Baire* if it can be represented as a symmetric difference

$$A = U \triangle M = (U \setminus M) \cup (M \setminus U),$$

where U is an open and M is a meager subset of X.

Theorem 9.37 *For any topological space X, the property of Baire subsets of X form a σ-field of subsets of X. It is the σ-field generated by the open subsets of X together with the meager sets.*

Recall that the field of *Borel subsets* of X is the σ-field generated by the open subsets of X. Thus, every Borel subset of X has the property of Baire, although typically not vice versa.

Recall also that for a subset B of some product $X \times Y$ and x in X, the
x-section, is the set
$$B_x = \{y \in Y : (x, y) \in B\},$$
while for $y \in Y$, the y-section of B is the set
$$B^y = \{x \in X : (x, y) \in B\}.$$

Theorem 9.38 (Kuratowski-Ulam) *Suppose B is a meager subset of some
product $X \times Y$ of two Polish spaces. Then the set*
$$\{x \in X : B_x \text{ is meager in } Y\}$$
is meager in X.

We assume the reader is familiar with the basic theory of the *Lebesgue
measure* λ_k in the finite-dimensional Euclidean space $\mathbb{R}^k$. For example, we
assume that the reader is familiar with the following characterization of sets
of Lebesgue measure-zero.

Lemma 9.39 *A subset A of $\mathbb{R}^k$ has Lebesgue measure-zero if and only if
there is a sequence (B_n) of Euclidean balls[6] such that $\sum_{n=0}^{\infty} \lambda_k(B_n) < \infty$
and $A \subseteq \bigcup_{m=0}^{\infty} \bigcup_{n=m}^{\infty} B_n$.*

Theorem 9.40 *A subset A of $\mathbb{R}^k$ is Lebesgue measurable if and only if can
be written as symmetric difference of some Borel subset and some measure-
zero subset of $\mathbb{R}^k$. Thus, the field of Lebesgue measurable subsets of $\mathbb{R}^k$ is the
σ-field generated by Borel and measure-zero subsets of $\mathbb{R}^k$.*

This shows that the Lebesgue measure is determined by its action on the
Borel sets and its outer regularity given in Lemma 9.39. Of the many other
properties of the Lebesgue measure, one can mention the fact that the corresponding *measure algebra* is separable, which amounts to the fact that
there is a sequence (B_n) of Lebesgue measurable sets such that for every
other Lebesgue measurable set A and every $\varepsilon > 0$, there is n such that
$\lambda_k(A \triangle B_n) < \varepsilon$. It is well known that any two uncountable Polish spaces X
and Y are *Borel isomorphic*, i.e., a bijection $\varphi : X \to Y$ that maps Borel
subsets of X into Borel subsets of Y. These maps can be used to transfer
the Lebesgue measure from $X = \mathbb{R}^k$ to any other uncountable Polish spaces
Y thus yielding a *Borel measure* on Y defined by $\mu(B) = \lambda_k(\varphi^{-1}(B))$. The
following fact shows that the converse is also true.

Theorem 9.41 *Let Y be a Polish space carrying a Borel measure μ such
that $\mu(Y) = 1$ and $\mu(\{y\}) = 0$ for all $y \in Y$. Then there exist Borel sets
$X_0 \subseteq [0, 1]$ and $Y_0 \subseteq Y$ and a Borel isomorphism $\varphi : X_0 \to Y_0$ such that
$\mu(X_0) = 1$ and $\lambda_1(Y_0) = 1$ and such that $\mu(B) = \lambda_1(\varphi^{-1}(B))$ for all Borel
sets $B \subseteq Y_0$.*

---

[6] Balls of $\mathbb{R}^k$ are sets of the form $B_\varepsilon(x) = \{x \in \mathbb{R}^k : (\sum_{i<k} |x_i - y_i|^2)^{\frac{1}{2}} < \varepsilon\}$ for some
$\varepsilon > 0$ and $y \in \mathbb{R}^k$.

There is, however, one important property of the Lebesgue measure not shared by all Borel measures on Polish spaces. This is the notion of *Lebesgue density* of a given measurable subset A of $\mathbb{R}^k$ at a given point $x \in \mathbb{R}^k$,

$$d_k(x, A) = \lim_{\delta \to 0^+} \frac{\lambda_k(B_\delta(x) \cap A)}{\lambda_k(B_\delta(x))}.$$

For a given measurable set $A \subseteq \mathbb{R}^k$, let

$$\phi(A) = \{x \in \mathbb{R}^k : d_k(x, A) = 1\}.$$

Theorem 9.42 (Lebesgue Density Theorem) *For every measurable subset A of $\mathbb{R}^k$, we have that $\lambda_k(A \triangle \phi(A)) = 0$.*

Another important property of the Lebegue measure (and in fact any complete measure space) is contained in the following famous result.

Theorem 9.43 (Fubini Theorem) *Suppose that B is a measure-zero subset of $\mathbb{R}^k$ and that $k = m + n$. Then*

$$\lambda_m(\{x \in \mathbb{R}^m : \lambda_n(B_x) \neq 0\}) = 0.$$

The following classical result (see [12] and [25]) is a variation on Fubini's theorem and it depends also on the regularity of the Lebesgue measure.

Theorem 9.44 (Brodski-Eggleston) *Let $k = m + n$, let $\epsilon > 0$, and let F be a closed subset of $\mathbb{R}^k$ such that $\lambda_n(F_x) > \epsilon$ for all $x \in \mathbb{R}^m$. Then there is a perfect subset P of $\mathbb{R}^m$ such that $\lambda_n\left(\bigcap_{x \in P} F_x\right) > \epsilon$.*

Corollary 9.45 *Let $k = m + n$ and F be a subset of $\mathbb{R}^k$ of positive Lebesgue measure. Then there is a perfect subset P of $\mathbb{R}^m$ such that $\lambda_n\left(\bigcap_{x \in P} F_x\right) > 0$.*

It is known that in Corollary 9.45 one cannot require that the perfect set P be of positive measure.

BOREL AND ANALYTIC SETS

We follow here standard references on this subject, such as [58] or [54]. Recall that a subset H of some metric space X is Borel if it belongs to the σ-field $\mathcal{B}(X)$ of subsets of X generated by open sets. Particularly well behaved is the family of all Borel subsets of some Polish space, as the following result shows.

Theorem 9.46 *Let $(X, \mathcal{T})$ be a Polish space and let H be a Borel subset of X. Then there is a Polish topology $\mathcal{T}_H \supseteq \mathcal{T}$ such that H is a clopen set relative to $\mathcal{T}_H$ but $\mathcal{B}(X, \mathcal{T}_H) = \mathcal{B}(X, \mathcal{T})$.*

Theorem 9.47 *Suppose $(X, \mathcal{T})$ is a Polish space and f is a Borel function from X into a separable metrizable space Y. Then there is a Polish topology $\mathcal{T}_f \supseteq \mathcal{T}$ such that $f : (X, \mathcal{T}_f) \to Y$ is continuous but $\mathcal{B}(X, \mathcal{T}_H) = \mathcal{B}(X, \mathcal{T})$.*

Definition 9.48 *Given a topological space X, we say that a function $f :$ $X \to \mathbb{R}$ is of Baire Class 1 if there is a sequence (f_n) of continuous real-valued functions defined on X such that f is a pointwise limit of (f_n), meaning that*

$$\lim f_n(x) = f(x) \text{ for all } x \in X.$$

We let $\mathcal{B}_1(X)$ denote the collection of all Baire Class 1 real-valued functions on X and we usually equip $\mathcal{B}_1(X)$ with the topology of pointwise convergence on X, i.e., the topology the Tychonov cube $\mathbb{R}^X$ induces on $\mathcal{B}_1(X)$.

The following fundamental result gives a useful characterization of this class of functions when X is assumed to be a Polish space.

Theorem 9.49 (Baire Characterization Theorem) *A real-valued function f defined on some Polish space X is of Baire Class 1 if and only if f has a point of continuity on any closed subset of X.*

A subset A if some metric space X is *analytic* if there is a Polish space Y and a continuous map $f : Y \to X$ such that $A = \text{range}(f)$. It is known that every Borel subset of X is analytic but that the converse is in general false. Recall also that there is a surjectively universal Polish space.

Theorem 9.50 *Every Polish space X is a continuous image of $\mathbb{N}^{\mathbb{N}}$.*

It follows that the Baire space $\mathbb{N}^{\mathbb{N}}$ has a special place in the class of Polish spaces. The following classical result (see, [54], (7.10)) shows the feature of $\mathbb{N}^{\mathbb{N}}$ that distinguishes it from Polish spaces such as $\mathbb{R}$.

Theorem 9.51 (Hurewicz) *A Polish space X is σ-compact if and only if it does not contain a closed subset homeomorphic to the Baire space $\mathbb{N}^{\mathbb{N}}$.*

It follows from Theorem 9.50 that for every analytic subset A of some metric space X there is a continuous map $f : \mathbb{N}^{\mathbb{N}} \to X$ such that $A = \text{range}(f)$. For $s \in \mathbb{N}^{<\infty}$, let

$$F_s = \overline{\{f(x) : x \supseteq s\}}.$$

Then we get a *Souslin scheme* F_s ($s \in \mathbb{N}^{<\infty}$) of closed subsets of X such that

$$A = \bigcup_{x \in \mathbb{N}^{\mathbb{N}}} \bigcap_{n \in \mathbb{N}} F_{x \restriction n}.$$

The set on the right-hand side is the result of the *Souslin operation* applied to the Souslin scheme and is usually denoted by $\mathcal{A}(F_s : s \in \mathbb{N}^{<\infty})$. The following perfect-set property of analytic sets is in general not shared with any other family of sets, such as, for example, the family of all coanalytic sets.

Theorem 9.52 (Souslin) *Every uncountable analytic set contains a compact subset homeomorphic to the Cantor space $2^{\mathbb{N}}$.*

The following fundamental property of analytic binary relations is also worth pointing out.

Theorem 9.53 (Boundedness Theorem for Analytic Relations) *Let R be a well-founded binary relation on some Polish space X. Then the well-founded structure (X, R) has countable rank.*

The following classical result shows another important property of analytic sets.

Theorem 9.54 (Christensen) *The following are equivalent for a separable metrizable space X :*

(a) *X is Polish,*

(b) *The family $\mathcal{K}(X)$ of all compact subsets of X is Tukey-reducible[7] to $\mathbb{N}^{\mathbb{N}}$,*

(c) *The family $\mathcal{K}(X)$ of all compact subsets of X equipped with the Vietoris topology is analytic.*

SEQUENCES IN NORMED SPACES

We follow here some standard references such as [64] or [42]. A non-negative real-valued function $\| \, . \, \|$ defined on some vector space X over $\mathbb{R}$ or $\mathbb{C}$ is called a *norm* if

(1) $\| x \| = 0$ if and only if $x = 0$,

(2) $\| \lambda x \| = |\lambda| \, \| x \|$,

(3) $\| x + y \| \leq \| x \| + \| y \|$.

A *normed space* is a vector space with the norm, $(X, \| \, . \, \|)$. A *Banach space* is a normed linear space $(X, \| \, . \, \|)$ that is complete in the associated metric defined by $\rho(x, y) = \| x - y \|$. By B_X we denote the *unit ball* of X, the set

$$B_X = \{ x \in X : \| x \| \leq 1 \},$$

and by S_X we denote the *unit sphere* of X, the set

$$S_X = \{ x \in X : \| x \| = 1 \}.$$

Two Banach spaces X and Y are *isomorphic* if there is a one-to-one onto linear operator $T : X \to Y$ such that both T and T^{-1} are *bounded* linear operators in the sense that $\sup\{\| T(x) \| \, | \, \| x \| \leq 1\} < \infty$ and $\sup\{\| T^{-1}(y) \| \, | \, \| y \| \leq 1\} < \infty$. A Banach space Y is isomorphic to a *quotient space* of a Banach space X if there is a bounded linear operator from X *onto* Y.

---

[7] Here we consider $\mathcal{K}(X)$ and $\mathbb{N}^{\mathbb{N}}$ directed sets ordered by the inclusion and the coordinatewise ordering, respectively.

Each Banach space X has a *dual space* X^* consisting of all bounded linear functionals on X, i.e., linear functionals f on X such that

$$\| f \| = \sup\{|f(x)| : \| x \| \le 1\} < \infty$$

Note that this defines a norm on X^*, making X^* into a Banach space. A Banach space X is said to be *reflexive* if $X = X^{**}$, or in other words, if every bounded linear functional f^* on X^* is represented by a point x from X, i.e., $f^*(g) = g(x)$ for all $g \in X^*$.

One example of a Banach space is the space $\mathcal{C}(K)$ of all continuous real-valued functions defined on some compact space K with the norm defined by

$$\| f \|_\infty = \sup\{f(x) : x \in K\}.$$

Note that the space ℓ_∞ of all bounded sequences of scalars with the same norm $\| \cdot \|_\infty$ is an example of a space from this class, i.e., $\ell_\infty = \mathcal{C}(\beta\mathbb{N})$. This gives us also the definition of $\ell_\infty(\Gamma)$ for any other index set Γ in place of $\mathbb{N}$. Other examples of Banach spaces are the spaces ℓ_p $(1 \le p < \infty)$ of sequences of real numbers such that

$$\| x \|_p = \left(\sum_{n=0}^\infty |x|^p\right)^{\frac{1}{p}} < \infty. \tag{9.123}$$

The space c_0 of all sequences $x = (x_n)$ converging to 0 with the norm $\| x \|_\infty = \sup\{|x_n| : n \in \mathbb{N}\}$ is another standard example. Note that the sequence spaces c_0 and ℓ_p $(1 \le p < \infty)$ contain the sequences of the form

$$e_n = (0, 0, \dots, 1, 0, 0, \dots),$$

where the digit 1 occurs at the nth position. They have a special property in each of these sequence spaces isolated by the following notion.

Definition 9.55 *A Schauder basis for a Banach space X is a sequence (v_i) of elements of X such that for every $x \in X$ there is a unique sequence (a_i) of scalars such that $x = \sum_{i=0}^\infty a_i v_i$.*

Given a Schauder basis (v_i) of X and $n \in \mathbb{N}$, we can define the projection $P_n(\sum_{i=0}^\infty a_i v_i) = \sum_{i=0}^{n-1} a_i v_i$. Then the P_n are well-defined linear operators whose norms $\| P_n \| = \sup\{\| P_n(x) \| : \| x \| \le 1\}$ are uniformly bounded. This amounts to the existence of a constant $C \ge 1$ such that

$$\left\| \sum_{i=0}^{m-1} a_i v_i \right\| \le C \left\| \sum_{i=0}^{n-1} a_i v_i \right\| \tag{9.124}$$

for every $m < n$ and every sequence $(a_i)_{i<n}$ of scalars. The minimal such C is called the *basic constant*. If a sequence (v_i) of elements of some Banach space X is a Shauder basis for its closed linear span,

$$Y = \overline{\operatorname{span}}\{v_i : i \in \mathbb{N}\},$$

then we call it a *Schauder basic sequence*, or simply a *basic sequence*. Referring to the spaces c_0 and ℓ_p $(1 \le p < \infty)$, one sees not only that the sequence (e_n) is a Shauder basis for each of these spaces but also that it is an *unconditional basis* in the following sense.

Definition 9.56 *A Schauder basis (v_n) of a Banach space X is unconditional if there is a constant $C \geq 1$ such that for every finite $E \subseteq F \subseteq \mathbb{N}$ and every sequence $(a_n)_{n \in F}$ of scalars $\| \sum_{n \in E} a_n v_n \| \leq C \cdot \| \sum_{n \in F} a_n x_n \|.$*

Here is one of the results that shows that this is an important notion.

Theorem 9.57 (James) *Suppose that X is an infinite-dimensional Banach space with an unconditional basis. Then either X is reflexive, or c_0 embeds into X, or ℓ_1 embeds into X.*

There is of course the corresponding notion for sequences indexed by an arbitrary index set.

Definition 9.58 *For a given a constant $C \geq 1$, a sequence $(x_i)_{i \in I}$ of elements of some Banach space E is said to be C-unconditional if for every pair E and F of non-empty finite subsets of the index set I with $E \subseteq F$ and every choice $(a_i)_{i \in F}$ of scalars, we have $\| \sum_{i \in E} a_i x_i \| \leq C \cdot \| \sum_{i \in F} a_i x_i \|.$*

While it is not true that every Banach space has a basis or an infinite unconditional basic sequence, we have the following results.

Theorem 9.59 (Banach) *Every infinite-dimensional Banach space has an infinite Schauder basic sequence.*

Theorem 9.60 (Johnson-Rosenthal) *Every separable infinite-dimensional Banach space has an infinite-dimensional quotient with a Shauder basis.*

It is still unknown if this theorem is true without the separability restriction but there are many sufficient conditions that guarantee this. For example, we have the following list of sufficient conditions in terms of the dual space X^* of X (see, [42]; 5.10 on p.205).

Theorem 9.61 (Johnson-Rosenthal) *Let X be an infinite-dimensional Banach space. Then X has an infinite-dimensional quotient with a Shauder basis if one of the following conditions holds:*

(1) *X^* contains a infinite-dimensional reflexive subspace.*

(2) *X^* contains an isomorphic copy of the space c_0.*

(3) *X^* contains an isomorphic copy of the space ℓ_1.*

Combining this with Theorem 9.57, we have the following interesting sufficient condition.

Theorem 9.62 (Rosenthal) *If the dual X^* of a Banach space X contains an infinite normalized unconditional basic sequence then X has an infinite-dimensional quotient with a Shauder basis.*

We mention another result of this sort, which, however, involves a topological condition.

Theorem 9.63 (Talagrand) *The following are equivalent for an infinite-dimensional Banach space X :*

(1) ℓ_∞ *is a quotient of X.*

(2) *The unit ball of X^* equipped with the topology[8] of pointwise convergence on X contains a homeomorphic copy of $\beta\mathbb{N}$.*

---

[8] Usually called the *weak* topology* of X^*.

Bibliography

[1] S. A. Argyros, P. Dodos, and V. Kanellopoulos. A classification of separable Rosenthal compacta and its applications. *Dissertationes Math. (Rozprawy Mat.)*, 449:1–52, 2008.

[2] S. A. Argyros, V. Felouzis, and V. Kanellopoulos. A proof of Halpern-Läuchli partition theorem. *European J. Combin.*, 23(1):1–10, 2002.

[3] S. A. Argyros and S. Todorcevic. *Ramsey methods in analysis.* Advanced Courses in Mathematics. CRM Barcelona. Birkhäuser Verlag, Basel, 2005.

[4] R. Baire. Sur les fonctions de variables reélles. *Annali di matem., pura ed appl.*, (3),3:1–123, 1899.

[5] B. Balcar, J. Pelant, and P. Simon. The space of ultrafilters on N covered by nowhere dense sets. *Fund. Math.*, 110:11–24, 1980.

[6] J. E. Baumgartner. A short proof of Hindman's theorem. *J. Combinatorial Theory Ser. A*, 17:384–386, 1974.

[7] J. E. Baumgartner. Partition relations for countable topological spaces. *J. Combin. Theory Ser. A*, 43(2):178–195, 1986.

[8] V. Bergelson, A. Blass, and N. Hindman. Partition theorems for spaces of variable words. *Proc. London Math. Soc. (3)*, 68(3):449–476, 1994.

[9] A. Blass. A partition theorem for perfect sets. *Proc. Amer. Math. Soc.*, 82(2):271–277, 1981.

[10] A. Blass. Selective ultrafilters and homogeneity. *Ann. Pure Appl. Logic*, 38:215–255, 1988.

[11] J. Bourgain, D. H. Fremlin, and M. Talagrand. Pointwise compact sets of Baire-measurable functions. *Amer. J. Math.*, 100(4):845–886, 1978.

[12] M.L. Brodski. On some properties of sets of positive measure. *Uspehi Matem. Nauk*, 4:136–138, 1949.

[13] T. J. Carlson. An infinitary version of the Graham-Leeb-Rothschild theorem. *J. Combin. Theory Ser. A*, 44(1):22–33, 1987.

[14] T. J. Carlson. Some unifying principles in Ramsey theory. *Discrete Math.*, 68:117–169, 1988.

[15] T. J. Carlson and S. G. Simpson. A dual form of Ramsey's theorem. *Adv. in Math.*, 53(3):265–290, 1984.

[16] T. J. Carlson and S. G. Simpson. Topological Ramsey theory. In *Mathematics of Ramsey theory*, volume 5 of *Algorithms Combin.*, pages 172–183. Springer, Berlin, 1990.

[17] W. W. Comfort. Ultrafilters: some old and some new results. *Bull. Amer. Math. Soc.*, 83:417–455, 1977.

[18] G. Debs. Effective properties in compact sets of Borel functions. *Mathematika*, 34(1):64–68, 1987.

[19] D. Devlin. Some partition theorems and ultrafilters on ω. Ph.D. Thesis, Dartmouth College, 1979.

[20] C. A. Di Prisco and J. M. Henle. Partitions of products. *J. Symbolic Logic*, 58(3):860–871, 1993.

[21] C. A. Di Prisco, J. Llopis, and S. Todorcevic. Borel partitions of products of finite sets and the Ackermann function. *J. Combin. Theory Ser. A*, 93(2):333–349, 2001.

[22] C. A. Di Prisco, J. Llopis, and S. Todorcevic. Parametrized partitions of products of finite sets. *Combinatorica*, 24(2):209–232, 2004.

[23] C. A. Di Prisco and S. Todorcevic. Souslin partitions of products of finite sets. *Adv. Math.*, 176(1):145–173, 2003.

[24] P. Dodos and V. Kanellopoulos. On pairs of definable orthogonal families. *Illinois Journal of Mathematics*, 52:181–201, 2008.

[25] H.G. Eggleston. Two measure properties of cartesian product sets. *Quart. J. Math. Oxford*, (2),5:108–115, 1954.

[26] A. Ehrenfeucht and A. Mostowski. Models of axiomatic theories admitting automorphisms. *Fund. Math.*, 43:50–68, 1956.

[27] E. Ellentuck. A new proof that analytic sets are Ramsey. *J. Symbolic Logic*, 39:163–165, 1974.

[28] R. Ellis. *Lectures on topological dynamics*. W. A. Benjamin, Inc., New York, 1969.

[29] P. Erdős and R. Rado. Combinatorial theorems on classifications of subsets of a given set. *Proc. London Math. Soc. (3)*, 2:417–439, 1952.

[30] P. Erdös and R. Rado. A combinatorial theorem. *J. London Math. Soc.*, 25:249–255, 1950.

[31] I. Farah. Semiselective coideals. *Mathematika*, 45(1):79–103, 1998.

[32] H. Furstenberg and Y. Katznelson. Idempotents in compact semigroups and Ramsey theory. *Israel J. Math.*, 68(3):257–270, 1989.

[33] F. Galvin. A generalization of Ramsey's theorem. *Notices of the Amer. Math. Soc.*, 15:548, 1968.

[34] F. Galvin. Partition theorems for the real line. *Notices of the Amer. Math. Soc.*, 15:660, 1968.

[35] F. Galvin and K. Prikry. Borel sets and Ramsey's theorem. *J. Symbolic Logic*, 38:193–198, 1973.

[36] K. Gödel. *The Consistency of the Continuum Hypothesis*. Annals of Mathematics Studies, no. 3. Princeton University Press, Princeton, N. J., 1940.

[37] W. T. Gowers. Lipschitz functions on classical spaces. *European J. Combin.*, 13(3):141–151, 1992.

[38] W. T. Gowers. An infinite Ramsey theorem and some Banach-space dichotomies. *Ann. of Math. (2)*, 156(3):797–833, 2002.

[39] R. L. Graham, K. Leeb, and B. L. Rothschild. Ramsey's theorem for a class of categories. *Advances in Math.*, 8:417–433, 1972.

[40] R. L. Graham and B. L. Rothschild. Ramsey's theorem for n-parameter sets. *Trans. Amer. Math. Soc.*, 159:257–292, 1971.

[41] R. L. Graham, B. L. Rothschild, and J. H. Spencer. *Ramsey theory*. Wiley-Interscience Series in Discrete Mathematics and Optimization. John Wiley & Sons Inc., New York, second edition, 1990. A Wiley-Interscience Publication.

[42] P. Hájek, V. Montesinos, J. Vanderwerff, and V. Zizler. *Biorthogonal systems in Banach spaces*. CMS Books in Mathematics/Ouvrages de Mathématiques de la SMC, 26. Springer, New York, 2008.

[43] A. W. Hales and R. I. Jewett. Regularity and positional games. *Trans. Amer. Math. Soc.*, 106:222–229, 1963.

[44] J. D. Halpern and H. Läuchli. A partition theorem. *Trans. Amer. Math. Soc.*, 124:360–367, 1966.

[45] J. D. Halpern and A. Lévy. The Boolean prime ideal theorem does not imply the axiom of choice. In *Axiomatic Set Theory (Proc. Sympos. Pure Math., Vol. XIII, Part I, Univ. California, Los Angeles, Calif., 1967)*, pages 83–134. Amer. Math. Soc., Providence, R.I., 1971.

[46] J. M. Henle. The consistency of one fixed omega. *J. Symbolic Logic*, 60:172–177, 1995.

[47] N. Hindman. Finite sums from sequences within cells of a partition of N. *J. Combinatorial Theory Ser. A*, 17:1–11, 1974.

[48] N. Hindman and R. McCutcheon. Partition theorems for left and right variable words. *Combinatorica*, 24(2):271–286, 2004.

[49] N. Hindman and D. Strauss. *Algebra in the Stone-Čech compactification*, volume 27 of *de Gruyter Expositions in Mathematics*. Walter de Gruyter & Co., Berlin, 1998. Theory and applications.

[50] V. M. Ivanova. On the theory of the space of subsets. *Doklady Akad. Nauk. SSSR*, 101:601–603, 1955.

[51] W. B. Johnson and H. P. Rosenthal. On ω^*-basic sequences and their applications to the study of Banach spaces. *Studia Math.*, 43:77–92, 1972.

[52] A. Kanamori and K. McAloon. On Gödel incompleteness and finite combinatorics. *Ann. Pure Appl. Logic*, 33(1):23–41, 1987.

[53] V. Kanellopoulos. Ramsey families of subtrees of the dyadic tree. *Trans. Amer. Math. Soc.*, 357(10):3865–3886, 2005.

[54] A. S. Kechris. *Classical descriptive set theory*, volume 156 of *Graduate Texts in Mathematics*. Springer-Verlag, New York, 1995.

[55] J. Ketonen and R. Solovay. Rapidly growing Ramsey functions. *Ann. of Math. (2)*, 113(2):267–314, 1981.

[56] O. Klein and O. Spinas. Canonical forms of Borel functions on the Milliken space. *Trans. Amer. Math. Soc.*, 357(12):4739–4769, 2005.

[57] A. Krawczyk. On the Rosenthal compacta and analytic sets. *Proc. Amer. Math. Soc.*, 115:1095–1100, 1992.

[58] K. Kuratowski. *Topology I*. Academic Press, New York, 1966.

[59] M. Laczkovich. A Ramsey theorem for measurable sets. *Proc. Amer. Math. Soc.*, 130(10):3085–3089, 2002.

[60] C. Laflamme, N. W. Sauer, and V. Vuksanovic. Canonical partitions of universal structures. *Combinatorica*, 26(2):183–205, 2006.

[61] R. Laver. Products of infinitely many perfect trees. *J. London Math. Soc. (2)*, 29(3):385–396, 1984.

[62] H. Lefmann. Canonical partition behaviour of Cantor spaces. In *Irregularities of partitions (Fertőd, 1986)*, volume 8 of *Algorithms Combin. Study Res. Texts*, pages 93–105. Springer, Berlin, 1989.

[63] H. Lefmann. Canonical partition relations for ascending families of finite sets. *Studia Sci. Math. Hungar.*, 31(4):361–374, 1996.

[64] J. Lindenstrauss and L. Tzafriri. *Classical Banach spaces. I.* Springer-Verlag, Berlin, 1977. Sequence spaces, Ergebnisse der Mathematik und ihrer Grenzgebiete, Vol. 92.

[65] J. Lopez-Abad. Canonical equivalence relations on nets of PS_{c_0}. *Discrete Math.*, 307(23):2943–2978, 2007.

[66] J. López Abad and S. Todorcevic. Partial unconditionality of weakly null sequences. *RACSAM Rev. R. Acad. Cienc. Exactas Fís. Nat. Ser. A Mat.*, 100(1-2):237–277, 2006.

[67] A. Louveau. Sur un article de S. Sirota. *Bull. Sci. Math. (2)*, 96:3–7, 1972.

[68] A. Louveau. Une démonstration topologique de théorèmes de Silver et Mathias. *Bull. Sci. Math. (2)*, 98(2):97–102, 1974.

[69] A. Louveau. Une méthode topologique pour l'étude de la propriété de Ramsey. *Israel J. Math.*, 23(2):97–116, 1976.

[70] A. Louveau, S. Shelah, and B. Veličković. Borel partitions of infinite subtrees of a perfect tree. *Ann. Pure Appl. Logic*, 63(3):271–281, 1993.

[71] E. Marczewski. Sur une classe de fonctions de sierpinski et la classe correspondante d'ensembles. *Fund. Math.*, 24:17–34, 1935.

[72] A. R. D. Mathias. On a generalization of Ramsey's theorem. *Notices of the Amer. Math. Soc.*, 15:931, 1968.

[73] A. R. D. Mathias. A remark on rare filters. In *Infinite and finite sets (Colloq., Keszthely, 1973; dedicated to P. Erdös on his 60th birthday), Vol. III*, pages 1095–1097. Colloq. Math. Soc. János Bolyai, Vol. 10. North-Holland, Amsterdam, 1975.

[74] A. R. D. Mathias. Happy families. *Ann. Math. Logic*, 12(1):59–111, 1977.

[75] J. G. Mijares. A notion of selective ultrafilter corresponding to topological Ramsey spaces. *MLQ Math. Log. Q.*, 53(3):255–267, 2007.

[76] A. W. Miller. Infinite combinatorics and definability. *Annals of Pure and Appl. Logic*, 41:179–203, 1980.

[77] K. R. Milliken. Ramsey's theorem with sums or unions. *J. Combinatorial Theory Ser. A*, 18:276–290, 1975.

[78] K. R. Milliken. A partition theorem for the infinite subtrees of a tree. *Trans. Amer. Math. Soc.*, 263(1):137–148, 1981.

[79] J. C. Morgan, II. On the general theory of point sets. II. *Real Anal. Exchange*, 12(1):377–386, 1986/87.

[80] J. Mycielski. Independent sets in topological algebras. *Fund. Math.*, 55:139–147, 1964.

[81] J. Mycielski. Algebraic independence and measure. *Fund. Math.*, 61:165–169, 1967.

[82] C. St. J. A. Nash-Williams. On well-quasi-ordering transfinite sequences. *Proc. Cambridge Phil. Soc.*, 61:33–39, 1965.

[83] E. Odell and H. P. Rosenthal. A double-dual characterization of separable Banach spaces containing l^1. *Israel J. Math.*, 20(3-4):375–384, 1975.

[84] J. C. Oxtoby. *Measure and category. A survey of the analogies between topological and measure spaces*, volume 2 of *Graduate Texts in Mathematics*. Springer-Verlag, New York, 1971.

[85] J. Pawlikowski. Parametrized Ellentuck theorem. *Topology Appl.*, 37(1):65–73, 1990.

[86] R. Pol. On pointwise and weak topology in function spaces. *University of Warsaw Preprint Series*, Preprint 4/84:1–53, 1984.

[87] M. Pouzet and N. Sauer. Edge partitions of the Rado graph. *Combinatorica*, 16:505–520, 1996.

[88] H. J. Prömel, S. G. Simpson, and B. Voigt. A dual form of Erdős-Rado's canonization theorem. *J. Combin. Theory Ser. A*, 42(2):159–178, 1986.

[89] H. J. Prömel and B. Voigt. Canonical partition theorems for parameter sets. *J. Combin. Theory Ser. A*, 35(3):309–327, 1983.

[90] H. J. Prömel and B. Voigt. Baire sets of k-parameter words are Ramsey. *Trans. Amer. Math. Soc.*, 291(1):189–201, 1985.

[91] H. J. Prömel and B. Voigt. Canonical forms of Borel-measurable mappings $\Delta\colon [\omega]^\omega \to \mathbf{R}$. *J. Combin. Theory Ser. A*, 40(2):409–417, 1985.

[92] P. Pudlák and V. Rödl. Partition theorems for systems of finite subsets of integers. *Discrete Math.*, 39(1):67–73, 1982.

[93] F.P. Ramsey. On a problem of formal logic. *Proc. London Math. Soc.*, 30:264–296, 1930.

[94] H. P. Rosenthal. Letter of December 7, 1998.

[95] H. P. Rosenthal. A characterization of Banach spaces containing l^1. *Proc. Nat. Acad. Sci. U.S.A.*, 71:2411–2413, 1974.

[96] H. P. Rosenthal. Pointwise compact subsets of the first Baire class. *Amer. J. Math.*, 99(2):362–378, 1977.

[97] A. Rosłanowski and S. Shelah. Norms on possibilities. I. *Mem. Amer. Math. Soc.*, 141(671):xii+167, 1999.

[98] H. L. Royden. *Real analysis*. Macmillan Publishing Company, New York, third edition, 1988.

[99] J. Silver. Every analytic set is Ramsey. *J. Symbolic Logic*, 35:60–64, 1970.

[100] S. M. Sirota. A product of topological groups, and extremal disconnectedness. *Mat. Sb. (N.S.)*, 79 (121):179–192, 1969.

[101] S. Solecki and S. Todorcevic. Cofinal types of topological directed orders. *Ann. Inst. Fourier (Grenoble)*, 54:1877–1911, 2004.

[102] O. Spinas. Ramsey and freeness properties of Polish planes. *Proc. London Math. Soc. (3)*, 82(1):31–63, 2001.

[103] J. Stern. A Ramsey theorem for trees, with an application to Banach spaces. *Israel J. Math.*, 29(2-3):179–188, 1978.

[104] A. D. Taylor. A canonical partition relation for finite subsets of ω. *J. Combinatorial Theory Ser. A*, 21(2):137–146, 1976.

[105] A. D. Taylor. Partitions of pairs of reals. *Fund. Math.*, 99(1):51–59, 1978.

[106] S. Todorcevic. Analytic gaps. *Fund. Math.*, 150(1):55–66, 1996.

[107] S. Todorcevic. Compact subsets of the first Baire class. *J. Amer. Math. Soc.*, 12(4):1179–1212, 1999.

[108] S. Todorcevic. A new quantitative analysis of some basic principles of the theory of functions of a real variable. *Bull. Cl. Sci. Math. Nat. Sci. Math.*, 26:133–144, 2001.

[109] S. Todorcevic and C. Uzcategui. Analytic k-spaces. *Topology Appl.*, 146/147:511–526, 2005.

[110] J. W. Tukey. *Convergence and Uniformity in Topology*. Annals of Mathematics Studies, no. 2. Princeton University Press, Princeton, N. J., 1940.

[111] B. Velickovic and W.H. Woodin. Complexity of reals in inner models of set theory. *Ann. Pure Appl. Logic*, 92:283–295, 1998.

[112] B. Voigt. Canonization theorems for finite affine and linear spaces. *Combinatorica*, 4(2-3):219–239, 1984.

[113] B. Voigt. Canonizing partition theorems: diversification, products, and iterated versions. *J. Combin. Theory Ser. A*, 40(2):349–376, 1985.

[114] B. Voigt. Extensions of the Graham-Leeb-Rothschild theorem. *J. Reine Angew. Math.*, 358:209–220, 1985.

[115] J. Zapletal. Preserving *P*-points in definable forcing. *Fund. Math.*, 204(2):145–154, 2009.

Subject Index

Index of Notation

285

Milton Keynes UK
Ingram Content Group UK Ltd.
UKHW020640290824
447545UK00007B/212